Lecture Notes in Computer Science   11057

Commenced Publication in 1973
Founding and Former Series Editors:
Gerhard Goos, Juris Hartmanis, and Jan van Leeuwen

More information about this series at http://www.springer.com/series/7409

Eva Méndez · Fabio Crestani
Cristina Ribeiro · Gabriel David
João Correia Lopes (Eds.)

Digital Libraries
for Open Knowledge

22nd International Conference on Theory and Practice
of Digital Libraries, TPDL 2018
Porto, Portugal, September 10–13, 2018
Proceedings

 Springer

Editors
Eva Méndez 🄳
University Carlos III
Madrid
Spain

Fabio Crestani 🄳
USI, Università della Svizzera Italiana
Lugano
Switzerland

Cristina Ribeiro 🄳
INESC TEC, Faculty of Engineering
University of Porto
Porto
Portugal

Gabriel David 🄳
INESC TEC, Faculty of Engineering
University of Porto
Porto
Portugal

João Correia Lopes 🄳
INESC TEC, Faculty of Engineering
University of Porto
Porto
Portugal

ISSN 0302-9743   ISSN 1611-3349   (electronic)
Lecture Notes in Computer Science
ISBN 978-3-030-00065-3   ISBN 978-3-030-00066-0   (eBook)
https://doi.org/10.1007/978-3-030-00066-0

Library of Congress Control Number: 2018952645

LNCS Sublibrary: SL3 – Information Systems and Applications, incl. Internet/Web, and HCI

This Springer imprint is published by the registered company Springer Nature Switzerland AG
The registered company address is: Gewerbestrasse 11, 6330 Cham, Switzerland

Preface

Open Access, Open Science, Open Education, Open Data, Open Culture... We are in the "Open Era" and really making knowledge open is the big challenge for digital libraries and other information infrastructures of the XXI century. The International Conference on Theory and Practice of Digital Libraries (TPDL) brings together researchers, developers, content providers, and users in digital libraries and digital content management. The 22nd TPDL took place in Porto, Portugal on September 10–13, 2018, jointly organized by INESC TEC and the Faculty of Engineering of the University of Porto.

The general theme of TPDL 2018 was "Digital Libraries for Open Knowledge". The year 2017 was considered "Year of Open" by the Open Education Consortium and TPDL 2018 is "the TPDL of Open". TPDL 2018 aimed to gather all the communities engaged in making knowledge more and more open, disseminating new ideas using the available technologies, standards and infrastructures, while reflecting on new challenges, policies, and other issues to make it happen. TPDL 2018 provided the community in computer and information science the opportunity to reflect, discuss, and contribute to the complex issues of making knowledge open, not only to users but also to re-users, among different information infrastructures and digital assets.

This volume of proceedings contains the papers presented at TPDL 2018. We received a total of 81 submissions, including 51 full papers, 17 short papers, plus 13 posters and demos. The Program Committee followed the highest academic standards to ensure the selection of papers to be presented at the conference. Of the full-paper submissions, 16 (31%) were accepted for long oral presentation. However, following the recommendations of reviewers, some selected full-paper submissions that included novel and interesting ideas were redirected for evaluation as potential short or poster papers. From the 17 short-paper submissions, plus the remaining long papers, 9 short papers (17%) were accepted for short oral presentation. Of the 13 poster and demo submissions, plus the full and short submissions not considered for presentation, 20 (36%) were accepted for poster and demo presentation. Each submission was reviewed by at least three Program Committee members and one Senior Program Committee member, with the two Program Chairs overseeing the reviewing process and prompting the necessary follow-up discussions to complete the selection process.

The conference was honored by three outstanding keynote speakers, covering important and current topics of the digital library field: Medha Devare, from the International Food Policy Research Institute, talked about "Leveraging Standards to Turn Data to Capabilities in Agriculture"; Natalia Manola, from the University of Athens, explored "Open Science in a Connected Society"; and Herbert Van de Sompel, from the Los Alamos National Laboratory, presented "A Web-Centric Pipeline for Archiving Scholarly Artifacts".

The program also included a doctoral consortium track, jointly organized with the co-located Dublin Core Metadata Initiative (DCMI) annual international conference, two tutorials on "Linked Data Generation from Digital Libraries" and "Research the Past Web Using Web Archives", and a hands-on session on "Europeana".

A set of workshops, also jointly organized with the DCMI annual international conference, allowed for more in-depth work on specialized topics or communities. Two half-day workshops took place: "Domain Specific Extensions for Machine-Actionable Data Management Plans" and a "Special Session on Metadata for Manufacturing". Four full-day workshops were also offered after the main conference: "18th European Networked Knowledge Organization Systems (NKOS)", "Web Archive — An Intro-duction to Web Archives for Humanities and Social Science Research", "Multi-domain Research Data Management: From Metadata Collection to Data Deposit", and "Internet of Things Workshop: Live Repositories of Streaming Data".

We would like to thank all our colleagues for trusting their papers to the conference, as well as our Program Committee and Senior Program Committee members for the precise and thorough work they put into reviewing the submissions.

September 2018

Eva Méndez
Fabio Crestani
Cristina Ribeiro
Gabriel David
João Correia Lopes

Organization

General Chairs

Cristina Ribeiro INESC TEC, FEUP, University of Porto, Portugal
Gabriel David INESC TEC, FEUP, University of Porto, Portugal

Program Co-chairs

Eva Méndez University Carlos III of Madrid, Spain
Fabio Crestani USI, Università della Svizzera Italiana, Switzerland

Doctoral Consortium Chairs

Ana Alice Baptista University of Minho, Portugal
José Borbinha INESC-ID, IST, University of Lisbon, Portugal

Poster and Demonstration Chairs

Antoine Isaac Europeana and Vrije Universiteit Amsterdam,
Netherlands
Maria Manuel Borges FLUC, University of Coimbra, Portugal

Tutorial Chairs

Andreas Rauber Vienna University of Technology, Austria
Irene Rodrigues DI, University of Évora, Portugal

Workshop Chairs

Daniel Gomes FCT: Arquivo.pt (Portuguese Web Archive), Portugal
Thomas Risse Frankfurt University Library J. C. Senckenberg,
Germany

Publication Chair

João Correia Lopes INESC TEC, FEUP, University of Porto, Portugal

Local Organization Chair

João Rocha da Silva INESC TEC, FEUP, University of Porto, Portugal

Program Committee

Senior Program Committee

Trond Aalberg	Norwegian University of Science and Technology, Norway
Maristella Agosti	University of Padua, Italy
Thomas Baker	DCMI
José Borbinha	University of Lisbon, Portugal
George Buchanan	University of Melbourne, Australia
Stavros Christodoulakis	Technical University of Crete, Greece
Panos Constantopoulos	Athens University of Economics and Business, Greece
Nicola Ferro	University of Padua, Italy
Edward Fox	Virginia Tech, USA
Norbert Fuhr	University of Duisburg-Essen, Germany
Richard Furuta	Texas A&M University, USA
Paul Groth	Elsevier Labs
Sarantos Kapidakis	Ionian University, Greece
Clifford Lynch	CNI, USA
Carlo Meghini	ISTI-CNR, Italy
Wolfgang Nejdl	L3S and University of Hanover, Germany
Michael Nelson	Old Dominion University, USA
Erich Neuhold	University of Vienna, Austria
Christos Papatheodorou	Ionian University, Greece
Edie Rasmussen	The University of British Columbia, Canada
Laurent Romary	Inria and HUB-ISDL, France
Mário J. Silva	University of Lisbon, Portugal
Hussein Suleman	University of Cape Town, South Africa
Costantino Thanos	ISTI-CNR, Italy
Giannis Tsakonas	University of Patras, Greece
Pertti Vakkari	University of Tampere, Finland
Herbert Van de Sompel	Los Alamos National Laboratory, USA

Program Committee

Hamed Alhoori	Northern Illinois University, USA
Robert B. Allen	Yonsei University, South Korea
Vangelis Banos	Aristotle University of Thessaloniki, Greece
Valentina Bartalesi	ISTI-CNR, Italy
Christoph Becker	University of Toronto, Canada
Alexandro Bia	Miguel Hernández University, Spain
Maria Bielikova	Slovak University of Technology in Bratislava, Slovakia
Maria Manuel Borges	University of Coimbra, Portugal

Séamus Lawless	Trinity College Dublin, Ireland
Hyowon Lee	Singapore University of Technology and Design, Singapore
Suzanne Little	Dublin City University, Ireland
João Magalhães	NOVA University of Lisbon, Portugal
Yannis Manolopoulos	Aristotle University of Thessaloniki, Greece
Zinaida Manžuch	Vilnius University, Lithuania
Bruno Martins	University of Lisbon, Portugal
Philipp Mayr	GESIS - Leibniz Institute for Social Sciences, Germany
Cezary Mazurek	Poznań Supercomputing and Networking Center, Poland
Robert H. Mcdonald	Indiana University, USA
Dana Mckay	University of Melbourne, Australia
Andras Micsik	Hungarian Academy of Sciences, Hungary
Agnieszka Mykowiecka	Polish Academy of Sciences, Poland
Heike Neuroth	Potsdam University of Applied Sciences, Germany
David Nichols	University of Waikato, New Zealand
Ragnar Nordlie	Oslo Metropolitan University, Norway
Kjetil Nørvåg	Norwegian University of Science and Technology, Norway
Nils Pharo	Oslo Metropolitan University, Norway
Francesco Piccialli	University of Naples Federico II, Italy
Dimitris Plexousakis	Foundation for Research and Technology Hellas, Greece
Andreas Rauber	Vienna University of Technology, Austria
Thomas Risse	Frankfurt University Library J. C. Senckenberg, Germany
Irene Rodrigues	Universidade de Évora, Portugal
Ian Ruthven	University of Strathclyde, UK
J. Alfredo Sánchez	Laboratorio Nacional de Informática Avanzada, Mexico
Heiko Schuldt	University of Basel, Switzerland
Michalis Sfakakis	Ionian University, Greece
Frank Shipman	Texas A&M University, USA
João Rocha da Silva	INESC TEC, Portugal
Gianmaria Silvello	University of Padua, Italy
Nicolas Spyratos	Paris-Sud University, France
Shigeo Sugimoto	University of Tsukuba, Japan
Tamara Sumner	University of Colorado Boulder, USA
Atsuhiro Takasu	National Institute of Informatics, Japan
Diana Trandabăț	University Alexandru Ioan Cuza of Iasi, Romania
Theodora Tsikrika	Centre for Research and Technology Hellas, Greece
Chrisa Tsinaraki	European Commission – Joint Research Center
Douglas Tudhope	University of South Wales, UK
Yannis Tzitzikas	University of Crete, Greece
Nicholas Vanderschantz	University of Waikato, New Zealand

Stefanos Vrochidis   Centre for Research and Technology Hellas, Greece
Michele Weigle   Old Dominion University, USA
Marcin Werla   Poznań Supercomputing and Networking Center,
                            Poland
Iris Xie   University of Wisconsin-Milwaukee, USA
Maja Žumer   University of Ljubljana, Slovenia

Additional Reviewers

André Carvalho   Anastasia Moumtzidou
Serafeim Chatzopoulos   André Mourão
Panagiotis Giannakeris   Panagiotis Papadakos
Tobias Gradl   David Semedo
Eleftherios Kalogeros   Thanasis Vergoulis

TPDL/DCMI Workshops

Domain Specific Extensions for Machine-Actionable Data Management Plans

João Cardoso   INESC-ID, Portugal
Tomasz Miksa   SBA Research, Austria
Paul Walk   Antleaf Ltd., UK

Special Session on Metadata for Manufacturing

Ana Alice Baptista   University of Minho, Portugal
João P. Mendonça   University of Minho, Portugal
Paula Monteiro   University of Minho, Portugal

18th European Networked Knowledge Organization Systems (NKOS)

Joseph A. Busch   Taxonomy Strategies, USA
Koraljka Golub   Linnaeus University, Sweden
Marge Hlava   Access Innovations, USA
Philipp Mayr   GESIS - Leibniz Institute for the Social Sciences,
                           Germany
Douglas Tudhope   University of South Wales, UK
Marcia L. Zeng   Kent State University, USA

Web Archive — An Introduction to Web Archives for Humanities and Social Science Research

Daniel Gomes   FCT - Foundation for Science and Technology,
                           Portugal
Jane Winters   University of London, UK

Multi-domain Research Data Management: From Metadata Collection to Data Deposit

Ângela Lomba   University of Porto, Portugal
João Aguiar Castro   INESC TEC, University of Porto, Portugal

Internet of Things Workshop: Live Repositories of Streaming Data

Artur Rocha   INESC TEC, Portugal
Alexandre Valente Sousa   Porto Digital, Instituto Universitário da Maia, Portugal
Joaquin Del Rio Fernandez Universidad Politècnica de Catalunya, Spain
Hylke van der Schaaf   Fraunhofer IOSB, Germany

TPDL/DCMI Tutorials

Linked Data Generation from Digital Libraries

Anastasia Dimou   Ghent University, Belgium
Pieter Heyvaert   Ghent University, Belgium
Ben Demeester   Ghent University, Belgium

Research the Past Web Using Web Archives

Daniel Gomes   FCT: Arquivo.pt, Portugal
Daniel Bicho   FCT: Arquivo.pt, Portugal
Fernando Melo   FCT: Arquivo.pt, Portugal

Europeana Hands-On Session

Hugo Manguinhas   Europeana
Antoine Isaac   Europeana and Vrije Universiteit Amsterdam, Netherlands

Sponsors

The Organizing Committee is very grateful to the following organizations for their support:

CNI	The Coalition for Networked Information
Porto.	Visit Porto and the North of Portugal
FLAD	Luso-American Development Foundation
LNCS	Springer Lecture Notes in Computer Science
U.PORTO	University of Porto, Faculty of Engineering
INESC TEC	Institute for Systems and Computer Engineering, Technology and Science

Keynotes

Conference Patron Message

Maria Fernanda Rollo (ID)

Portuguese Secretary of State of Science, Technology and Higher Education

What if everybody, everybody indeed, could have full access to knowledge? And what if everybody, regardless of her/his individual circumstances, could access what is written across the world? The purpose of the book would therefore be accomplished; the highest goal for knowledge, its plenitude of access and universality.

What used to be a dream, is now a path that we pursue through the affirmation and expansion of the digital libraries.

Together with the will of people and the commitment of diverse entities, science and technology can accomplish this goal: by unlocking new areas of knowledge, by challenging impossibilities, by overcoming potential difficulties and by tearing down those which are less likely to be obstacles; by betting on the future, preserving and claiming for the property of knowledge production.

Between the dream and utopia, digital libraries provide a real opportunity for access and universal "appropriation" of knowledge. The right to education and training, is still one of the most difficult challenges to overcome and nonetheless one of the most fundamental for the creation of a fairer society.

I believe that the revolutionary demand of knowledge raised up by technology will also grow, designed for a common well-being and founded on democratisation of access to knowledge. However, society, as a whole, will have to help and to assume that purpose as its own.

I truly believe in the justice of the principles and the purposes as in the solidary, driving and contagious forces of all actors, scientists, politicians and social agents, those who inspire and proclaim the movement of open knowledge for all.

Knowledge, including science, is a common good, belonging to everyone, and everyone must benefit of it. Its universality, coupled with the capacity for change, provides it with the ability to cross political, cultural and psychological boundaries towards sustainable development.

The preservation, dissemination and sharing of knowledge, making it accessible and beneficial to all, is an essential way to build a fairer, more democratic society. It is our responsibility for the generations to come and for a better wellbeing.

Asserting Open Science/Knowledge for all, to which my country is devoted, represents a new commitment with society to the production and access, as to the sharing and usage of science and knowledge in general and in its contribution for a sustainable development and for the building of a fairer society.

Knowledge sharing, access to knowledge, open access and open science, provided properly and in equitable manners, is an opportunity to share results and data at the North-South and South-South levels, an unprecedented opportunity that can stimulate inclusion, enable local researchers, regions, and society in general in the countries that make up the world to be effectively included in the knowledge community — to have access and an integral part of global knowledge.

The digital libraries are effective means and essential catalysts for the acceleration and achievement of this goal.

Short-Bio

Professor Maria Fernanda Rollo, the Secretary of State for Science, Technology and Higher Education, has kindly accepted to be the Patron of TPDL 2018. Professor Maria Fernanda Rollo is a researcher in Contemporary History, with a vast contribution to Digital Humanities. Her work as a Secretary of State is actively promoting all aspects of Open Knowledge and Open Science.

Leveraging Standards to Turn Data to Capabilities in Agriculture

Medha Devare (iD)

CGIAR Consortium: Montpellier, Languedoc-Roussillon, France

CGIAR is a global research partnership of 15 Centers primarily located in developing countries, working in the agricultural research for development sector. Research at these Centers is focused on poverty reduction, enhancing food and nutrition security, and improving natural resource management to address key development challenges. It is conducted in close collaboration with local partner entities, including national and regional research institutes, civil society organizations, academia, development organizations, and the private sector. Thus, the CGIAR system is charged with tackling challenges at a variety of scales from the local to the global; however, research outputs are often not easily discoverable and research data often resides on individual laptops, not being well annotated or stored to be accessible and usable by the wider scientific community.

Innovating in this space and enhancing research impact increasingly depends upon enabling the discovery of, unrestricted access to, and effective reuse of the publications and data generated as primary research outputs by Center scientists. Accelerating innovation and impact to effectively address global agricultural challenges also requires that data be easily aggregated and integrated, which in turn necessitates interoperability. In this context, open is inadequate, and the concept of FAIR (Findable, Accessible, Interoperable, Reusable) has proven more useful. CGIAR Centers have made strong progress implementing publication and data repositories that meet minimum interoperability standards; however, work is still needed to enable consistent and seamless information discovery, integration, and interoperability across outputs. For datasets, this generally means annotation using standards such as controlled vocabularies and ontologies.

The Centers are therefore working to create an enabling environment to enhance access to research outputs, propelled by funder requirements and a system-wide Open Access and Data Management Policy implemented in 2013 (CGIAR, 2013). Guidance and the impetus for operationalization is being provided via the CGIAR Big Data Platform for Agriculture, and its Global Agricultural Research Data Innovation and Acceleration Network (GARDIAN). GARDIAN is intended to provide seamless, semantically-linked access to CGIAR publications and data, to demonstrate the full value of CGIAR research, enable new analyses and discovery, and enhance impact.

There are several areas in which standards and harmonized approaches are being leverages to achieve FAIRness at CGIAR, some of which are outlined below.

Data sourcing, handling. Research at CGIAR Centers focuses on different commodities, agro-ecologies, disciplinary domains, geographies and scales, resulting in varied data streams—some born digital, often characterized by large size and speed of generation, and frequent updates. Data ranges from agronomic trial data collected by field technicians in a variety of ways and formats, through input and output market information and socioeconomic data on technology adoption and enabling drivers, to weather data and high-throughput sequencing and phenotypic information and satellite images. These datasets cannot all be treated in the same manner; the curation and quality control needs differ significantly, for instance necessitating somewhat customized approaches depending on the data type. Yet, to address key challenges, they must be discoverable, downloadable, reusable, and able to be aggregated where relevant. As a first step towards these goals, Centers have agreed on and mapped repository schemas to a common Dublin Core based set of required metadata elements (the CG Core Metadata Schema v.1.0).

Enhancing interoperability. Interoperability is critical to providing meaning and context to CGIAR's varied data streams and enabling integration between linked content types (e.g., related data and publications) and across related data types (e.g. an agronomic data set and related socioeconomic data).

CGIAR's approach to interoperability and data harmonization focuses on the use of standard vocabularies (AGROVOC/GACS), and strong reliance on ontologies developed across CGIAR (efforts such as the Crop Ontology, the Agronomy Ontology - AgrO, the in-development socioeconomic ontology - SociO), and other entities (ENVO, UO, PO, etc.)

Discovery framework. Recognizing the need to democratize agricultural research information and make it accessible to partners — particularly those in developing countries – CGIAR's aspirations focus on enabling data discovery, integration, and analysis via an online, semantically-enables infrastructure. This tool, built under the auspices of the Big Data platform, harvests metadata from CGIAR Center repositories, and includes the ability to relatively seamlessly leverage it with existing and new analytical and mapping tools. While there is no blueprint for building such an ecosystem in the agriculture domain, there are successful models to learn and draw from. Of particular interest are the functionalities demonstrated by the biomedical community via the National Center for Biotechnology Information (NCBI) suite of databases and tools, with attendant innovations for translational medicine and human health. CGIAR efforts to enable similar functionalities to NCBI's are underlain by strong and enduring stakeholder engagement and capacity building.

Harmonizing data privacy and security approaches as appropriate. Concern regarding data privacy and security is becoming increasingly significant with recent breaches of individual privacy and the GDPR. Any CGIAR repositories and harvesters of data need to provide assurance of data anonymity with respect to personally identifiable information, yet this presents a conundrum when spatial information is so integral to the ability to provide locally actionable options to farming communities. Related to these issues is the concern around ethics, particularly with respect to surveys. The Big Data Platform is therefore focusing on facilitating the creation of and continued support for Institutional Review Boards (IRBs) or their equivalent at Centers, including via guidelines on ethical data collection and handling. Lastly, whether agricultural data is closed or open, it needs to be securely held in the face of such threats as hacking and unanticipated loss.

It is important to recognize that without incentives and a culture that encourages and rewards best practices in managing research outputs, technical attempts to promote the use of standards and enable FAIR resources will meet with limited success, at best. Among some factors influencing these goals: Clarity on incentives (e.g., from funding agency incentives to data contributors understanding the benefits of sharing data) and easy processes, workflows and tools to make data FAIR, with continued support for stakeholders. Researchers need to be accountable for making their outputs FAIR (e.g., through contractual obligation, annual performance evaluation and recognition, funder policies etc.) Only through a multi-faceted approach that recognizes and addresses systemic and individual constraints in both the cultural and technical domain will CGIAR succeed in leveraging its research outputs to fuel innovation and impact, and transform agricultural research for development.

Short-Bio

Medha Devare is Senior Research Fellow with the International Food Policy Research Institute (IFPRI) and leads its Big Data Platform efforts to organize data across the CGIAR System's 15 Centers. She has led CGIAR food security projects in South Asia, and its Open Access/Open Data Initiative. Medha also has expertise in data management and semantic web tools; while at Cornell University she was instrumental in the development of VIVO, a semantic web application for representing scholarship.

Open Science in a Connected Society

Natalia Manola🆔

Communication and Knowledge Technologies, Athena Research Center
in Information, Greece

Open science comes on the heels of the fourth paradigm of science, which is based on data-intensive scientific discovery, and represents a new paradigm shift, affecting the entire research lifecycle and all aspects of science execution, collaboration, communication, innovation. From supporting and using (big) data infrastructures for data archiving and analysis, to continuously sharing with peers all types of research results at any stage of the research endeavor and to communicating them to the broad public or commercial audiences, openness moves science away from being a concern exclusively of researchers and research performing organisations and brings it to center stage of our connected society, requiring the engagement of a much wider range of stakeholders: digital and research infrastructures, policy decision makers, funders, industry, and the public itself.

Although the new paradigm of science is shifting towards openness, participation, transparency, and social impact, it is still unclear how to measure and assess these qualities. This presentation focuses on the way the scientific endeavor is assessed and how one may shape up science policies to address societal challenges, as science is becoming an integral part of the wider socio-economic environment. It discusses how one may measure the impact science has on innovation, the economy, and society in general, and how the need for such measurement influences the collection, stewardship, preservation, access, and analysis of digital assets. It argues that an open transfer of both codified and tacit knowledge lies at the core of impact creation and calls for a consistently holistic systematic approach to research. In particular, it includes codified knowledge in the form of traditional publications and datasets, but also formal intellectual property (patents, copyright, etc.) and soft intellectual property (e.g., open software, databases or research methodologies), as well as tacit knowledge in the form of skills, expertise, techniques, and complex cumulative knowledge, conceptual models, and terminology.

Putting the spotlight on (open) data collection and analysis, this presentation further illustrates a use case based on the collaboration between OpenAIRE (www.openaire.eu) and the Data4Impact project (www.data4impact.eu) on the use of an open scholarly communication graph, combined with text mining, topic modeling, machine learning, and citation based approaches to trace and classify the societal impact of research funded by the European Commission.

Short-Bio

Natalia Manola is a research associate in the University of Athens, Department of Informatics and Telecommunications and the "Athena" Technology and Innovation Research Center. She holds a Physics degree from the University of Athens, and an MS in Electrical and Computing Engineering from the University of Wisconsin at Madison. She has several years of employment as a Software Engineer and Architect employed in the Bioinformatics sector. From 2009 she has served as the managing director of OpenAIRE (www.openaire.eu), a pan European e-Infrastructure supporting open access in all scientific results, and OpenMinTeD (www.openminted.eu) an infrastructure on text and data mining.

She has served in EC' Future Emerging Technology (FET) advisory group 2013–2017 and is currently a member of EC's Open Science Policy Platform, a high level Expert Group to provide advice about the development and implementation of open science policy in Europe. Her research interests include the topics of e-Infrastructures development and management, scientific data management, data curation and validation, text and data mining, complex data visualization, and research analytics.

A Web-Centric Pipeline for Archiving Scholarly Artifacts

Herbert Van de Sompel ⓘ

Los Alamos National Laboratory, USA

Scholars are increasingly using a wide variety of online portals to conduct aspects of their research and to convey research results. These portals exist outside of the established scholarly publishing system and can be dedicated to scholarly use, such as myexperiment.org, or general purpose, such as GitHub and SlideShare. The combination of productivity features and global exposure offered by these portals is attractive to researchers and they happily deposit scholarly artifacts there. Most often, institutions are not even aware of the existence of these artifacts created by their researchers. More importantly, no infrastructure exists to systematically and comprehensively archive them, and the platforms that host them rarely provide archival guarantees; many times quite the opposite.

Initiatives such as LOCKSS and Portico offer approaches to automatically archive the output of the established scholarly publishing system. Platforms like Figshare and Zenodo allow scholars to upload scholarly artifacts created elsewhere. They are appealing from an open science perspective and researchers like the citable DOIs that are provided for contributions. But these platforms don't offer a comprehensive archive for scholarly artifacts since not all scholars use them, and the ones that do are selective regarding their contributions.

The Scholarly Orphans project funded by the Andrew W. Mellon Foundation, explores how these scholarly artifacts could automatically be archived. Because of the scale of the problem — the number of platforms and artifacts involved — the project starts from a web-centric resource capture paradigm inspired by current web archiving practice. Because the artifacts are often created by researchers affiliated with an institution, the project focuses on tools for institutions to discover, capture, and archive these artifacts. The Scholarly Orphans team has started devising a prototype of an automatic pipeline that covers all three functions. Trackers monitor the APIs of productivity portals for new contributions by an institution's researchers. The Memento Tracer framework generates web captures of these contributions. Its novel capturing approach allows generating high-quality captures at scale. The captures are subsequently submitted to a — potentially cross-institutional — web archive that leverages IPFS technology and supports the Memento "Time Travel for the Web" protocol. All components communicate using Linked Data Notifications carrying ActivityStreams2 payloads.

Without adequate infrastructure, scholarly artifacts will vanish from the web in much the same way regular web resources do. The Scholarly Orphans project team hopes that its work will help raise awareness regarding the problem and contribute to finding a sustainable and scalable solution for systematically archiving web-based scholarly artifacts. This talk will be the first public communication about the team's experimental pipeline for archiving scholarly artifacts.

Short-Bio

Herbert Van de Sompel is an Information Scientist at the Los Alamos National Laboratory and, for 15 years, has led the Prototyping Team. The Team does research regarding various aspects of scholarly communication in the digital age, including information infrastructure, interoperability, and digital preservation. Herbert has played a major role in creating the Open Archives Initiative Protocol for Metadata Harvesting, the Open Archives Initiative Object Reuse & Exchange specifications, the OpenURL Framework for Context-Sensitive Services, the SFX linking server, the bX scholarly recommender service, info URI, Web Annotation, ResourceSync, Memento "time travel for the Web", Robust Links, and Signposting the Scholarly Web. He graduated in Mathematics and Computer Science at Ghent University, Belgium, and holds a Ph.D. in Communication Science from the same university.

Contents

Tutorials

Metadata

Content-Based Quality Estimation for Automatic Subject Indexing of Short Texts Under Precision and Recall Constraints

Martin Toepfer[1]([✉]) and Christin Seifert[2]([iD])

[1] ZBW – Leibniz Information Centre for Economics, Kiel, Germany
m.toepfer@zbw.eu
[2] University of Twente, Enschede, The Netherlands
c.seifert@utwente.nl

Abstract. Digital libraries strive for integration of automatic subject indexing methods into operative information retrieval systems, yet integration is prevented by misleading and incomplete semantic annotations. For this reason, we investigate approaches to detect documents where quality criteria are met. In contrast to mainstream methods, our approach, named Qualle, estimates quality at the document-level rather than the concept-level. Qualle is implemented as a combination of different machine learning models into a deep, multi-layered regression architecture that comprises a variety of content-based indicators, in particular label set size calibration. We evaluated the approach on very short texts from law and economics, investigating the impact of different feature groups on recall estimation. Our results show that Qualle effectively determined subsets of previously unseen data where considerable gains in document-level recall can be achieved, while upholding precision at the same time. Such filtering can therefore be used to control compliance with data quality standards in practice. Qualle allows to make trade-offs between indexing quality and collection coverage, and it can complement semi-automatic indexing to process large datasets more efficiently.

Keywords: Quality estimation · Automatic subject indexing
Document-level constraints · Multi-label classification · Short-text

1 Introduction

Semantic annotations from automatic subject indexing can improve information retrieval (IR) by query expansion, however, classification performance is a critical factor to gain the benefits [21]. Research across disciplines advanced multi-label text classification over the last decades [2,8–11,17,20,22], yet several challenges remain. Just to give an example, precision@5 = 52% [9] has recently been reported for a dataset in the legal domain [10], which means that on average

© Springer Nature Switzerland AG 2018
E. Méndez et al. (Eds.): TPDL 2018, LNCS 11057, pp. 3–15, 2018.
https://doi.org/10.1007/978-3-030-00066-0_1

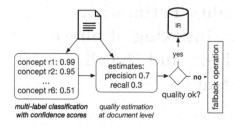
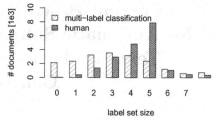

Fig. 1. Schematic overview of the main application context. Document-level quality estimation enables filtering of automatic subject indexing results.

Fig. 2. Illustration of low document-level recall by comparing distributions of label set size (human vs. multi-label classification) [Dataset: EURLEX].

per document only half of the five top-ranked subjects matched human annotations. Applying state-of-the-art algorithms only according to averaged f_1 scores is therefore not enough. It is furthermore mandatory to separate the wheat from the chaff since institutional quality requirements often put severe constraints on precision [1] as well as document-level recall. Compliance with quality standards is typically realized with semi-automatic workflows (e.g. [8]). Peeking into every document is, however, infeasible for very large datasets. Realizing compliance in a more efficient automatic manner is therefore desired, yet automatic subject indexing and multi-label text classification miss essential research in this regard. We aim to fill this gap since rapidly growing databases make integration of autonomous processes into operative IR systems indispensable. Semi-automatic tools could be utilised as a fall-back operation (cf. Fig. 1 on the right).

As depicted in Fig. 1, we consider scores at the *concept-level* (individual subject term assignments) and the *document-level* (set of subject term assignments). Most automatic indexing methods[1] provide a score for each concept [8,12,17,22], hence allowing to exclude individual concept predictions that might be incorrect (see box on the bottom left of Fig. 1). Such precision-oriented concept-level filtering removes single assignments from documents, which consequently lowers the average number of subject terms per document and typically impairs document-level recall. As exemplified in Fig. 2, subsequent assessment of document-level quality is difficult; the plain number of assigned concepts to a document is not a satisfying indicator, since human indexers use a wide range of label set sizes. In fact, uncertainty in document-level recall is an inherent and inevitable phenomenon of text classification when only a few preconditions are met, as we will outline in Sect. 3.1. For this reason, complementing concept-level confidence scores with document-level quality estimates is crucial.

In summary, the contributions of this work are the following:

- We provide a brief conceptual analysis of confidence and quality estimation for automatic subject indexing.

---

[1] For brevity, the term *subject* may be omitted in subject indexing, subject indexer, ..., respectively.

- We propose a quality estimation approach, termed *Qualle*, that combines multiple content-based features in a multi-layered regression architecture.
- We show the impact of different feature groups and the effectiveness of *Qualle* for quality estimation and filtering in an empirical study.

The empirical study is centered around the following questions:

Q1: Do predictions of recall and actual recall correlate with each other?
Q2: How accurate are the recall estimates?
Q3: Which of the feature groups contribute most to recall prediction?
Q4: What are the effects of filtering based on recall estimates on the percentage of documents passing the filter (collection coverage) as well as document-level precision?

The following sections address related work, quality estimation (analysis and our approach), experiments and conclusions.

2   Related Work

Confidence scores are an integral part of many machine learning (ML) approaches for multi-label text classification [7,17]. For instance, rule-learning typically computes a confidence score for each rule, dividing the number of times the rule correctly infers a class label by the number of times the rule matches in total. Naive-Bayes approaches use Bayes' Rule to derive conditional probabilities. Flexible techniques have been developed to perform probability calibration [23]. Thus, systems using multi-label classification (MLC) machine learning methods for subject indexing often provide confidence scores for each subject heading. Medelyan and Witten [12] used decision trees to compute confidence scores for dictionary matches. Huang et al. [8] similarly applied a learning-to-rank approach on MeSH term recommendation based on candidates from k-nearest-neighbors. In general, binary relevance (BR) approaches also provide probabilities for each concept, for instance by application of probability calibration techniques (e.g. [22]). Tang et al. [19] proposed a BR system which additionally creates a distinct model to determine the number of relevant concepts per document, calibrating label set size. In summary, the scores provided by the above mentioned systems are limited to concept-level confidence, that is, referring to individual subjects.

In the context of classifier combination, Bennett et al. [2] proposed *reliability-indicator variables* for model selection. They identified four types of indicator variables and showed their utility. In contrast to their work, we focus on different objectives. We apply such features (reliability indicators) for quality estimation, which in particular comprises estimation of recall. By contrast, *precision-constrained* situations have recently been studied by Bennett et al. [1]. *Confidence in predictions and classifiers* has recently gained attention in the context of *transparent machine learning* (e.g. [15]). Contrary to transparent machine learning, quality estimation does not aim to improve interpretability, and it thus may be realized by black box ML models. Nevertheless, quality estimates may be relevant for humans to gain trust in ML.

Confidence estimation has been studied in different *application domains*, and it has been noted that different levels of confidence scores are relevant. For instance, Culotta and McCallum [3] distinguished between field confidence and record confidence (entire record is labeled correctly) in information extraction. They compared different scoring methods and also trained a classifier that discriminates correct and incorrect instances for fields and records, respectively.

3   Quality Estimation

Our approach to quality estimation (Sect. 3.2) stems from an analysis of common practice, as described in the following (Sect. 3.1).

3.1   Analysis

In the past, quality of automatic subject indexing has been assessed in different ways that have individual drawbacks. Traditionally, library and information scientists examined indexing quality, effectiveness, and consistency [16]. Quality assessment that requires human judgements is, however, costly, which can be a severe issue on large and diverse datasets. For this reason, evaluations of automatic subject indexing often just rely on consistency with singly annotated human indexing, yielding metrics which are known as precision and recall. As described in Sect. 2, common indexing approaches provide confidence scores for each class, denoting posterior probabilities $p(y_j = 1|\mathcal{D})$, where y_j refers to a single concept of the controlled vocabulary, thus they are referred to as *concept-level confidence* in this work. Statistical associative approaches derive confidence scores based on dependencies between terms and class labels from examples. As a consequence, the performance of these methods largely depends on the availability of appropriate training examples and the stability of term and concept distributions, whereas lexical methods require vocabularies that exhaustively cover the domain. When concept drift occurs, that is, if observed terms and the set of relevant concepts differ between training data and new data, both types of indexing approaches considerably decrease in performance [20]. Interestingly, since these algorithms merely learn to assign recognized subjects of the controlled vocabulary, they will silently miss to assign relevant subjects not covered by the controlled vocabulary, and moreover they are unable to recognize and represent the loss in document-level content representation. It is further plausible that these issues are more pronounced when only titles of documents are processed, since for title-based indexing the complete subject content is compressed into only a few words which makes understanding of each single word more crucial compared to processing full texts. As the evolution of terms and concepts is an inherent property of language (cf. e.g. [18]), accurate recognition of insufficient exhaustivity is essential in the long term. It must be assumed that uncertainty in recall is an inherent and inevitable phenomenon of automatic indexing and multi-label text classification in general. For these reasons, in order to guarantee quality, indexing systems must gain knowledge relating to classifier reliability based on additional representations (cf. [2]), exploiting information such as out-of-vocabulary term occurrences and document length, just to give an example. Therefore, we propose to directly address document-level quality instead of concept-level confidence.

3.2 Qualle: Content-Based Quality Estimation

Multi-label classification methods can be tuned by regularization and configuration of thresholds to satisfy constraints on precision. Hence, the main challenge for our approach on quality estimation, *Qualle*, is to estimate document-level recall. As a solution, we propose the architecture which is exemplified in Fig. 3. It can be seen that Qualle has a multi-layered design that in particular builds upon distinct machine learning models. Learning in deep layers is used to create feature representations for top-level quality regression. The input layer (Fig. 3: top) shows a fictitious title of a document to be indexed, which is then represented by multiple features. The content is processed by a multi-label text classification method, producing a set of concepts and corresponding confidence values, e.g., $\pi(c_{29638-6}) = 0.98$.[2] Moreover, a multi-output regression module offers expectations regarding the proper number of concepts for a document (label set size calibration, in short: label calibration, LC). If possible, Qualle considers distinct semantic categories, for instance geographic names ($\hat{L}_{\text{Geo}}$) or economics subject terms ($\hat{L}_{\text{Econ}}$), commodities, and much more. In the given example, the phrase "three European countries" clearly points out that it would be reasonable to assume three geographic names when access to the full text is possible, however, without particularly specifying which ones to choose. The input is not precise enough. In Fig. 3, the LC module of the multi-layered architecture estimates $\hat{L}_{\text{Geo}} = 2.7$ correspondingly. Drawing connections between the predicted concept set L^* and the estimated numbers of concepts $\hat{L}$ can signalize recall issues. For instance, $|\hat{L}_{\text{Geo}} - L^*_{\text{Geo}}| = 2.7$ indicates that the proposed index terms probably miss more than two geographic names. Such reasoning is not covered by ordinary statistical text categorization methods. In addition, basic reliability indicators are included as features, such as content length ($\#\_\text{Char}$), individual term indicators (e.g.: $\text{TERM}_{\text{analysis}}$), the number of out-of-vocabulary terms ($\#\_\text{W\_OOV}$), or different types of aggregations (Π) of the confidence scores of the assigned concepts. Finally, quality aspects are estimated using regression models that leverage the complex features derived in the deep architecture.

Development of the feature groups (Fig. 3: V, C, Π, LC) was driven by conceptual considerations. In particular, we wanted the features to represent: *imprecise input* (e.g., "three European countries": inherent ambiguity), *lack of input information* (e.g., title with fewer than 4 words: information is scarce), as well as *lack of knowledge* (e.g., "On Expected Effects of Brexit on European law":

---

[2] The concept identifier 29638-6 refers to the concept "Low-interest-rate policy".

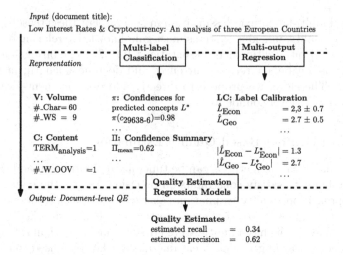

Fig. 3. Multi-layered regression architecture for quality estimation (example).

information is present but can not be interpreted, if the term "Brexit" has not been observed before).

In general, the architecture of Qualle is a framework which, for example, allows to apply arbitrary regression methods for quality estimation. Since the number of completely correct records in automatic subject indexing is extremely low, we do not consider re-ranking by MaxEnt, which has been investigated for record-based confidence estimation in information extraction [3]. In this paper, we focus our analysis of Qualle on document-level recall. In addition, basic indicators have been considered for document-level precision estimation, that is, the mean (Π_{mean}), product, median and minimum of the confidence values of the assigned concepts.

More details on the implemented configurations are given in Sect. 4.1.

4 Experiments

The experimental study is centered around the four research questions (Q1-Q4) that have been announced at the end of Sect. 1. These questions are relevant in practice for different reasons. Ranking documents by document-level recall (Q1) allows to separate high-recall documents from low-recall documents. Accurate estimates (Q2) allow to control filtering with meaningful constraints. Applicability of the filtering approach would, however, be prevented if either document-level precision was decreased considerably or if the number of documents passing the filter was too low (Q4). The following paragraphs first describe the setup and then turn to the results and their discussion.

4.1 Setup

We evaluate the approach in two domains. We first perform a basic experiment on legal texts, addressing questions Q1 and Q2. Subsequently, we go into details regarding economics literature, treating questions Q1–Q4.

The adequacy of quality estimation is measured in two ways. Since perfect quality estimates follow their corresponding actual counterparts linearly, we consider the Pearson product-moment *correlation* coefficient ρ for (Q1)[3]. A strong correlation between predicted and true quality allows to order documents correctly, that is, corresponding to the true performance. ρ has been used in related studies [3]. For measuring the exactness of estimated recall values (Q2), we consider the *mean squared error* (MSE). To gain knowledge about the utility of the feature groups (Q3), we perform a systematic analysis of different configurations. Feature groups are removed separately from the complete set of features (ablation study), and measurements are also collected for each feature group alone (isolation study). Question Q4 was addressed by evaluating different thresholds on estimated recall and measuring average true precision and recall over the corresponding selected documents. In addition, the $coverage = \frac{|\{\mathcal{D}_{\text{selected}}\}|}{N}$ was measured, with N being the total number of documents and $\mathcal{D}_{\text{selected}}$ the selected subset of the whole dataset. We also report the relative recall *gain* (RG) on theses subsets. The accuracy of initial multi-label classification is reported briefly for comparability, using metrics as described in Sect. 3.

Regarding law, we employ EURLEX [10] to address Q1 and Q2. EURLEX has 19,314 documents, each having 5.31 EUROVOC[4] subject terms on average. For further details, refer to [10] and the website of the dataset[5]. Please note that our experiments only use the titles rather than the full text of the documents and that different train/test splits were used. Regarding economics, we use three datasets, which comprise roughly 20,000 (T20k), 60,000 (T60k), and 400,000 documents (T400k), respectively. Each document is associated with several descriptors (e.g., 5.89 on average for T400k) from the STW Thesaurus for Economics (STW)[6]. Both, the STW and EUROVOC, comprise thousands of concepts, yielding challenging multi-label classification tasks.

For each dataset, we perform cross validation with 5-folds. And for each of those 5 runs, we apply nested cross validation runs, likewise with 5 folds used for parameter optimization and learning of quality estimation, that is, each training set is subdivided into *dev-train* and *dev-test* splits. For validation, a new model is trained from random samples of the same size as one of the dev-train splits. Consequently, training and prediction of MLC as well as LC are carried out $5 \cdot (5 + 1) = 30$ times for each dataset. Quality estimation is evaluated on the corresponding *eval-test* data folds.

For *multi-label text classification*, we chose binary relevance logistic regression (BRLR) optimized with stochastic gradient descent (cf. [1, 22]).

---

[3] If only ranking is relevant, rank-based correlation coefficients should be considered.
[4] http://eurovoc.europa.eu/, accessed: 31.12.2017.
[5] http://www.ke.tu-darmstadt.de/resources/eurlex, accessed 31.12.2017.
[6] http://zbw.eu/stw/version/latest/about.en.html, accessed: 09.01.2018.

Regarding *reliability indicator variables,* the EURLEX study relies on just two features: the estimated number of concepts for the document, and the difference to the number of actually predicted concepts for the document by BRLR. For the detailed study on economics documents, all feature groups were employed (Sect. 3.2). Label calibration has been realized with tree-based methods (EURLEX: ExtraTreesRegressor [6], Economics: GradientBoostingRegressor [5]). Only the total number of concepts per document is considered for EURLEX. The economics experiments compute label calibration estimates for the seven top categories of the STW. For EURLEX and economics, $\#$_Char, $\#$_WS and $TERM_i$ have been used as features for label calibration.

Several regression methods implemented in scikit-learn [14] were considered for recall regression. For EURLEX, rather basic models like LinearRegression and DecisionTreeRegression are tested, as well as ensemble machine learning methods, namely, ExtraTrees [6], GradientBoosting [5], and AdaBoostRegressor [4]. Regarding the more detailed experiments on economics, only the two regression methods that performed best on EURLEX were investigated. Extensive grid searches over the configuration parameters are left for future work.

4.2   Results

EURLEX. From the different regression models, LinearRegression produced the lowest correlation coefficient ($\rho = .214 \pm .026$) between predicted recall and true recall. AdaBoostRegressor reached the highest correlation coefficient ($\rho = .590 \pm .013$) and the lowest mean squared error (MSE $= 0.067 \pm 0.002$). Only AdaBoostRegressor and GradientBoostingRegressor achieved correlation coefficients greater than .500. Although being worse than the AdaBoostRegressor on average, the results for the ExtraTreesRegressor were more balanced.

Economics. Comparing the two selected regression methods, we found that the best configurations of GradientBoosting dominated the best configurations of AdaBoost on all datasets and with respect to both metrics (ρ, MSE). Thus, Adaboost has been excluded from further analysis.

Table 1 offers the numbers for ablation and isolation of feature groups. For each collection, the complete set of features (first row corresponding to each collection) is always among the top configurations, where differences are not greater than the sum of their standard deviations. For all collections, the largest decrease in performance is recognized when the group of features related to label calibration is removed. In accordance, this feature group yields the strongest individual results, and it can achieve performance close to the complete set of features on some configurations (T400k). Volume features, including length of the document, was found to be the lowest ranking group and has little impact when removed from the complete set of features. In nearly all cases of configurations, more data yields higher correlation coefficients, however, not necessarily lower mean squared error. In the following, we focus on reporting results regarding T400k. Figures for T20k and T60k were similar.

Table 1. Feature analysis for economics with GradientBoosting. ✓: presence of feature group. Δ: Difference in relation to complete set of features. †: Absolute difference to condition with all features is greater than the sum of their *sd*.

Configuration	V	C	LC	Π	$\rho\pm$ std	Δ_ρ	MSE $\pm$ std	Δ_{MSE}
T20k	✓	✓	✓	✓	0.597±0.014	-0.0%	0.039±0.001	-0.0%
T20k (ablation)		✓	✓	✓	0.596±0.014	-0.2%	0.040±0.001	0.2%
T20k (ablation)	✓		✓	✓	0.595±0.015	-0.3%	0.039±0.001	-0.6%
T20k (ablation)	✓	✓	✓		0.583±0.015	-2.3%	0.040±0.001	1.8%
T20k (ablation)	✓	✓		✓	0.384±0.005	-35.6%†	0.050±0.001	26.5%†
T20k (isolation)			✓		0.569±0.014	-4.7%†	0.041±0.001	2.6%
T20k (isolation)				✓	0.362±0.007	-39.3%†	0.051±0.001	28.0%†
T20k (isolation)		✓			0.196±0.013	-67.1%†	0.056±0.001	41.1%†
T20k (isolation)	✓				0.128±0.008	-78.6%†	0.056±0.001	43.0%†
T60k	✓	✓	✓	✓	0.617±0.011	-0.0%	0.043±0.000	-0.0%
T60k (ablation)		✓	✓	✓	0.615±0.010	-0.3%	0.044±0.000	0.3%
T60k (ablation)	✓		✓	✓	0.602±0.009	-2.5%	0.044±0.001	1.8%
T60k (ablation)	✓	✓	✓		0.600±0.010	-2.8%	0.044±0.000	2.4%†
T60k (ablation)	✓	✓		✓	0.420±0.009	-31.9%†	0.055±0.001	26.1%†
T60k (isolation)			✓		0.574±0.005	-6.9%†	0.046±0.001	5.4%†
T60k (isolation)				✓	0.391±0.011	-36.6%†	0.056±0.001	28.7%†
T60k (isolation)		✓			0.216±0.017	-64.9%†	0.062±0.001	43.9%†
T60k (isolation)	✓				0.069±0.009	-88.8%†	0.064±0.001	48.2%†
T400k	✓	✓	✓	✓	0.648±0.002	-0.0%	0.042±0.000	-0.0%
T400k (ablation)	✓	✓	✓		0.649±0.001	0.1%	0.042±0.000	-0.1%
T400k (ablation)		✓	✓	✓	0.648±0.001	0.0%	0.042±0.000	0.2%
T400k (ablation)	✓		✓	✓	0.644±0.002	-0.6%†	0.042±0.000	0.8%†
T400k (ablation)	✓	✓		✓	0.528±0.002	-18.5%†	0.050±0.000	19.5%†
T400k (isolation)			✓		0.640±0.001	-1.3%†	0.043±0.000	1.1%†
T400k (isolation)				✓	0.511±0.002	-21.2%†	0.051±0.000	21.6%†
T400k (isolation)		✓			0.225±0.003	-65.3%†	0.064±0.000	51.6%†
T400k (isolation)	✓				0.122±0.002	-81.1%†	0.065±0.000	55.3%†

Figure 4(a) depicts recall estimation results for T400k. The plot illustrates the degree of linear relation and also reveals the distributions of estimated and true recall values. Most of the documents have a true recall that is less than 60%. Regarding the scoring functions for document-level precision, the product of concept-level confidence scores exhibited the highest correlations for T20k and T60k, however, still staying below .500. On T400k, all scoring functions were very close to each other, and their correlation coefficients were above .500. Figure 4(b) depicts results for the product of concept confidence values.

Figure 5 finally reveals key findings of our study, that is, how different thresholds on estimated recall affect properties of the resulting document selections.

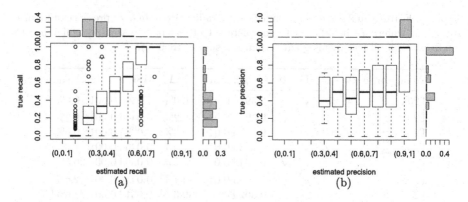

Fig. 4. Quality estimates and true values (Economics: T400k). (a) Recall estimation by Qualle, (b) Precision score by product of concept confidence values. Marginal distributions (bin count/total count) are shown on the top and on the right, respectively.

The plot shows coverage, as well as mean document-level true recall and true precision. For instance, constraining estimated recall to be at least 30% achieves a gain RG=44% of true recall on roughly half of T400k. Figure 5 confirms that putting constraints on estimated recall leads to real improvements regarding document-level recall. Notably, precision on the selected subsets remained the same or even increased when putting harder constraints on estimated recall. Coverage gradually falls when the threshold is raised.

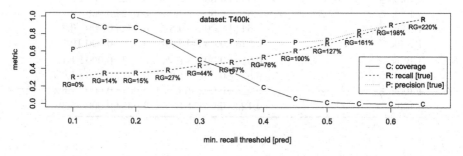

Fig. 5. Quality vs. coverage (collection: T400k): coverage, mean document-level true recall and true precision for different predicted recall thresholds. RG: relative gain in document-level recall on selected subset compared to the full dataset.

Multi-label Classification. BRLR performed comparable to related studies, for instance, with sample-based avg. $f_1 = 0.361$, precision $= 0.528$, recall $= 0.327$ on T20k, and $f_1 = 49.1\%$ (micro avg.) on EURLEX.

4.3 Discussion

The basic set of features used on EURLEX reached respectable correlations ($\rho > .500$) between predicted and true recall (Q1) only for the sophisticated machine learning methods AdaBoost and GradientBoosting. Differences in the balance of predictions should be considered for applications, just like the notable amount of variance that remains around the predicted values (Q2). In summary, the EURLEX study already shows the feasibility of the approach, in accordance with the economics experiments which provide deeper insights.

The outcomes on the economics datasets, especially Fig. 5, show that the proposed quality estimation approach can successfully identify subsets of document collections where soft constraints on precision as well as recall are met (Q4). Finally, it remains a decision depending on the application context to make trade-offs according to multi-criteria objectives, which notably comprise coverage. Regarding recall, the ranking and accuracy of predictions are sufficient enough (Q1, Q2). Interestingly, precision was not affected negatively (cf. Fig. 5). Based on Table 1, applications should consider the full set of features, which belongs to the top performing configurations in all cases and outperformed individual feature groups. Label calibration information is found to be a strong individual predictor. It is the most relevant reliability indicator (Q3) compared to the volume, content, and concept-confidence related feature groups. The mean squared errors of predictions indicate that considerable vagueness remains (Q2). Possibly, it may be caused by the errors in concept assignments, which influence the label calibration related features.

Our results highlight the inevitable difficulties (cf. Sect. 3.1) in multi-label text classification, namely, suffering from low document-level recall when the model misses knowledge (either dictionary entries or training examples), or when the observed input is inherently ambiguous. Quality estimation enables to handle such issues by controlling, that is, making trade-offs between quality and coverage. Since the proposed approach is not bound to specific MLC or regression methods, further progress in this regard can be integrated and is assumed to improve coverage. Another direction for future work is to consider alternative quality metrics that take semantic relations into account (see e.g., [11,13]).

5 Conclusion

In order to assure data quality in operative information retrieval systems with large and diverse datasets, we investigated an important yet less addressed research topic, namely quality estimation of automatic subject indexing with a focus on the document level. Our experimental results on two domains spanning over collections of different sizes show that the proposed multi-layered architecture is effective and thus enables quality control in settings where high standards have to be met. The approach allows to define different thresholds, which resulted in considerable gains of document-level recall, while upholding precision at the same time. Label calibration was the most relevant reliability indicator.

References

1. Bennett, P.N., Chickering, D.M., Meek, C., Zhu, X.: Algorithms for active classifier selection: maximizing recall with precision constraints. In: Proceedings of WSDM 2017, pp. 711–719. ACM (2017)
2. Bennett, P.N., Dumais, S.T., Horvitz, E.: Probabilistic combination of text classifiers using reliability indicators: models and results. In: Proceedings of SIGIR 2002, pp. 207–214. ACM (2002). https://doi.org/10.1145/564376.564413
3. Culotta, A., McCallum, A.: Confidence estimation for information extraction. In: Proceedings of HLT-NAACL 2004: Short Papers, pp. 109–112. ACL (2004)
4. Drucker, H.: Improving regressors using boosting techniques. In: Proceedings of ICML 1997, pp. 107–115. Morgan Kaufmann (1997)
5. Friedman, J.H.: Stochastic gradient boosting. Comput. Stat. Data Anal. **38**(4), 367–378 (2002). https://doi.org/10.1016/s0167-9473(01)00065-2
6. Geurts, P., Ernst, D., Wehenkel, L.: Extremely randomized trees. Mach. Learn. **63**(1), 3–42 (2006). https://doi.org/10.1007/s10994-006-6226-1
7. Gibaja, E., Ventura, S.: A tutorial on multilabel learning. ACM Comput. Surv. **47**(3), 52:1–52:38 (2015). https://doi.org/10.1145/2716262
8. Huang, M., Névéol, A., Lu, Z.: Recommending MeSH terms for annotating biomedical articles. JAMIA **18**(5), 660–667 (2011)
9. Liu, J., Chang, W., Wu, Y., Yang, Y.: Deep learning for extreme multi-label text classification. In: Proceedings of SIGIR 2017, pp. 115–124. ACM (2017)
10. Loza Mencía, E., Fürnkranz, J.: Efficient multilabel classification algorithms for large-scale problems in the legal domain. In: Francesconi, E., Montemagni, S., Peters, W., Tiscornia, D. (eds.) Semantic Processing of Legal Texts. LNCS (LNAI), vol. 6036, pp. 192–215. Springer, Heidelberg (2010). https://doi.org/10.1007/978-3-642-12837-0_11
11. Medelyan, O., Witten, I.H.: Measuring inter-indexer consistency using a thesaurus. In: Proceedings of JCDL 2006, pp. 274–275. ACM (2006)
12. Medelyan, O., Witten, I.H.: Domain-independent automatic keyphrase indexing with small training sets. JASIST **59**(7), 1026–1040 (2008)
13. Neveol, A., Zeng, K., Bodenreider, O.: Besides precision & recall: exploring alternative approaches to evaluating an automatic indexing tool for MEDLINE. In: AMIA Annual Symposium Proceedings, pp. 589–593 (2006)
14. Pedregosa, F., et al.: Scikit-learn: machine learning in python. J. Mach. Learn. Res. **12**, 2825–2830 (2011)
15. Ribeiro, M.T., Singh, S., Guestrin, C.: Why should I trust you?: explaining the predictions of any classifier. In: Proceedings of SIGKDD 2016, pp. 1135–1144. ACM (2016)
16. Rolling, L.N.: Indexing consistency, quality and efficiency. Inf. Process. Manag. **17**(2), 69–76 (1981)
17. Sebastiani, F.: Machine learning in automated text categorization. ACM Comput. Surv. **34**(1), 1–47 (2002)
18. Tahmasebi, N., Risse, T.: On the uses of word sense change for research in the digital humanities. In: Kamps, J., Tsakonas, G., Manolopoulos, Y., Iliadis, L., Karydis, I. (eds.) TPDL 2017. LNCS, vol. 10450, pp. 246–257. Springer, Cham (2017). https://doi.org/10.1007/978-3-319-67008-9_20
19. Tang, L., Rajan, S., Narayanan, V.K.: Large scale multi-label classification via metalabeler. In: Proceedings of WWW 2009, pp. 211–220. ACM (2009)

20. Toepfer, M., Seifert, C.: Descriptor-invariant fusion architectures for automatic subject indexing. In: Proceedings of JCDL 2017, pp. 31–40. IEEE Computer Society (2017)
21. Trieschnigg, D., Pezik, P., Lee, V., de Jong, F., Kraaij, W., Rebholz-Schuhmann, D.: MeSH up: effective MeSH text classification for improved document retrieval. Bioinformatics **25**(11), 1412–1418 (2009)
22. Wilbur, W.J., Kim, W.: Stochastic gradient descent and the prediction of MeSH for PubMed records. In: AMIA Annual Symposium Proceedings 2014, pp. 1198–1207 (2014)
23. Zadrozny, B., Elkan, C.: Transforming classifier scores into accurate multiclass probability estimates. In: Proceedings of SIGKDD 2002, pp. 694–699. ACM (2002)

Metadata Synthesis and Updates on Collections Harvested Using the Open Archive Initiative Protocol for Metadata Harvesting

Sarantos Kapidakis[✉]

Laboratory on Digital Libraries and Electronic Publishing,
Department of Archive, Library and Museum Sciences,
Faculty of Information Science and Informatics, Ionian University,
72, Ioannou Theotoki Str., 49100 Corfu, Greece
sarantos@ionio.gr

Abstract. *Harvesting* tasks gather information to a central repository. We studied the metadata returned from 744179 harvesting tasks from 2120 harvesting services in 529 harvesting rounds during a period of two years. To achieve that, we initiated nearly 1,500,000 tasks, because a significant part of the Open Archive Initiative harvesting services never worked or have ceased working while many other services fail occasionally. We studied the synthesis (elements and verbosity of values) of the harvested metadata, and how it evolved over time. We found that most services utilize almost all Dublin Core elements, but there are services with minimal descriptions. Most services have very minimal updates and, overall, the harvested metadata is slowly improving over time with "description" and "relation" improving the most. Our results help us to better understand how and when the metadata are improved and have more realistic expectations about the quality of the metadata when we design harvesting or information systems that rely on them.

Keywords: Metadata · Dublin core elements · OAI-PMH · Harvesting
Reliability · Quality · Patterns · Evolution

1 Introduction and Related Work

An established protocol for exchange of metadata is the Open Archive Initiative Protocol for Metadata Harvesting[1] (OAI-PMH). All metadata providers act as OAI-PMH servers that accept requests to provide their metadata. A central node acts as OAI-PMH client that issues requests to many OAI-PMH servers, and is using the collected metadata records to construct a central repository, which will be preferred for searching. The central node will also regularly update its metadata records with new and changed ones and therefore the OAI-PMH communication should be repeated regularly, with a new task each time.

Metadata harvesting is used very often, to incorporate the resources of small or big providers to large collections. The metadata harvesters, like Science Digital Library and

---

[1] http://www.openarchives.org/OAI/2.0/openarchivesprotocol.htm.

© Springer Nature Switzerland AG 2018
E. Méndez et al. (Eds.): TPDL 2018, LNCS 11057, pp. 16–31, 2018.
https://doi.org/10.1007/978-3-030-00066-0_2

Europeana, accumulate metadata from many collections (or sources) through the appropriate services, belonging to metadata providers mostly memory institutions, by automatically contacting their services and storing the retrieved metadata locally. Their goal is to enable searching on the huge quantity of heterogeneous content, using only their locally store content. Metadata harvesting is very common nowadays and is based on the Open Archives Initiative Protocol for Metadata Harvesting. As examples, we mention the Directory of Open Access Repositories (OpenDOAR) that provides an authoritative directory of academic open access repositories, and the OAIster database of OCLC with millions of digital resources from thousands of providers.

In [13] Lagoze et al. discuss the National Science Digital Library development and explains why OAI-PMH based systems are not relatively easy to automate and administer with low people cost, as one would expect from the simplicity of the technology. It is interesting to investigate the deficiencies of the procedure. In [17] Ward analyses the provenance and the record distribution of 100 Data Providers registered with the Open Archives Initiative, of which 18 provided no data and 5 others were responding only sometimes. He also examines how the Dublin Core elements are used and shows that Dublin Core is not used to its fullest extent. We will present some of his results on Table 2, side by side with our similar results.

Additional quality approaches are applied on OAI-PMH aggregated metadata: Bui and Park in [2] provide quality assessments for the National Science Digital Library metadata repository, studying the uneven distribution of the one million records and the number of occurrences of each Dublin Core element in these. Another approach to metadata quality evaluation is applied to the open language archives community (OLAC) in [5] by Hughes that is using many OLAC controlled vocabularies. Ochoa and Duval in [15] perform automatic evaluation of metadata quality in digital repositories for the ARIADNE project, using humans to review the quality metric for the metadata that was based on textual information content metric values.

Other metadata quality approaches are based on metadata errors and validation, e.g. appropriate type of values and obligatory values. Hillmann and Phipps in [4] are additionally using statistics on the number of use of metadata elements in their application profiles, based on Dublin Core. For Beall in [1] the metadata quality deals with errors in metadata, proposing a taxonomy of data quality errors, (such as typographical, conversion, etc.), a statistical analysis on the types of errors and a strategy for error management.

The quality of the content, is important to the successful use of the content and the satisfaction from the service. The evaluation and quality of metadata is examined as one dimension of the digital library evaluation frameworks and systems in the related literature, like [3, 14, 16, 18]. Fuhr et al. in [3] and Vullo et al. in [16] propose a quality framework for digital libraries that deal with quality parameters, but do not provide any practical way to measure it, or to depict it to any quantitative metric, that can be used in decision making. In [6] Kapidakis presents quality metrics and a quality measurement tool, and applied them to compare the quality in Europeana and other collections, that are using the OAI-PMH protocol to aggregate metadata. It can also be applied here to estimate the quality of our examined services.

National or large established institutions consistently try to offer their metadata and data reliably and current and to keep the quality of their services as high as possible, but local and smaller institutions often do not have the necessary resources for consistent quality services – sometimes not even for creating metadata, or for digitizing their objects. In small institutions, the reliability and quality issues are more prominent, and decisions often should also take the quality of the services under consideration.

In order to study the metadata we first have to collect them from the systems that carry them. The big diversity of computer services, their different requirements, designs and interfaces and also network problems and user-side malfunctions make it hard to reliably contact a service. The regular tasks that request metadata records run unattended, and the system administrators assume they are successful most of the time. If a small number of harvesting tasks fail occasionally, probably due to temporary network errors, they only affect the central node temporarily. The harvesting mechanism is normally established and then scheduled to run forever, but it is observed that after some time a significant part of these services stopped working permanently.

The reliability of the services is important for ensuring current information, and an obstacle to measuring metadata quality. When the resources are not available, the corresponding user requests are not satisfied, affecting the quality of the service. In [7] Kapidakis further studies the responsiveness of the same OAI-PMH services, and the evolution of the metadata quality over 3 harvesting rounds between 2011 and 2013. In [8, 9] Kapidakis examines how solid the metadata harvesting procedure is, by exploiting harvesting results to conclude on the availability of their metadata, and how it evolves over these harvesting rounds. The unavailability of the metadata may be either OAI-PMH specific or due to the networking environment. In [10] Kapidakis examined when a harvesting task, which includes many stages of information exchange and any one of them may fail, is considered successful. He found that the service failures are quite a lot, and many unexpected situations occur. The list of working services was decreasing every month almost constantly, and less than half of the initial services continued working at the end. The remaining services seem to work without human supervision and any problems are difficult to be detected or corrected. Since almost half of the harvesting tasks fail or return unusual results, contacting humans to get more information about these service cannot be applied globally.

In [11, 12] Kapidakis examined the operation and the outcome messages of information services and tried to find clues that will help predicting the consistency of the behavior. He studied the different ways of successful and failed task termination, the reported messages and the information they provide for the success of the current or the future harvesting tasks, the number of records returned, and the required response time and tried to discover relations among them. To do that, he gathered a lot of information by performing a large number of harvesting tasks and rounds and examined in detail the harvesting failures from their warning messages. In this work we concentrate on the successful harvesting tasks and examine the returned records – we do not examine on the outcome messages or the errors of the harvesting tasks.

Some of the problems of OAI-PMH will probably be improved by using its successor protocol, ResourceSync[2], but there is not yet a large enough installation base, to

---

[2] http://www.openarchives.org/rs/1.1/resourcesync.

make a fare comparison. Even then, issues related to the content of the metadata or to server unavailability will not be improved.

The rest of the paper is organized as follows: In Sect. 2 we describe our methodology and how we selected our services and used the software we made to create our dataset, and we examine general characteristics of the harvested metadata, to be aware to the differences among different services. In Sect. 3 we study the synthesis of the metadata that the harvesting tasks returned for the different services, giving attention to the 15 Dublin Core elements. In Sect. 4 we study the differences of the metadata among different harvesting rounds, to see the long term effect of the updates to the metadata. In Sect. 5 we measure the increase in words of the individual Dublin Core elements and on Sect. 6 we conclude. In all cases, we do not present our full data, but we emphasize the most interesting observations.

2 Collection of the Information

Good quality metadata are important for many operations, such as for searching a collection, for presenting records or the search results, for searching across collections, for identifying the records and even for detecting duplicate record s and merging them. The metadata schema in use also affects the detail of the description, and the quality of the described metadata. In this work we will only concentrate on computer services that use the OAI-PMH and we study the harvested metadata in Dublin Core.

A big challenge was to harvest the metadata needed for such a study. They have to be collected over a long period of time, and they constitute a huge volume. Additionally, the sources, the data and the involved procedures change over time, creating additional challenges.

To harvest and study the provided metadata, we created an OAI-PMH client using the oaipy library and used it over several harvesting rounds, where on each one we asked each service from a list of OAI-PMH services for a similar task: to provide 1000 (valid – non deleted) metadata records. Such tasks are common for the OAI-PMH services, which periodically satisfy harvesting requests for the new or updated records, and involve the exchange of many OAI-PMH requests and responses, starting from the a negotiation phase for the supported OAI-PMH features of the two sides.

The sources listed in the official OAI-PMH Registered Data Providers[3] site was used as the list of services that we used. We used the initial list with 2138 entries, as on January of 2016, for our first 201 rounds, and then an updated list with 2870 entries for the next 328 rounds. The two lists of services had 1562 entries in common, while the first list has 576 entries that were seized later on, when 1308 new entries were added. As [8–10] has shown, an noticeable number of services stop working every month, thus an regular update on the list of services should be in order.

Our sequential execution of all these record harvesting tasks from the corresponding specific services normally takes much more than 24 h to complete. Sometimes the tasks time out resulting to abnormal termination of the task: we set a timeout

---

[3] https://www.openarchives.org/Register/BrowseSites.

deadline to 1 h for each task, and interrupted any incomplete task just afterwards, so that it will not last forever.

We repeat a new harvesting round with a task for each service in constant intervals, asking the exact same requests every 36 h for a period of 9 months for the first 201 rounds, and every 24 h for a period of 6 months for the remaining 215 rounds, initially, and for 113 more rounds, after some technical updates. The rounds are close enough to each other to avoid significant changes on the services. We selected this interval so that we do not overload the services and our client machine.

Ideally, a task will complete normally, returning all requested metadata records – 1000, or all available records if fewer are only available on that service. Other behaviors are also possible - and of interest when studying the behavior of each service. A task may return less records, or even 0. A harvesting task includes many stages of information exchange, and each one of them may fail – but with different consequences each time. Additionally, a task may not declare a normal completion, but report a warning message indicating a problem, with some supplemental information detail. These two situations are not mutually exclusive: a task may declare normal completion and return no records, or a task may report a warning message and still return records – sometimes even 1000 records!

According to [10], when we had to briefly characterise the complex procedures of a task as successful or not, we will call it successful if it returned any metadata records. We can then process and study these metadata. We still consider responses with fewer than the requested records as successful, as we can still process their returned metadata.

Overall, the amount of information collected from these services was huge, therefore difficult to store, process and manage, especially when software updates adopted new data output format. Most of our collected evidence cannot fit in this space, therefore only its essential details are described below.

A significant part of the conducted Open Archive Initiative harvesting services never responded or have ceased working while many other services fail occasionally. As a result, 756028 tasks, only about half of our initiated harvesting tasks, succeeded by returning any records, and we could only process them. We also decided to only consider the 744179 successful task for the services that resulted to at least 100 successful tasks in all our harvesting rounds, to increase the consistency of our data. This resulted to tasks from only 2120 from the conducted OAI-PMH services.

The examined services had on the average 351 successful tasks each (out of the 529 rounds) and provided on the average 555.2 metadata records (out of the 1000 asked during each task). Each harvesting task always asks for the first 1000 records of the service. Nevertheless, the returned records are not the same every time: some records may have changed, replaced or deleted. Among the first 1000 valid records we also get references (but not enough metadata) from a number of invalid (usually replaced/deleted) records, which we ignore. On the average we get information about 62.74 such invalid records, while the maximum we got was 10117 invalid records in a harvesting task.

The metadata records that we study are the ones that the information sources provide in Dublin Core, even though they may contain even richer metadata, for local use. When aggregating data through OAI-PMH, the native metadata scheme is

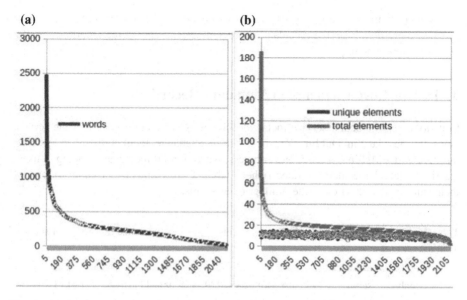

Fig. 1. (a) The number of words in the description, for every service. (b) The number of unique and repeated Dublin Core elements per average record, for every service.

normally not visible, and only the resulted aggregated metadata is available to everyone, and this metadata schema (usually in Dublin Core) has to be considered.

Each service provided records with its own synthesis, which is more appropriate for the described objects. More detailed descriptions, with more elements and lengthier values, are normally preferred, but require mode effort to create them. Figure 1 demonstrates some aspects of the different metadata synthesis of the services. Figure 1 (a) depicts the average length of the record description, i.e. the number of words per record, for each service. This can be considered as a very simple quality metric for the record of the service, and we use it as such here, while more a elaborate quality metrics can be found in [6]. We see that few collections have very high quality, and also few collections have very low quality.

Similarly, in Fig. 1(b) we can see the average number of the 15 Dublin Core elements in use and also the total number of distinct Dublin Core element declarations, for each service. Again, we can see a big variation, although the distinct Dublin Core element declarations cannot be more than 15:

One service provided records containing on average 2479 words and had on the average 187 Dublin Core element declarations. All other services provided less metadata, usually a lot less. This was more than 10 times the average service record: The harvested records had on average 10.77 out of the 15 Dublin Core elements. Some of them were repeated, and had on the average 17.89 element declarations, containing on average a total of 237 words.

We can also measure that each Dublin Core element was described on the average in 21 words while it was repeated on the average 1.66 times, among all tasks. In the service with the maximum such values, each Dublin Core element contained on the average record in 187 words, distributed into 17 distinct declarations. Later on we will examine more details on the metadata record synthesis, breaking the record to specific Dublin Core elements.

3   Dublin Core Elements in Metadata Records

In many cases the actual service metadata scheme is not Dublin Core and the service maps its metadata in Dublin Core in order to exchange them with other systems, through OAI-PMH in our case. Nevertheless, when a good metadata mapping is used, the richer actual metadata produce richer Dublin Core elements, in either extend of their values (in words) or in the number of elements.

Table 1. Number of distinct Dublin Core Elements in use and the number of services that adopt them

Number of elements	15	14	13	12	11	10	9	8	7	6	5	4	2	1
Number of services	179	668	408	281	239	176	90	42	25	6	3	1	1	1

In Table 1 we can see the number of distinct Dublin Core Elements in use in any record of the service and the number of services that adopt that many. We observe that 179 services use all 15 Dublin Core metadata elements, while the majority of the 2120 services use 14, or a few less elements. Very few services use 1–6 elements: 1 service is only using "identifier", 1 service is only using "identifier" and "title" and 1 service is only using "identifier", "title", "type" and "publisher", creating quite minimal metadata records.

At a first glance we see that most of the services try to maintain detailed metadata, that include most elements, while some of the elements may not apply for specific services.

In Table 2 we can see the 15 Dublin Core metadata elements and how often they occur in any of the harvested records. The column 2, "services", present how many of the 2120 studied services ever use the element in the first column. The most common elements, "identifier" and "title" are only absent from 4 services each – but not the same 4 services.

It was a surprise that "identifier" is not present in all services: it is useful for technical reasons (usually containing a URL) and applies to all collections and can be constructed easily. Furthermore, the services without "identifier" were not among the ones with the fewer metadata elements: 1 of them contained all other 14 elements, 1 of them contained most (10) other elements, and the other 2 contained about half the elements (7 and 8 respectively).

The element "title" usually applies to most collections. The services that did not contain "title" contain usually very few elements: Only 1 service of them contained 9 elements, while 2 services contained only 5 and the last one only one element ("identifier"). When "title" is absent, "creator" and "contributor" are absent too – but this may happen by coincidence.

Table 2. Dublin Core elements from 2120 collections: presence in services, frequency in the records and size of their values. Columns 4 and 7 are similar data in 82 services harvested by Ward in [17].

Element	Services	% 2120	% 82	Unique/avg	Total/2120	Total/82	Total/max	Words/avg	Words/max
title	2116	99.8	98.8	1.00	1.18	0.95	4.59	12.63	77.27
identifier	2116	99.8	91.5	1.00	1.86	1.42	25.98	4.73	241.05
date	2094	98.8	92.7	0.98	1.26	0.92	6.98	1.31	14.14
creator	2090	98.6	95.1	0.95	1.98	1.78	50.30	8.73	451.14
description	2076	97.9	72.0	0.84	1.12	0.51	45.40	167.44	1402.42
type	2052	96.8	87.8	0.98	1.87	0.88	6.04	2.50	14.24
publisher	2016	95.1	50.0	0.89	0.96	0.26	4.00	4.71	32.38
subject	1907	90.0	82.9	0.75	2.35	0.54	168.38	10.72	322.14
language	1836	86.6	52.4	0.93	0.97	0.16	6.03	1.06	11.48
format	1805	85.1	47.6	0.90	1.08	0.15	23.51	1.42	24.70
rights	1640	77.4	43.9	0.72	1.04	0.34	4.05	38.90	1008
relation	1579	74.5	19.5	0.80	1.73	0.05	44.85	20.55	1490.36
source	1542	72.7	36.6	0.92	2.18	0.04	6.23	14.34	316.93
contributor	1260	59.4	39.0	0.23	0.38	0.04	6.12	7.57	81.94
coverage	364	17.2	19.5	0.39	0.66	0.22	6.03	4.84	52.21
Any element				10.77	17.89	8.26	186.16	237.02	2478.57

Five other Dublin Core elements are used in over 2000 services and some others in almost 2000. The element "coverage" is the most rare in these services, been present only in 364 of them.

Table 2 also shows the average occurrences of each element in the 2120 services. Column 2 shows the number of services that each Dublin Core element is present. Column 3 shows the percentage of the 2120 services that the element is present, and column 4 shows the same percentage for the 82 services that Ward in [17] had harvested. From these two columns we can observe that the usage of most elements (except "coverage") has been increased in the current services – and in many elements this increase is really big (e.g. "relation" and "publisher").

Column 5 counts the average number of times that an element is present (any number of times) in any record, which can be at most 1. It only considers services that the element is present, and computes the ratio of the records that actually contain the element. We observe that elements "title" and "identifier" are present in all records, "date" and many others are present into most records while "contributor" is present in less than a quarter of the records.

Column 6 counts the average number of declarations of an element in any record (in the services that it is present), considering all its repetitions. Element "contributor" seems to be the element used less times – it may not be relevant anyway. Element "language" seems to be the element repeated less times, on average, because from its 0.97 declarations, span over 0.93 of the records. Element "subject" seems to have the most declarations, 2.35 per record, and considering it is present only on 0.75 of the records, it has an average of 3.13 declarations in the records it is present.

Column 7 is like column 6, but for the 82 services that Ward harvested in [17]. We observe that the records are clearly improving: the average number of elements per record more than doubled (and is now 17.89 compared to 8.26) and the average number of each individual element has increased. The increase in the four most usual elements ("title", "identifier", "date", "creator") is small, while in the less usual elements is much bigger. Therefore, the records of the newer collections/services seem to better utilise all Dublin Core elements and to have much richer descriptions.

Column 8 shows the maximum, over all services, of the number of declarations of an element on the average service record, i.e. of the average number of declarations over the tasks of the service in all harvesting rounds. To avoid non-representative service records, here we consider the average service record, the average values over all harvested records in all harvesting tasks. We observe that some tasks contain element repetitions well beyond the average task, and the elements with the most repetitions are "subject", "description" and "creator".

Columns 9 and 10 examine the length of all the record declarations for each element, in words. They both consider the average number of words over all records in all service tasks. Column 9 presents its average and column 10 its maximum value over all services. The elements with the shortest values are "language", "date" and "format". The longest values are found in "description", followed by "rights" and "relation". The maximum value can be many times higher, especially in "relation", "identifier" and "creator".

Long element values are usually considered a plus, but for some elements like "rights" they are normally just used to describe in more detail a specific licence, and are usually repeated intact in all (or many) records. Similarly, longer values in "identifier" do not actually provide more information.

The average service record has declarations for 10.77 different elements and 17.89 declarations in total, with a maximum of 186.16 declarations. The average service record contains 237 words, with a maximum of 2479 words over all services (see also Fig. 1(a) above).

Other metrics, like the standard deviation, lead to similar results: there is a huge variation of the average task record among tasks. Most services provide records with many (if not all) Dublin Core elements, and few elements with a small number of repetitions. But there are tasks with very low or high values on these metrics, resulting to very poor or very rich metadata records. The high number of element repetitions may originate from automatically mapping the records from already existing rich descriptions.

4 Updates of Metadata

Our tasks take place over many rounds, that may be performed in short intervals, but overall they expand over a significant period of time. On the previous section we used the average service record, considering all harvesting tasks, but among the harvesting rounds some records can change, although we do not expect such changes to affect very much the average record. Therefore, we will study how the tasks of the same service evolve over time, using the data from our harvesting rounds, in order to see (a) how the

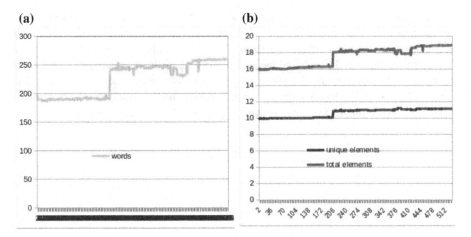

Fig. 2. (a) The number of words in the average record for every round. (b) The number of unique and repeated DC elements per average record, for every round.

updates affect the records, and if they improve them and (b) if the updates affect the records significantly, or the basic record characteristics and metrics remain mostly the same.

All records in a specific service are expected to have a similar degree of detail in the description, as they all comply with their creation policy. Still, we always ask the same records from each task of the same service, and we consider the average record. Any changes that we find during tasks in different rounds may be due to (a) maintenance updates in some of the harvested records, (b) permanent replacements of some records with others: if records are deleted they are permanently replaced by the next records and (c) by temporary communication errors that prevent some of the records to be considered on this specific harvesting round.

We use the average record per harvesting round, average over all services that return records during the round, to study the record evolution for all services during the harvesting rounds. In addition to the reasons above, a permanent change to the participation services in the harvesting rounds can change the average record we use. But this situation happens when we decide so, which will only be very few times: one.

Figure 2 depicts how some aspects of metadata evolve over time, over our 529 harvesting rounds. For each round, we can see the average length of the record content in Fig. 2(a), showing the number of words per record (considering all elements), and the average number of the 15 Dublin Core elements in use and also the total number of distinct Dublin Core element declarations (i.e. including repetitive declarations) in Fig. 2(b).

Any increase in these aspects contributes to the improvement in the quality of the records. We observe that all these aspects present a slow but steady increase over time, therefore the quality of the metadata increases. We also immediately observe a big boost that happened at the same time for all metrics: on the round that we updated the list of our services. This occurs because the newly added services that were added had better (in number of words/elements) metadata than those of the services they replaced.

We expect the record updates to also depend on the services themselves, their policy and maintenance procedures. Figure 3 presents the average increase in the record description (in words) for each service, from its first to its last harvesting round. This is not affected by changes to the included harvested services, as the harvesting tasks for each service are considered separately. The average increase over all sources is 1.98 words but with standard deviation 56.86, showing that the services really update on a different way. In fact, the update ranges from 952 words increase to 549.83 words decrease.

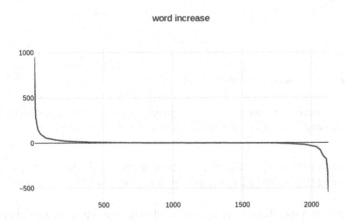

Fig. 3. The number of words in the description in the average record, for every service.

This unusual average decrease in words cannot be an intentional effort to reduce the information in the records. It mostly indicates that the average record may change a lot during the updates and even changes of that size changes may come from normal record maintenance procedures. The updates of individual services may not lead us to secure conclusions and we should consider the overall increase over all services.

Looking closer into Fig. 3 we can also see that 779 services have an average increase of more than 1 word, 541 services have an average decrease of more than 1 word and 800 services have a variation of less than 1 word. From these 800, 351 services had no word difference at all from their first to their last harvesting round: they should have no maintenance at all during this time.

Apart from that, we can see that the metadata are slowly improving on all above aspects anyway. The updates to the collections seem to improve on the length and granularity of the description.

5 Updates in Individual Elements

The updates in record metadata are not the same for all elements. In Table 3 we can see basic properties of the change for each Dublin Core element. Column 6 presents the average increase (in words) between the first and last harvesting rounds, and column 7 presents the percentage that this increase represents from its average value (in column 2).

Table 3. Statistics for the size in words of the individual Dublin Core metadata elements

Element	Average	sdev	Min	Max	Last-First	Grow%
description	164.46	10.99	146.78	178.58	21.01	12.78%
rights	39.44	2.28	30.78	43.35	−1.19	−3.02%
relation	14.2	8.04	3.83	23.45	19.62	138.17%
source	14.17	1.54	11.71	16.45	4.46	31.47%
title	12.49	0.57	11.63	13.19	1.3	10.41%
subject	10.58	0.45	9.86	11.25	1.03	9.74%
creator	8.19	1.11	6.56	9.53	2.14	26.13%
contributor	7.23	0.45	6.4	9.16	1.26	17.43%
identifier	5.25	0.48	4.48	6.08	−0.92	−17.52%
coverage	4.82	0.16	4.3	5.27	−0.13	−2.70%
publisher	4.74	0.17	4.41	4.99	0.39	8.23%
type	2.47	0.2	2.17	2.8	0.46	18.62%
format	1.49	0.12	1.37	1.68	−0.27	−18.12%
date	1.35	0.06	1.27	1.45	−0.12	−8.89%
language	1.05	0.01	1.03	1.06	0.02	1.90%

The wordiest element, "description", also has the highest increase in words. The significant increase in the values in the element "relation" agrees with the larger number of relations that are described (as seen in Table 4). The rest of the elements have a much smaller increase in description words, or even a small decrease, that can be due to random/statistical error (e.g. a service not returning records on a harvesting round, temporarily affecting the results).

Elements like "description", "source", "creator", "contributor", "title" and "subject" show a constant increase, while other elements like "language", "date", "format" and "coverage" show small (and sometimes negative) increase. This seems reasonable: most declarations of "language", "format", "type" or "date" need constant space.

The element "relation" has the highest percentage of increase in words – but this increase is not smooth and seems to vary very much, as can be observed from its standard deviation (column 3) and also Fig. 4. The sudden change and recovery to the previous level of the words of an element (e.g. "relation" and in a smaller degree to others such as "source") may be explained if a service that includes many more than average such declarations becomes unavailable for some harvesting rounds.

We can see similar increase if we consider the number of repeated DC elements, that represent the increase in the number of element declarations, and also if we consider the number of unique DC elements, that represent the enrichment of the records with new DC elements.

In Table 4 we can see how the element declarations are growing. For example, the element "description" has on average 1.12 declarations on each record (column 2), with standard deviation 0.05 (column 3). The maximum across all services average number of declarations is 1.18 (column 4). The average increase of declarations, from the first to the last harvesting task, is 0.13 (column 5) which corresponds to an 11.61% (column

6) increase in declarations per record. The element "description" is also present on 82% of the services (column 7), and during our harvesting rounds it was added on 9% of the services (column 8).

Table 4. Statistics for the number of declarations for the individual Dublin Core metadata elements.

Element	Declarations					Presence	
	Average	sdev	Max	Last-First	Grow%	Average%	Last-First %
description	1.12	0.05	1.18	0.13	11.61	82	9
rights	0.98	0.16	1.15	0.38	38.78	70	16
relation	1.47	0.37	1.89	0.94	63.95	78	13
source	2.02	0.39	2.50	1.04	51.49	90	6
title	1.17	0.02	1.20	0.04	3.42	100	0
subject	2.43	0.24	2.78	−0.52	−21.40	75	0
creator	1.95	0.09	2.09	0.23	11.79	95	3
contributor	0.43	0.06	0.52	−0.12	−27.91	26	−6
identifier	1.89	0.04	1.98	−0.06	−3.17	100	1
coverage	0.69	0.03	0.75	−0.05	−7.25	43	−6
publisher	0.96	0.02	0.98	0.03	3.13	88	5
type	1.83	0.18	2.07	0.43	23.50	98	2
format	1.09	0.01	1.13	0.03	2.75	90	6
date	1.30	0.05	1.37	−0.08	−6.15	98	1
language	0.97	0.02	1.00	0.06	6.19	93	5

We observe that the element "relation" had the highest increase in declarations (63.95%), followed by elements "source" (51.49%) and "rights" (38.78%). The element "rights" was added to most services (16%) followed by elements "relation" (13%) and "description" (9%).

Element "title" had a 3.42 increase in declarations and element "subject" had a 21.40% decrease in declarations, while the number of services they are present did not change. The element "coverage" had a decrease into both declarations and services it is present.

Elements "title" and "identifier" are present on the 100% of the services, while element "title" was added to the 1% of the services during the harvesting rounds.

Figure 4 shows the increase effect of the updates for the 15 elements round by round. The scale for some elements ("description" and "rights") is differently adjusted, so that all elements can be shown in one picture, and cross-element comparisons cannot be derived from this picture. The order of the elements in the legend match the size of their values, so that we can distinguish which colored line corresponds to which element. But the clear distinction is not necessary, as most elements have a similar distribution in the 522 rounds.

In particular we observe that elements like "description", "source", "creator", "contributor", "title" and "subject" show a constant increase, while other elements like "language", "date", "format" and "coverage" show small (and sometimes negative) increase. This seems reasonable: most declarations of "language", "format", "type" or "date" need limited space for their and cannot benefit from a more detailed description.

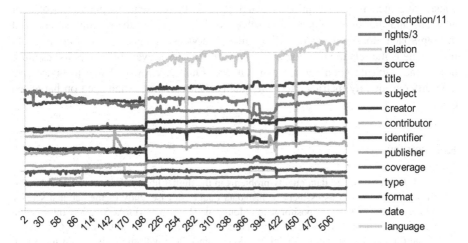

Fig. 4. The number of words in the values for each element, over the 522 harvesting rounds. The order of the elements in the legend match the size of their values. The scale for "description" and "rights" has been adjusted, so all elements can be seen together.

6   Conclusions and Future Work

When we use harvested metadata, we have no knowledge of how good they are and if they comply with any quality threshold. Our analysis can be used to investigate the quality of the participating services and as a tool to service maintainers and mainly aggregators to detect some problems in them and correct them. We cannot comment on the accuracy and correctness of the metadata descriptions, but we can examine their length, element synthesis and update patterns, and investigate the most unusual situations.

The metadata on the OAI-PMH harvested services seem quite complete, utilizing most Dublin Core elements, despite of been much different in record size. On some exceptions, commonly used elements like "identifier" or "title" were absent and the "worst" service provided only "identifier". We conclude it is possible to enforce quality requirements for the services to be in the list of sources we used. The quality requirements that will improve the searching are achieved by most services and are therefore feasible.

If we should express in one phrase the observations we made, we would say that the metadata improves in a slow rate over time, with "description" been improved the most and "relation" following. The improvement over time is not the same for all elements and many elements get no measurable improvement. But over the years (compared to the data from 2003), the improvement is significant.

In the future, we could examine into more detail specific Dublin Core elements that have mostly free language values distinguishing them from the elements containing URLs or controlled values. Additionally, someone could try to explain the unusual findings and/or no responding services by examining the individual services, either by considering their responses in more detail, such as the contents of the values of their elements and the digital objects themselves, or by contacting their administrator. But this approach does not scale well to many services. Another interesting issue to investigate is the relation of good metadata with the server availability and its response time – that indicate good service maintenance. Finally, we would like to investigate how different quality metrics affect the quality measurement, when used on harvested data.

References

1. Beall, J.: Metadata and data quality problems in the digital library. J. Digit. Inf. **6**(3) (2005)
2. Bui, Y., Park, J.: An assessment of metadata quality: a case study of the national science digital library metadata repository. In: Moukdad, H. (ed.) CAIS/ACSI 2006 Information Science Revisited: Approaches to Innovation. Proceedings of the 2005 Annual Conference of the Canadian Association for Information Science held with the Congress of the Social Sciences and Humanities of Canada at York University, Toronto, Ontario (2005)
3. Fuhr, N.: Evaluation of digital libraries. Int. J. Digit. Libr. **8**(1), 21–38 (2007)
4. Hillmann, D.I., Phipps, J.: Application profiles: exposing and enforcing metadata quality. In: Proceedings of the International Conference on Dublin Core and Metadata Applications, Singapore (2007)
5. Hughes, B.: Metadata quality evaluation: experience from the open language archives community. In: Chen, Z., Chen, H., Miao, Q., Fu, Y., Fox, E., Lim, E.-p. (eds.) ICADL 2004. LNCS, vol. 3334, pp. 320–329. Springer, Heidelberg (2004). https://doi.org/10.1007/978-3-540-30544-6_34
6. Kapidakis, S.: Comparing metadata quality in the europeana context. In: Proceedings of the 5th ACM International Conference on PErvasive Technologies Related to Assistive Environments (PETRA 2012), Heraklion, Greece, 6–8 June 2012, ACM International Conference Proceeding Series, vol. 661 (2012)
7. Kapidakis, S.: Rating quality in metadata harvesting. In: Proceedings of the 8th ACM international conference on PErvasive Technologies Related to Assistive Environments (PETRA 2015), Corfu, Greece, 1–3 July 2015, ACM International Conference Proceeding Series (2015). ISBN 978-1-4503-3452-5
8. Kapidakis, S.: Exploring metadata providers reliability and update behavior. In: Fuhr, N., Kovács, L., Risse, T., Nejdl, W. (eds.) TPDL 2016. LNCS, vol. 9819, pp. 417–425. Springer, Cham (2016). https://doi.org/10.1007/978-3-319-43997-6_36
9. Kapidakis, S.: Exploring the consistent behavior of information services. In: 20th International Conference on Circuits, Systems, Communications and Computers (CSCC 2016), Corfu, 14–17 July 2016 (2016)
10. Kapidakis, S.: When a metadata provider task is successful. In: Kamps, J., Tsakonas, G., Manolopoulos, Y., Iliadis, L., Karydis, I. (eds.) TPDL 2017. LNCS, vol. 10450, pp. 544–552. Springer, Cham (2017). https://doi.org/10.1007/978-3-319-67008-9_44

11. Kapidakis, S.: Unexpected errors from metadata OAI-PMH providers. In: 10th Qualitative and Quantitative Methods in Libraries International Conference, QQML 2018, Chania, Greece, 22–25 May 2018 (2018)

12. Kapidakis, S.: Error analysis on harvesting data over the internet. In: 11th ACM International Conference on PErvasive Technologies Related to Assistive Environments, PETRA 2018, Corfu, 26–29 June 2018 (2018)

13. Lagoze, C., Krafft, D., Cornwell, T., Dushay, N., Eckstrom, D., Saylor, J.: Metadata aggregation and "automated digital libraries": a retrospective on the NSDL experience. In: Proceedings of the 6th ACM/IEEE-CS Joint Conference on Digital libraries (JCDL 2006), pp. 230–239 (2006)

14. Moreira, B.L., Goncalves, M.A., Laender, A.H.F., Fox, E.A.: Automatic evaluation of digital libraries with 5SQual. J. Informetr. **3**(2), 102–123 (2009)

15. Ochoa, X., Duval, E.: Automatic evaluation of metadata quality in digital repositories. Int. J. Digit. Libr. **10**(2/3), 67–91 (2009)

16. Vullo, G., et al.: Quality interoperability within digital libraries: the DL.org perspective. In: 2nd DL.org Workshop in Conjunction with ECDL 2010, 9–10 September 2010, Glasgow, UK (2010)

17. Ward, J.: A quantitative analysis of unqualified Dublin core metadata element set usage within data providers registered with the open archives initiative. In: Proceedings of the 3rd ACM/IEEE-CS Joint Conference on Digital Libraries (JCDL 2003), pp. 315–317 (2003). ISBN 0-7695-1939-3

18. Zhang, Y.: Developing a holistic model for digital library evaluation. J. Am. Soc. Inf. Sci. Technol. **61**(1), 88–110 (2010)

Metadata Enrichment
of Multi-disciplinary Digital Library:
A Semantic-Based Approach

Hussein T. Al-Natsheh[1,2,3]($\boxtimes$) (iD), Lucie Martinet[2,4] (iD), Fabrice Muhlenbach[1,5] (iD),
Fabien Rico[1,6] (iD), and Djamel Abdelkader Zighed[1,2] (iD)

[1] Université de Lyon, Lyon, France
[2] Lyon 2, ERIC EA 3083, 5 Avenue Pierre Mendès France, 69676 Bron Cedex, France
[3] CNRS, MSH-LSE USR 2005, 14 Avenue Berthelot, 69363 Lyon Cedex 07, France
`hussein.al-natsheh@cnrs.fr`
[4] CESI EXIA/LINEACT, 19 Avenue Guy de Collongue, 69130 Écully, France
[5] UJM-Saint-Etienne, CNRS, Lab. Hubert Curien UMR 5516,
42023 Saint Etienne, France
[6] Lyon 1, ERIC EA 3083, 5 Avenue Pierre Mendès France, 69676 Bron Cedex, France

Abstract. In the scientific digital libraries, some papers from different research communities can be described by community-dependent keywords even if they share a semantically similar topic. Articles that are not tagged with enough keyword variations are poorly indexed in any information retrieval system which limits potentially fruitful exchanges between scientific disciplines. In this paper, we introduce a novel experimentally designed pipeline for multi-label semantic-based tagging developed for open-access metadata digital libraries. The approach starts by learning from a standard scientific categorization and a sample of topic tagged articles to find semantically relevant articles and enrich its metadata accordingly. Our proposed pipeline aims to enable researchers reaching articles from various disciplines that tend to use different terminologies. It allows retrieving semantically relevant articles given a limited known variation of search terms. In addition to achieving an accuracy that is higher than an expanded query based method using a topic synonym set extracted from a semantic network, our experiments also show a higher computational scalability versus other comparable techniques. We created a new benchmark extracted from the open-access metadata of a scientific digital library and published it along with the experiment code to allow further research in the topic.

Keywords: Semantic tagging · Digital libraries · Topic modeling
Multi-label classification · Metadata enrichment

1 Introduction

The activity of researchers has been disrupted by ever greater access to online scientific libraries –in particular due to the presence of open access digital libraries.

© Springer Nature Switzerland AG 2018
E. Méndez et al. (Eds.): TPDL 2018, LNCS 11057, pp. 32–43, 2018.
https://doi.org/10.1007/978-3-030-00066-0_3

Typically when a researcher enters a query for finding interesting papers into the search engine of such a digital library it is done with a few keywords. The match between the keywords entered and those used to describe the relevant scientific documents in these digital libraries may be limited if the terms used are not the same. Every researcher belongs to a community with whom she or he shares common knowledge and vocabulary. However, when the latter wishes to extend the bibliographic exploration beyond her/his community in order to gather information that leads him/her to new knowledge, it is necessary to remove several scientific and technical obstacles like the size of digital libraries, the heterogeneity of data and the complexity of natural language.

Researchers working in a multi-disciplinary and cross-disciplinary context should have the ability of discovering related interesting articles regardless of the limited keyword variations they know. They are not expected to have a prior knowledge of all vocabulary sets used by all other related scientific disciplines. Most often, semantic networks [6] are a good answer to the problems of linguistic variations in non-thematic digital libraries by finding synonyms or common lexical fields. However, In the scientific research context, using general language semantic network might not be sufficient when it comes to very specific scientific and technical jargons. Such terms also have the challenge of usage evolution over time in which having an updated semantic network counting for new scientific terms would be very expensive to achieve. Another solution could be brought by the word embedding approach [11]. This technique makes it possible to find semantically similar terms. Nevertheless, this approach presents some problems. It is not obvious to determine the number of terms that must be taken into account to be considered semantically close to the initial term. In addition, this technique does not work well when it comes to a concept composed of several terms rather than a single one. Another strategy is to make a manual enrichment of the digital libraries with metadata in order to facilitate the access to the semantic content of the documents. Such metadata can be other keywords, tags, topic names but there is a lack of a standard taxonomy and they are penalized by the subjectivity of the people involved in this manual annotation process [1].

In this paper we present an approach combining two different semantic information sources: the first one is provided by the synonym set of a semantic network and the second one from the semantic representation of a vectorial projection of the research articles of the scientific digital library. The latter takes advantage of learning from already tagged articles to enrich the metadata of other similar articles with relevant predicted tags. Our experiments show that the average F1 measure is increased by 11% in comparison with a baseline approach that only utilizes semantic networks. The paper is organized as follows: the next section (Sect. 2) provides an overview of related work. In Sect. 3 we introduce our pipeline of multi-label semantic-based tagging followed by a detailed evaluation in Sects. 4 and 5. Finally, Sect. 6 concludes the paper and gives an outlook on future work.

2   State of the Art

According to the language, a concept can be described by a single term or by an expression composed of multiple words. Therefore the same concept may have different representations in different natural languages or even in the same language in the case of different disciplines. This causes an information retrieval challenge when the researcher does not know all the term variations of the scientific concept he is interested in. Enriching the metadata of articles with semantically relevant keywords facilitates the access of scientific articles regardless of the search term used in the search engine. Such semantically relevant terms could be extracted thanks to lexical databases (e.g., *WordNet* [12]) or knowledge bases (e.g., *BabelNet* [13], *DBpedia* [8], or *YAGO* [10]). Another solution is to use word embedding techniques [5] for finding semantically similar terminologies. Nevertheless, it is difficult in this approach to identify precisely the closeness of the terms in the projection and then if two terms have still close meanings.

When the set of terms is hierarchically organized, it composes a taxonomy. A *faceted* or *dynamic taxonomy* is a set of taxonomies, each one describing the domain of interest from a different point of view [16]. Recent research in this area has shown that it improves the interrogation of scientific digital libraries to find specific elements, e.g., for finding chemical substances in pharmaceutical digital libraries [18].

The use of *Latent Dirichlet Allocation* (LDA) [3] for assigning documents to topics is an interesting strategy in this problem and it has shown that it helps the search process in scientific digital libraries by integrating the semantics of topic-specific entities [14]. For prediction problems, the unsupervised approach of LDA has been adapted to a supervised one by adding an approximate maximum-likelihood procedure to the process [2]. Using LDA for topic tagging however has a fundamental challenge in mapping the user defined topics with the LDA's latent topics. We can find a few variations of LDA trying to solve this mapping challenge. For example, *Labeled LDA* technique [15] is kind of a supervised version of LDA that utilize the user define topic. Semi-supervised LDA approaches are also interesting solutions for being able to discover new classes in unlabeled data in addition to assigning appropriate unlabeled data instances to existing categories. In particular, we can mention the use of weights of word distribution in *WWDLDA* [19], or an interval semi-supervised approach [4]. However, in the case of a real application to millions of documents, such as a digital library with collections of scientific articles covering many disciplines, over a large number of years, even recent evolutionary approaches of LDA require the use of computationally powerful systems, like the use of a computer cluster [9], which is a complex and costly solution.

3   Model Pipeline

The new model we propose can be resumed following a pipeline of 4 main components as illustrated in Fig. 1. In this section we will describe each of this components.

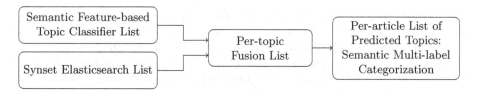

Fig. 1. High-level illustration of the model pipeline. The *Semantic Feature-based Topic Classifier* phase is used to generate *Top N* articles ranked by the probability of topic belonging. Another ranked list is generated by querying the synonym set (synset) of the topic using a text-based search engine which is presented in *Synset Elasticsearch* phase. A *Per-topic Fusion List* is then generated using a special mean rank approach in which only *Top a × N* are considered where *a* is experimentally determined. Finally, each article is tagged by a list of topics that was categorized with in the *Fusion list*.

3.1   Semantic Feature-Based Topic Classifier

This is computationally a big component that itself includes a pipeline of data transformation and a multi-label classification steps. The main phases of it are described as the following:

Extract Semantic Features. Starting from a multi-disciplinary scientific digital library with an open-access metadata, we extract a big number of articles, i.e., millions in which researchers want to explore. The retrieved data from the metadata of these articles are mainly the *title* and the *abstract*. These two fields will then be concatenated in order to be considered as the textual representation of the article in addition to a unique *identifier*. These set of articles will be denoted as *Corpus*. A TF–IDF weighted bag-of-word vectorization is then applied to transform the *Corpus* into a sparse vector space. This vectorized representation is then semantically transformed into a dense semantic feature vector space, typically 100–600 vector size. The result of this stage is an $(N \times M)$ matrix, where N is the semantic feature vector size and M is the number of articles. It must be accompanied with a dictionary that maps the article unique identifier of the article to the row index of the matrix.

Topic Classifier. For each topic name, i.e., scientific category name or a key-phrase of a scientific topic, we generate a *dataset* of *positive* and *negative* examples. The *positive* examples are obtained using a text-based search engine, e.g. *Elasticsearch*, which is a widely used search engine web service built on Apache Lucene, as the resulted articles that have *topic name* matches in *title* OR *abstract*. The negative examples, however, are randomly selected articles from the *Corpus* but with no matches with the *topic name* in any of the metadata text fields. Using this *dataset*, we build a kind of *One-vs-All* topic classifier. This classifier must have the ability of providing the predicted probability value of belonging to the topic, i.e. the class.

Probability-Based Multi-label Classification. Each of the obtained *One-vs-All* topic classifiers are then used in a multi-label classification task where each article in *Corpus* will have a probability value of belonging to the topic. This could be thought of as a kind of *fuzzy clustering* or *supervised topic modeling* where the article can be assigned to more than one topic but with a probability of belonging. The result of this stage is a top 100 K ranked list of articles per topic with the probability value as the ranking score.

3.2   Synset Elasticsearch

This component is computationally simple but has a great value in the pipeline. It is a kind of query expansion where the query space is increased by finding synonyms and supersets of query terms. So, it also requires a text-based search engine, e.g., *Elasticsearch*. We first need a semantic network or a lexicon database, e.g., WordNet, that can provide a set of synonyms of a giving concept name. For each topic in the set of topics, we generate a set of topic name synonyms, that is denoted by *Synset* (synonym set). Using *Elasticsearch* we then generate a ranked list of articles that have matches in their metadata with any of the synonyms in the topic *Synset*. So, the output of this component is a ranked list of articles per topic. As in Sect. 3.1, this output could be considered as a multi-label classification output but with ranking information rather than a probability score.

3.3   Fusion and Multi-label Categorization

This final stage constitutes the main contribution part of this experimentally designed pipeline. It uses an introduced ranked list fusion criteria of combining the 2 rankings of an article A which are the rank in the *Synset Elasticseach* list denoted by s_A and the rank in the semantic feature-based topic classifier list, denoted by r_A. If an article is present both in the 2 lists, we use a special version of *Mean Rank* score ($t_A = \frac{s_A + r_A}{2}$). Otherwise, the default score value of the article is given by equation ($t_A = r_A \times |S|$) where $|S|$ is the size of the *Synset Elasticseach* list.

The rank score of the *Fusion List* will be finally used to re-rank the articles to generate a new ranked list with a list size that ranges from the $max(|S|, |R|)$ and $|S| + |R|$ where $|R|$ is the size of the semantic feature-based topic classifier list. However, in our model we define a hyper-parameter a that determines the size of the *Fusion* list as in equation ($|F| = a \times |S|$). The hyper-parameter a will be experimentally determined based on multi-label classification statistics and evaluation that would be presented in Sect. 4.

The output of this component, and also the whole pipeline, is a list of articles with their predicted list of topics, i.e. scientific category names. Such list is obtained by applying a *lists inversion* process that takes as input all the per topic *Fusion* lists and generates a per article list of topics for all articles presented in any of the *Fusion* lists. The obtained list of predicted topics per article are optionally presented with a score value that reflects the ranking of the article in the *Fusion* list of the topic. That score could be used to set an additional hyperparameter replacing a which would be a score threshold that determines if the topic would be added to the set of predicted topic tags of the article. However, a simple and efficient version, as would be shown in Sect. 4, would only relay of the ranking information but having in place the design parameter a.

4   Experiments

4.1   Data Description

Scientific Paper Metadata from ISTEX Digital Library. The dataset used for running the experiments is extracted from *ISTEX*[1], a French open-access metadata scientific digital library [17]. This digital library is the result of the *Digital Republic Bill*, a law project of the French Republic discussed from 2014, one of whose aims is a "wider data and knowledge dissemination"[2].

ISTEX digital library contains 21 million documents from 21 scientific literature corpora in all disciplines, more than 9 thousands journals and 300 thousands ebooks published between 1473 and 2015 (in April 2018).

Private publishers (e.g., Wiley, Springer, Elsevier, Emerald...) did not leave access to their entire catalog of publications, that is why the publication access does not cover the most recent publications. In addition, because the contracts were signed with the French Ministry of Higher Education and Research, even if anybody can access to the general information about the publications with ISTEX platform (title, names of the authors and full references of the publication, and also metadata in MODS or JSON format), the global access is limited to the French universities, engineering schools, or public research centers: documents in full text (in PDF, TEI, or plain text format), XML metadata and other enrichments (e.g., bibliographical references in TEI format and other useful tools and criteria for automatic indexing).

For our experiments, we considered only a subpart of ISTEX corpus: the articles must be published during the last twenty years, written in English and related to sufficient metadata, including their title, abstract, keywords and subjects.

Scientific Topic from Web of Science. For each scientific article, we also use a list of tags extracted from the collection of *Web of Science*[3] which contains more than 250 flattened topics. These flattened topics are obtained as follows: when a topic is a sub-topic of another one, we can aggregate to the subcategory terms those of the parent category (e.g., [computer science, artificial intelligence] or [computer science, network]). Some of the topics are composition of topics, like "art and humanities."

---

[1] Excellence Initiative of Scientific and Technical Information https://www.istex.fr/.

[2] https://www.republique-numerique.fr/pages/in-english.

[3] https://images.webofknowledge.com/images/help/WOS/hp_subject_category_terms _tasca.html.

The selected 33 topics are: [Artificial Intelligence; Biomaterials; biophysics; Ceramics; Condensed Matter; Emergency Medicine; Immunology; Infectious Diseases; Information Systems; Literature; Mechanics; Microscopy; Mycology; Neuroimaging; Nursing; Oncology; Ophthalmology; Pathology; Pediatrics; Philosophy; Physiology; Psychiatry; Psychology; Rehabilitation; Religion; Respiratory System; Robotics; Sociology; Substance Abuse; Surgery; Thermodynamics; Toxicology; Transplantation].

In our experiments, to facilitate the analysis of the results without bias due to lexical pretreatment, we work only with topics containing neither punctuation nor linkage words. Moreover, we have kept in our experiences only *Web of Science* topics with enough articles (in ISTEX digital library) for having a significant positive subset of documents not used for the learning part (at least 100 scientific articles). The topics, which can be single words (as "thermodynamic") or a concatenation of words (as "artificial intelligence"), should be known in the semantic network to benefit of a consequent synonyms list. In our work, we present the results obtained with 33 topics, which are English single words or the concatenation of several words.

Synonym Sets from BabelNet. In our experiments, we produce a semantic enrichment by using a list of synonyms for each concept, also known as "synset" (for "synonym set"). To build our *synset* list, we need a semantic network. After some preliminary tests on several semantic networks, we chose *BalbelNet* [13] which gave better results. A sample synset from *BabelNet* for the topic *Mycology* is [Mycology, fungology, History of mycology, Micology, Mycological, Mycologists, Study of fungi].

Supervised LDA. Based on the state-of-the-art review as described in Sect. 2, we started by developing a model based on LDA. We defined a supervised version of the LDA (*sLDA*) where we the number of topics was set to 33 topics. Each topic was guided by boosting the terms of the topic synonym set obtained from *BabelNet* where the boosting values were [1, 10, 20, 30]. The dataset for experimenting this model were extracted from ISTEX scientific corpus by using *Elasticsearch* getting all articles that have at least one match of any of the 33 topics in any of these metadata fields: *Title, abstract, subjects* or *keywords*. However, the text used to build the *sLDA* were limited to the *title* and the *abstract*. The evaluation of the *sLDA* model will then be performed on a test set that is constructed from the *keywords* and the *subjects* fields.

4.2 Experimental Process

Initially, we defined an accuracy indicator that is based on the count of tagged articles with a list of prediction topics that has at least one label intersection with ground truth. This indicator will be denoted as *At least one common label* metric. The other measure including label cardinality, Hamming loss and Jaccard index could be found in the literature[4].

In order to build an experiment of our proposed pipeline, we need to experimentally determine some hyper-parameters of it as follows:

Semantic Feature-Based Topic Classifier: We limit our text representation of the article to its title and abstract, which are available metadata. Comparing Paragraph vector [7] and Randomized truncated SVD [7] based on a metric that maximizes the inner cosine similarity of articles from the same topics and minimizes it for a randomly selected articles, we choose SVD decomposition of the TF–IDF weighted bag of words and bi-grams resulting in 150 features for more than 4 millions articles. As for the topic classifier, also by comparative evaluation, we select *Random Forest Classifier*, tuning certain design parameters, and use it to rank the scientific corpus. We consider the top 100 K articles of each topic classifier to be used in the fusion step.

Synonym Set Elasticsearch: Reviewing many available semantic networks, we found that BabelNet was the most comprehensive one combining many other networks [13]. So, we use it to extract a set of synonyms, i.e., a *synset* for each topic. This synset is then used to query the search engine of ISTEX which is built on Elasticsearch server. As would be shown in Sect. 5. This technique will be used as the experiment baseline.

Fusion and Per Multi-label Categorization: The main design parameter of this phase is the size of the ranked list that is achieved by setting it to the double size of the *Synset Elasticsearch* list.

5 Results and Discussion

First, we run an experiment on *sLDA* as described in Sect. 4. The result of this designed experiment was very disappointing based on the evaluation metrics. The best performing *sLDA* model, that was with a boosting value of 30, resulted in the following evaluation: *F1 measure* = 0.02828, *At-least-one-common-label* = 0.0443, *Jaccard index* = 0.0219 and *Hamming loss* = 0.0798. Comparing to using our pipeline with $a = 2$ having *F1 measure* of the 33 topics was 0.6032. So, *sLDA* was obviously not a good candidate to be used as a baseline. However, it was an additional motivation for designing and proposing our pipeline. After dropping *sLDA* from further experiments due to the very low evaluation results, we have added 2 more topics to the set of the 33 topics totaling to 35 topics. The 2 additional topics were [International Relations; Biodiversity Conservation]. We have also added more examples to the test set counting for an additional ISTEX metadata field called *categories:wos* that is actually does not exists in all the

---

[4] https://en.wikipedia.org/wiki/Multi-label_classification.

Table 1. Evaluation results based on the evaluation metrics *Recall* and *At least one common label* denoted here as the *Common-Match* metric. The table also shows the size of the intersection between the method results and the test set that was used in computing the evaluation metric, denoted here as *Intersection*. The value of *Intersection* might also be a good indicator of the method being able to tag more articles.

Method	Intersection	Common-Match	Recall
Synset	22,192	0.5284	0.5285
Fusion1	22,123	0.5736	0.5735
Fusion2	41,642	0.6375	0.6374
Fusion3	56,114	**0.6470**	**0.6473**
Fusion4	**67,625**	**0.6470**	0.6464

articles but was still considered as a good source for increasing the test examples in our published benchmark.

We define 5 methods for the experiment. One is a method of *Synset Elasticsearch*, denoted here by *Synset* which will be the baseline of benchmark. The other 4 methods are variations of our proposed pipeline but with variant values of the design parameter $a = [1, 2, 3, 4]$. The pipeline methods are then denoted respectively with the value of a as *Fusion1, Fusion2, Fusion3* and *Fusion4*. The results of the multi-label classification evaluation metrics, described in Sect. 4.2, are shown in Table 1 and Fig. 2.

While the evaluation metric values in Table 1 recommend higher a values, 3 or 4 with no significant value difference, we can see from Fig. 2 that the best value is $a = 2$ based on *Precision, F1 measure, Jaccard index* and *Hamming loss*. This means that if we increase the size of the fusion ranked list more than the double of the size of the Synset method, we will start loosing accuracy. Another indicator that we should limit the size of the Fusion list is Fig. 2a that shows that if we increase the size of the Fusion list, the difference of the *Label Cardinality* between the predicted results and the compared test set will increase. This difference is a negative effect that should be minimized, otherwise, the model will tend to predict too much labels that would be more probably irrelevant to the article.

Due to the fact that the test set was not generated manually but by filtering on a set of scientific category terms in relevant metadata fields, we believe that it is an incomplete ground truth. However, we think it is very suitable to compare models as a guidance for designing an efficient one because the test labels are correct even incomplete. Accordingly, we tried to perform some error analysis where we found that in most of the cases, the extra suggested category names are either actual correct topic having the article a multi-disciplinary one or topics from very similar and related topic. For example, a medical article from ISTEX[5] is tagged with the category name ['Transplantation'] in the test set. The predicted topics by our method was ['Mycology', 'Transplantation'] resulting into 0.5 precision value. However, when we read the abstract of that article, we find

---

[5] https://api.istex.fr/document/23A2BC6E23BE8DE9971290A5E869F1FA4A5E49E4.

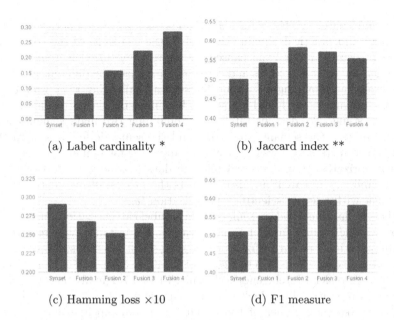

(a) Label cardinality *

(b) Jaccard index **

(c) Hamming loss ×10

(d) F1 measure

Fig. 2. Results of *label cardinality difference, Jaccard index, Hamming loss* and *F1 measure* evaluation metrics. While Synset is the method that uses synonyms of the category name as a query in Elasticsearch, Fusion 1, 2, 3 and 4 represent respectively the values of the pipeline design parameters $a = [1, 2, 3, 4]$ that determine the number of annotated articles per topic as an integer multiple of the size of *Synset Elasticsearch* list. *: Difference value with the label cardinality of the compared test set of each of the methods. **: Equivalent to *Precision* in our case of a test set label cardinality = 1.

that it talks about *dematiaceous fungi* which is actually a 'Mycology' topic. So, in many cases where there is at least one common tag, the other tags are actually the aimed discovered knowledge rather than a false prediction. In another example, the model predicted the tags 'Psychology', 'Sociology' in addition to 'Religion' resulting in 0.3333 precision while they are actually relevant predicted tags when we read the abstract of the article[6] that also talks about *social networks*. The complete list of results –where these cases could be verified– are published as well as all the experimental data and reproducibility code[7].

6 Conclusion and Future Work

Governments, public organizations and even the private sector have recently invested in developing multi-disciplinary open-access scientific digital libraries. However, these huge scientific repositories are facing many information retrieval issues. Nevertheless, this opens opportunities for text-mining based solutions

---

[6] https://api.istex.fr/document/BA63065CCE8B0520F36B7DA90CF26F2DEF6CED7F.
[7] https://github.com/ERICUdL/stst.

that can automate cognitive efforts in data curation. In this paper, we proposed an efficient and practical pipeline that solves the challenge of the community-dependent tags and the issue caused by aggregating articles from heterogeneous scientific topic ontologies and category names used by different publishers. We believe that providing a solution for such a challenging issue would foster trans-disciplinary research and innovation by enhancing the corpus information retrieval systems. We demonstrated that combining two main semantic information sources – the semantic networks and the semantic features of the text of the article metadata – was a successful approach for semantic based multi-label categorization. Our proposed pipeline does not only enable for a better trans-disciplinary research but also supports the process of metadata semantic enrichment with relevant scientific categorization tags.

Other available methods in semantic multi-label categorization, such as LDA, are not suitable in this context for many reasons. For instance, they require powerful computational resources for processing big scientific corpus. Moreover, they need a pre-processing step to detect concepts that are composed of more than one word (e.g., "Artificial Intelligence"). Finally, LDA is originally an unsupervised machine learning model in which it is problematic to define some undetermined parameters like the number of topics. Our proposed pipeline, however, overcomes all of these limitations and provides efficient results. Towards improving the query expansion component of the pipeline (Synset Elasticsearch), we are planning to study the impact of using extra information from *BabelNet* semantic network other than only the synonym sets. In particular, we want to include the neighboring concept names as well as the category names of the concept. We expect that such term semantic expansion will improve the performance of the method.

Acknowledgment. We would like to thank ARC6 program (http://www.arc6-tic. rhonealpes.fr/larc-6/) of the Region Auvergne-Rhône-Alpes that funds the current PhD studies of the first author and thank ISTEX project.

References

1. Abrizah, A., Zainab, A.N., Kiran, K., Raj, R.G.: LIS journals scientific impact and subject categorization: a comparison between web of science and scopus. Scientometrics **94**(2), 721–740 (2013). https://doi.org/10.1007/s11192-012-0813-7
2. Blei, D.M., McAuliffe, J.D.: Supervised topic models. In: Platt, J.C., Koller, D., Singer, Y., Roweis, S.T. (eds.) Advances in Neural Information Processing Systems 20, Proceedings of the Twenty-First Annual Conference on Neural Information Processing Systems, Vancouver, British Columbia, Canada, 3–6 December 2007, pp. 121–128. Curran Associates, Inc. (2007)
3. Blei, D.M., Ng, A.Y., Jordan, M.I.: Latent Dirichlet allocation. J. Mach. Learn. Res. **3**, 993–1022 (2003)
4. Bodrunova, S., Koltsov, S., Koltsova, O., Nikolenko, S., Shimorina, A.: Interval semi-supervised LDA: classifying needles in a haystack. In: Castro, F., Gelbukh, A., González, M. (eds.) MICAI 2013. LNCS (LNAI), vol. 8265, pp. 265–274. Springer, Heidelberg (2013). https://doi.org/10.1007/978-3-642-45114-0_21

5. Bojanowski, P., Grave, E., Joulin, A., Mikolov, T.: Enriching word vectors with subword information. TACL **5**, 135–146 (2017)
6. Borgida, A., Sowa, J.F.: Principles of semantic networks - Explorations in the representation of knowledge. The Morgan Kaufmann Series in Representation and Reasoning. Morgan Kaufmann, Burlington (1991)
7. Halko, N., Martinsson, P.G., Tropp, J.A.: Finding structure with randomness: probabilistic algorithms for constructing approximate matrix decompositions. SIAM Rev. **53**(2), 217–288 (2011)
8. Lehmann, J., et al.: DBpedia - a large-scale, multilingual knowledge base extracted from Wikipedia. Semant. Web **6**(2), 167–195 (2015). https://doi.org/10.3233/SW-140134
9. Liang, F., Yang, Y., Bradley, J.: Large scale topic modeling: improvements to LDA on Apache Spark, September 2015. https://tinyurl.com/y7xfqnze
10. Mahdisoltani, F., Biega, J., Suchanek, F.M.: YAGO3: a knowledge base from multilingual wikipedias. In: CIDR 2015, Asilomar, CA, USA, 4–7 January 2015 (2015). www.cidrdb.org
11. Mikolov, T., Sutskever, I., Chen, K., Corrado, G.S., Dean, J.: Distributed representations of words and phrases and their compositionality. In: Burges, C.J.C., Bottou, L., Ghahramani, Z., Weinberger, K.Q. (eds.) Advances in Neural Information Processing Systems 26: 27th Annual Conference on Neural Information Processing Systems 2013. Proceedings of a Meeting held, 5–8 December 2013, Lake Tahoe, Nevada, United States, pp. 3111–3119 (2013)
12. Miller, G.A.: WordNet: a lexical database for English. Commun. ACM (CACM) **38**(11), 39–41 (1995). https://doi.org/10.1145/219717.219748
13. Navigli, R., Ponzetto, S.P.: BabelNet: the automatic construction, evaluation and application of a wide-coverage multilingual semantic network. Artif. Intell. **193**, 217–250 (2012)
14. Pinto, J.M.G., Balke, W.: Demystifying the semantics of relevant objects in scholarly collections: a probabilistic approach. In: Proceedings of the 15th ACM/IEEE-CE Joint Conference on Digital Libraries, Knoxville, TN, USA, 21–25 June 2015, pp. 157–164. ACM (2015). https://doi.org/10.1145/2756406.2756923
15. Ramage, D., Hall, D.L.W., Nallapati, R., Manning, C.D.: Labeled LDA: a supervised topic model for credit attribution in multi-labeled corpora. In: Proceedings of the 2009 Conference on Empirical Methods in Natural Language Processing, EMNLP 2009, 6–7 August 2009, Singapore, A meeting of SIGDAT, a Special Interest Group of the ACL, pp. 248–256. ACL (2009)
16. Sacco, G.M., Tzitzikas, Y. (eds.): Dynamic Taxonomies and Faceted Search: Theory, Practice, and Experience. The Information Retrieval Series. Springer, Berlin, Heidelberg (2009). https://doi.org/10.1007/978-3-642-02359-0
17. Scientific and Technical Information Department - CNRS: White Paper - Open Science in a Digital Republic. OpenEdition Press, Marseille (2016). https://doi.org/10.4000/books.oep.1635
18. Wawrzinek, J., Balke, W.-T.: Semantic facettation in pharmaceutical collections using deep learning for active substance contextualization. In: Choemprayong, S., Crestani, F., Cunningham, S.J. (eds.) ICADL 2017. LNCS, vol. 10647, pp. 41–53. Springer, Cham (2017). https://doi.org/10.1007/978-3-319-70232-2_4
19. Zhou, P., Wei, J., Qin, Y.: A semi-supervised text clustering algorithm with word distribution weights. In: Proceedings of the 2013 the International Conference on Education Technology and Information System (ICETIS 2013). Advances in Intelligent Systems Research, 21–22 June 2013, Sanya, China, pp. 1024–1028. Atlantis Press (2013). https://doi.org/10.2991/icetis-13.2013.235

Entity Disambiguation

Harnessing Historical Corrections to Build Test Collections for Named Entity Disambiguation

Florian Reitz[✉][iD]

Schloss Dagstuhl LZI, dblp group, Wadern, Germany
florian.reitz@dagstuhl.de

Abstract. Matching mentions of persons to the actual persons (the name disambiguation problem) is central for many digital library applications. Scientists have been working on algorithms to create this matching for decades without finding a universal solution. One problem is that test collections for this problem are often small and specific to a certain collection. In this work, we present an approach that can create large test collections from historical metadata with minimal extra cost. We apply this approach to the dblp collection to generate two freely available test collections. One collection focuses on the properties of name-related defects (such as similarities of synonymous names) and one on the evaluation of disambiguation algorithms.

Keywords: Name disambiguation · Historical metadata · dblp

1 Introduction

Digital libraries store lists of names which refer to real world persons (e.g., the authors of a document). Many applications require a map between these names and the real world persons they refer to. E.g., projects create author profiles which list all publications of a person. These profiles can be used by users who look for works of a specific person. They are also the basis of attempts to measure the performance of researchers and institutions. Mapping author mentions and persons is difficult. The name itself is not well-suited to refer to a person and many metadata records provide limited additional information such as email and institution name. Therefore, many profiles are defective. That is: they list publications from different persons (a homonym defect) or publications of one person are listed in different profiles (a synonym defect). Correct author disambiguation in bibliographic data has been the subject of intensive research for decades. For an overview on algorithmic approaches, see the survey by Ferreira et al. [3]. For manual strategies, see the paper by Elliot [1].

Many approaches concentrate on reclustering the existing data. I.e., the algorithm is provided with all mentions of persons and clusters these mentions into profiles. An advantage of this approach is that potentially wrong profiles can

© Springer Nature Switzerland AG 2018
E. Méndez et al. (Eds.): TPDL 2018, LNCS 11057, pp. 47–58, 2018.
https://doi.org/10.1007/978-3-030-00066-0_4

be ignored. The problem is that reclustering ignores disambiguation work which has already been invested into the collection. In a living collection, a significant amount of manual and automatic work has been invested in the correctness of data. With an increasing number of open data projects, we can expect more disambiguations by users (e.g., authors use ORCID to manage their publication profiles). We will also see collections getting larger as sharing and incorporating data will become easier. For a large and volatile collection, reclustering might be algorithmically unfeasible. An alternative disambiguation task is to identify profiles which are likely defective. These profiles can then be corrected automatically or checked by staff or in a crowdsourcing framework. One problem of developing algorithms for this task is the lack of suitable test collections for evaluation. Traditional test collections, such as the set provided by Han et al. [4], consist of mentions with the same name without any known author profiles. This is not useful for the defect detection task. In addition, there are the following problems: (1) Classic test collections are small. The largest collections discussed have several thousand mentions, while collections like dblp list several million mentions. This makes classic collections unusable for evaluation of running time and other resource requirements such as main memory. (2) Classic test collections cannot be used to study properties of defective profiles such as network relations of synonym profiles. Their properties can reveal new approaches to match mentions and persons. They can also show relevant differences between collections. Known defects can also be used to train detection algorithms. In this work, we describe two alternative test collections which try to overcome these problems.

Creating a classic test collection is expensive, mainly because it requires manual cleaning of author profiles. However, for a large digital library, we can expect that a number of defects have already been corrected. We extract these corrections from historical data and use them as examples of defective data. Since the defects have been corrected, we also know a (partial) solution to the defect. Based on the defects, we build two test collections. Our goal is to provide as much contextual information for each defect as possible. One of the collections focuses on individual defects. The other test collection focuses on the defect detection task in a large collection itself. Harnessing historical corrections has several advantages: (1) The collections we obtain are large compared to traditional test collections and created with minimal additional cost. (2) Unlike classic test collections, our approach is well-suited to study the properties of defects. This can lead to a better understanding of quality problems and can be used when designing new disambiguation algorithms. As defect corrections can be triggered in may different ways, we obtain a large variety of defects. (3) The framework we present can be used for all digital libraries that provide historical metadata. This might provide us with specific test collections which can be used to adjust algorithms to the properties of individual collections. The main contributions of this work are: (1) We present a framework to create test collections for the defect detection task from historical data. (2) We use the framework to create an open test collection based on the dblp collection.

After discussing related work, we describe how to build test collections from historical defect corrections (Sect. 3). In Sect. 4, we apply the framework to dblp and discuss possible applications of the test collections.

2 Related Work

The usual approach to build a test collection is to select a small portion of data from a collection and clean it thoroughly. This requires manual work which leads to small test collections. E.g., the often used test collection by Han [4] consists of about 8,400 mentions (which roughly equals publications), the KISTI collection by Kang et al. [6] consists of about 32,000 mentions. For comparison, the dblp collection listed about 10 million mentions in March 2018. For an overview, see the work of Müller et al. [9]. Most test collections consist of two sets, the challenge which is presented to the algorithms and the solution which contains the correct clustering of mentions into profiles. Algorithms are judged by how close they can approximate the solution. E.g., for Han et al., the challenge consists of publications from authors with common names (such as *C. Chen*). The authors first name is abbreviated to increase the difficulty. Data that could not be manually disambiguated was discarded. Most test collections provide the basic bibliographic metadata such as title, name of coauthors and publication year. This creates compact test collections but provides very little context information. E.g., the collections only contain partial information on the coauthors because most of their publications are not part of the test collection.

Since manual disambiguation for test collections is expensive, there have been several attempts to harness work which has already been invested into a collection. Reuther [12] compared two states of dblp from different years to see if publications had been reassigned between author profiles. Reuther gathered the publications of these profiles for a test collection that focuses on corrections of synonym defects. We will extend this approach by also considering other types of defects. Momeni and Mayr [8] built a test collection based on homonym profiles in dblp. When dblp notices that a name is used by several authors, the author mention is appended by a number. E.g., *Wei Wang 0001,...*, *Wei Wang 0135*. Momeni and Mayr built a challenge by removing the suffixes. The full name (including suffix) is used as solution. Like Reuther, this approach is limited to a single defect type (here: the homonym). Momeni and Mayr are also limited cases where authors have the exact same name, which excludes many real world problems. Müller et al. [9] describe how a test collection can be built by comparing the manual disambiguation work of different projects. For all these approaches, the data presentation is record-based as in the classic test collections. We will extend these approaches by adding contextual information which can be used by disambiguation algorithms. We also provide individual defect cases which can be studies to understand defects.

3 Extracting Test Collections from Historical Metadata

Assume that publications for a person *John Doe* are listed in author profiles *J. Doe* and *John Doe*. If that defect is uncovered, it will most likely be corrected. Many collections attempt author disambiguation when data is added. In the example, this did not work which indicates that this defect is not trivial which makes it interesting for defect detection algorithms. We use the state of the collection before the correction as challenge to see if an algorithm can detect the presence of a problem and possibly propose a solution. Using historical corrections has a number of advantages: (1) The corrections can come from a number of different sources which can include defects with different properties. E.g., for dblp, a significant amount of defects are reported by users [11]. (2) At the moment of the correction, data was available that might not be available today. In 2014, Shin et al. [14] reconsidered the data of Han et al. [4] from 2005 and determined that more than 22% of their gold standard mappings between mention and profiles could not be confirmed with external data. In 2005 that verification was probably possible (web pages went off line, publishers became defunct ...). We will now describe an extraction approach for historical defects and how to build test collections on top of them. In Sect. 4, we show how this approach can be applied to the dblp bibliography to create test collections.

3.1 Identifying Corrections in Historical Data

For our approach, we need suitable historical metadata. Locating this data turned out to be difficult. If a project provides historical data, it is often necessary to use secondary data sources like backup files. For these data, we cannot expect to capture every correction separately. Instead, we obtain observations of the data. The points of observation depend on the underlying data, e.g., the times the backups were created. If observations are far apart and edits frequent, we might not be able to extract individual corrections. Figure 1 shows an example with edits and observations. In this case, we obtain four states, $A, \ldots, D$ even though there are more edits. E.g., edits 3, 4 and 5 might be merged into one observed correction. For each dataset, we need to determine if the observation allows a reasonable correction analysis. E.g., for dblp, we used a collection [5] that has nightly observations. This granularity allowed for reliable correction extraction. We also considered weekly snapshots of the Internet Movie Database (IMDB)[1]. This granularity made interpreting the data difficult.

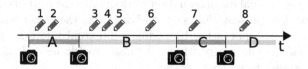

Fig. 1. Observer-based framework for historical metadata.

---

[1] ftp://ftp.fu-berlin.de/pub/misc/movies/database/frozendata.

If there is sufficient historical data, we can extract corrections. Most digital libraries provide interpretations of their person mentions. We call such an interpretation a **profile**. A profile contains the mentions (publications) that the digital library thinks are created by the person represented by the profile. The interpretation can be based on the name directly (as in dblp) or based on an identifier assigned to the mention. We require that the interpretation is contained in the historical metadata. I.e., we can reconstruct historical profiles. As explained above, some profiles will be defective (i.e., they deviate from the real person's work list). Let $t_1 < t_2$ be two time points of observation and let $p\langle t \rangle$ be the set of mentions that is assigned to profile p at time t. We can observe two types of relations between profiles from different observations:

Definition 1. *Let p_1, p_2 be two profiles. We call p_1 **reference predecessor** of p_2 if $\exists\, m \in p_1\langle t_1 \rangle : m \in p_2\langle t_2 \rangle$. We call p_2 **reference successor** of p_1. We call p_1 **consistent predecessor** of p_2 if $p_1\langle t_1 \rangle \subseteq p_2\langle t_2 \rangle$.*

There are two candidates for a defect correction:

1. A profile p has two or more reference predecessors.
2. A profile p has two or more reference successors.

In case (1), p was represented by multiple profiles before. If we assume that p is correct now, the successors were synonyms. Similarly, in case (2) we observe the correction of a homonym defect.

We can categorize modifications to profiles as follows:

Definition 2. *Let p be a profile and $t_1 < t_2$ two time points of observation. Let $P := p_1, \ldots, p_k$ be the reference predecessors of p with respect to t_1 and t_2. We call P a **merge group** if $k > 1$ and $\forall 1 \le s \le k : p_s\langle t_2 \rangle = \emptyset \vee p_s = p$.*

Between time t_1 and t_2, mentions in $p_1, \ldots, p_k$ were reassigned to p. These profiles, except p itself, do not have any mention left. We can consider a similar correction for homonyms:

Definition 3. *With p, t_1 and t_2 as above. Let $P := p_1, \ldots, p_k$ be the reference successors of p with respect to t_1 and t_2. We call P a **split group** if $k > 1$ and $\forall 1 \le s \le k : p_s\langle t_1 \rangle = \emptyset \vee p_s = p$.*

For a merge, we demand that the merged profiles are no longer referenced in the library. Similarly, we demand for a split that *new* profiles do not contain a mention at t_1. In addition to that, we need to consider a combination of merge and split, a **distribute**. In this case, a mention is moved from one profile to another without creating an empty profile. Distributes are different from merges and splits as both profiles are represented before and after the correction. An algorithm which aims to correct this defect must determine which mentions are to be reassigned. If both profiles exist before the correction, the algorithm can use their properties to determine if a mention needs reassignment. This might be easier than detecting completely merged mentions.

Merge and split groups can combine multiple corrections. This can create artifacts with the observation framework. Assume that there are two merge corrections $p_1, p_2 \rightarrow p_1$ and $p_1, p_3 \rightarrow p_1$. If these operations occur between two observations, we obtain a merge group $p_1, p_2, p_3 \rightarrow p_1$. If the observation occurs between the two corrections, we obtain two merge groups. To avoid splitting groups that are related, we group merge and split groups if they occur in close temporal proximity and have at least one common profile.

3.2   Structure of Test Collections

The extracted corrections can now be transformed into test collections. Classic test collections list the records of the disambiguated authors. This creates compact collections. However, it provides very little context information. For our test collections, we have the following goals:

1. The collections should allow to study the properties of defects. In particular, they should provide the context information which is commonly used by disambiguation algorithms such as parts of the coauthor network.
2. The collections should facilitate the development of algorithms that search for defects in an existing name reference interpretation. As opposed to collections that aim at algorithms that completely recluster author mentions.
3. The collections should support a performance-based evaluation (running time, memory requirement . . .).
4. The collections should be of manageable size.

To meet these goals, we create two collections that both differ from classic collections: A case-based collection that lists the individual corrections as small graphs. An embedded collection that integrates the detected defects into the total collection. We now discuss the general structure of the two collections.

Case-Based Collection. The case-based collection consists of isolated test cases that are directly derived from the observed corrections. For each correction, we provide two files. One file contains the state of the digital library directly before the correction, the other file contains the state right after the correction. The primary purpose of the case-based collection is to study the properties of defects. This requires that a certain context is provided. E.g., many disambiguation algorithms use common coauthors as evidence that there is a relation between two mentions. Classic test collections provide this information but they give no information about the relations between the coauthors. Consider Fig. 2 with synonymous profiles p_1, p_2, p_3, coauthors $c_1, \ldots, c_5$ and journals j_1 and j_2. p_1 and p_2 are strongly related by two common coauthors and a journal. p_3 is not in a direct relation to p_1 and p_2. The black solid lines represent the data available from a classic test collection. The dashed lines represent contextual data that is not in the test collection. c_2 and c_5 collaborated. However, that relation is defined by publications outside the test collection. Studying these indirect relations might help to develop a better disambiguation algorithm.

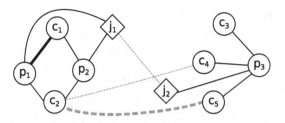

Fig. 2. Relations between metadata entities as a graph.

Obviously, we cannot provide the complete metadata context (e.g., the complete coauthor graph) for each test case. To provide at least local context, we code the test case as a graph. Consider again Fig. 2. Assume that p_1, p_2, p_3 profiles are part of a merge group. We create a graph as follows: We add nodes for all corrected profiles (p_1, p_2, p_3). We call these nodes primary nodes. We add a node for each entity that is in relation to a primary node (e.g., the coauthors). The set of available entities depends on the underlying digital library. Other entities might be conferences/journals or common topics. We then add an edge for each known relation between these nodes. The context is provided by the edges between the nodes that are no primary nodes. The types of relations depend on the data in the digital library. The edges can be weighted which makes it possible to convey the strength of a relation without massively inflating the files.

We encode the graphs in XML. For edges and nodes, we can provide properties in (key, value) form. The following example shows a document-type node that has title and publication year data. A similar notation is used for edges.

```
<node label="DOCUMENT" id="doc1">
    <property key="year" value="1999"/>
    <property key="title" value="The Ultrasonic Navigating."/>
</node>
```

Embedded Collection. The defects of the case-based collection are too small to analyze the running time performance of an algorithm. They also do not provide a full context. Some disambiguation algorithms require a full coauthor graph [2,15], which is not available from the local context of the individual cases. The embedded collection can solve these problems. It consists of two components: (1) A full copy of the collection's metadata at a certain point, provided as metadata records. (2) An annotation of detected defects in this version which are corrected later. This means that algorithms need to process the full collection (to test the running time performance) but have also access to all data (which removed limits imposed by partial data of classical test collections).

Since we provide the full version of the metadata, it is not possible to use a dense observation framework. I.e., for detecting defect corrections, we need to compare states of the collection which are some time apart (e.g., a full year). This will create a sufficient number of defects to be annotated. However, the long periods between the states of the dataset makes overlapping corrections more likely. Assume that we observe a distribute operation between author profiles p_1 and p_2 (publications are moved between these profiles). Further assume that profile p_3 is merged into profile p_1. For a dense observation framework, there are many different ways in which these operations can be performed. In a slightly different situation, we might have observed a distribute between p_2 and p_3 and a merge of p_1 into p_3. For a sparse observation framework, these corrections will most likely be merged together. This does not affect the presence of a defect (which should be detected by the algorithm) but makes the embedded collection unsuited to analyze individual corrections. The metadata of the collection can be provided in any way, e.g., as metadata records. Unlike the case-based collection, this might make importing the data easier for some approaches. The annotations are provided as simple XML-Files. The example below shows a small split case. doc1, doc2, p1 and p2 are identifiers taken from the underlying collection.

```
<source>
   <profile authorid="p1">
      <signature pkey="doc1" pos="1" surface="B. Doe"/>
      <signature pkey="doc2" pos="0" surface="B. Doe"/>
   </profile>
</source>
   <target>
      <profile authorid="p1">
         <signature pkey="doc1" pos="1" surface="Bob A. Doe"/>
      </profile>
      <profile authorid="p2">
         <signature pkey="doc2" pos="0" surface="Bob B. Doe"/>
      </profile>
   </target>
```

3.3 Biases and Limitations

The test collections we present here are different from the classic test collections as they do not provide a full gold standard. This means: (1) They provide examples of errors but have no examples of guaranteed correct data which could be used to detect false positives. (2) The corrections might be partial. See below for an example. In Sect. 4.2, we will very briefly discuss scenarios in which the collections can be used. It is important to note that these collections will not replace classic test collections but complement them. E.g., to study defects or to evaluate running time. Apart from the evaluation method itself, our approach has intrinsic biases which cannot be fully mitigated. In this section, we will discuss the most relevant points. Each of these threats to validity must be considered before undertaking a study based on historical defect corrections.

Assumption: Corrections improve data quality. We assume that a correction replaces defective data values with correct values. Obviously, there is no guarantee for that as the changes related to the correction can also introduce errors.

The likelihood of introducing new errors depends on the data curation process of the individual projects. Some projects use trained teams while others rely on direct or indirect user contribution. On the other hand, user contribution might be vandalism. In any case, we will obtain a number of partially or fully defective corrections.

Assumption: Corrections completely remove defects. A correction might remove a defect only partially. Assume that one profile contains publications from authors A, B and C. A correction (a split) might extract the publications of A but leave the publications of B and C behind. The original profile is still a homonym. If we build a test collection based on partial corrections, we must allow for a case where an algorithm finds the whole correction. This means that there is no gold standard solution to our test collection as there is for classical test collections. We need to define specific evaluation metrics to handle this situation. In case of a study of defect properties, we must also consider that some corrections are only partial.

Assumption: The corrected defects are representative of the set of all defects. Our approach is biased by the way defects are detected in the underlying dataset. Assume that a project applies a process which is good at finding defects with property $\mathcal{A}$ but can barely handle defects with property $\mathcal{B}$. In this case, defects with property $\mathcal{A}$ would be overrepresented and many defects with property $\mathcal{B}$ would be missing. It is also possible that a project is aware of a defect but does not fix it because it has a low priority. Again, it is unclear how community contribution can mitigate this problem. For all studies, we must assume that error classes exist that are significantly underrepresented.

4 Test Collections Based on Dblp

We apply the framework described above to the dblp bibliography[2]. The results are published under an open license [10]. The dblp project gathers metadata for publications in computer science and related fields. In March 2018, it contained 4.1 million publications and 2 million profiles. The dblp project creates nightly backups of its data which are combined into a historical data file [5]. This file can be used to trace modifications to the metadata records between June 1999 and March 2018.

4.1 Application of the Framework

The dblp project has two mechanisms to match author mentions with observed entities. (1) The name itself. The name might be appended with a numeric suffix such as *Wei Wang 0050*. (2) Authority records which map names to person entities. The authority records are part of the historical data. I.e., we can track changes to the authority data as well.

---

[2] https://dblp.org.

We use three different types of entities for the graphs of the case-based collection: *Document, Person* and *Venue* (journal or conference series) The primary function of *Document* is to provide the standard metadata such as title and year of publication. We model six different relations. *Created/Contributed* (Person → Document, unweighted): The person is author/editor of that document/proceedings. *Co-Created/Co-Contributed* (Person ↔ Person, weighted by number of common papers): The persons are authors/editors of at least one common paper. *Created-At/Contributed-At* (Person → Venue, weighted by number of papers): The person is author of a paper that appeared at the venue/editor of a proceedings of the venue. We decided to model editorship and authorship separately as they might have different implications for an algorithm. Coauthorship usually implies cooperation while being coeditors (e.g., of a proceedings) can simply mean that the authors are active in the same field. Weights are computed for the last date before the correction was observed. I.e., the weights represent the data which would have been available to an algorithm at that time. We provide all properties for the documents that are listed in dblp. However, we use the most recent data instead of the data available at the point of correction. The main reason is to provide current weblinks to the publication pages of publishers. Today, these links are mostly resolved via DOI. An algorithm can use the links to get additional information from the web. The case-based collection contains 138,532 merges, 16,532 splits and 55,362 distributes.

For the embedded test collection, we considered the state of dblp at the beginning of a year. Table 1 lists the number of corrections for some combinations of different dates. We do not consider states of dblp from before 2013 as the collection was small at that time and the number of possible corrections is negligible. The number of corrections is small compared to the case-based collection. The primary reasons are (1) short-lived defects that were introduced to the collection and corrected between the observations are missing. (2) As discussed above, we might merge multiple corrections into one.

Table 1. Number of identified corrections for different observation frameworks.

Observation dates	Split	Merge	Distribute	All
2013, 2017	2.207	19,175	5,346	26,728
2015, 2017	1.536	13,393	3,968	18,897
2017, 2018	978	8,608	2,666	12,252

4.2   Possible Applications

As stated above, both test collections do not provide full solutions of the name disambiguation task. Therefore, classic evaluations such as cluster alignment cannot be used to evaluate the approaches. However, the embedded collection can be used to test running time performance and the general ability to detect known defects in a collection. A simple evaluation strategy would be: (1) Use classic test collections to obtain precision/recall/cluster alignment in a fully solved scenario. (2) Check if the algorithm can handle the size of the embedded test collection. (3) Measure how many defects in the embedded collection are detected. This will filter out slow approaches and provide insides if the qualitative performance from the classic test collection translates to the embedded collection.

The case-based collection can be used to study properties of defects. As an example, we considered how suitable names are for simple blocking approaches. Blocking is a preprocessing step which partitions the into manageable bins. The idea is that similar names are placed in the same bin [4,7]. Some approaches use a similar idea to compute similarity between mentions [13]. Blocking is mostly recall-based so we can use the case-based collection to measure the performance. We consider two variations of name-based blocking: (1) Only the last name part is considered. E.g., from *John Doe* we use *Doe*. (2) The last name and the initial of the first name is used. Middle names are ignored. From *John A. Doe* we use *J. Doe*. We use both approaches with and without considering case. For merge and distribute cases, we compute how many name pairs are placed in the same block (the hit rate). Table 2 shows the result for dblp, compared to results from a test collection we built on IMDB. The data for dblp are from an older version of the test collection which covers the period 1999–2015.

Both blocking approaches are designed with the name abbreviation problem in mind. The approaches perform well for dblp with hit rates between 76.51% and 79.10%. This is due to the large number of abbreviated names in academic publications. While a hit rate of 79.1% is far from optional – of all name pairs 21% do not end in the same block – it might be acceptable. However, the results for IMDB are much worse, indicating that blocking strategies that work well for one project are not suited for other libraries.

Table 2. Comparison of abbreviation-based similarity. The table shows percentages of pairs which are considered similar.

Project	Pairs tested	Consider case		Ignore case	
		Initial + Last	Last	Initial + Last	Last
DBLP merge+dist	128,048	76.51%	78.56%	77.10%	79.10%
IMDB merge+dist	29,218	46.24%	56.64%	47.15%	57.57%

5   Conclusion

In this work, we described how historical defect corrections can be extracted and processed into test collections for the name disambiguation task. The collections do not permit classical evaluation but provide insights into the nature of defects and allow evaluation of aspects which have been difficult to test so far. At the moment, it is still difficult to find usable historical data for most collections. We hope that with an increasing number of open collections, this problem will be mitigated. At that point, it will be possible to create individual test collections. Using different collections will provide more stable algorithms that do not depend on properties of the underlying data.

Acknowledgements. The research in this paper is funded by the Leibniz Competition, grant no. LZI-SAW-2015-2. The author thanks Oliver Hoffmann for providing the data on which the dblp test collection is built and Marcel R. Ackermann for helpful discussions and suggestions.

References

1. Elliot, S.: Survey of author name disambiguation: 2004 to 2010 (2010). http://digitalcommons.unl.edu/libphilprac/473. Accessed Apr 2018
2. Fan, X., Wang, J., Pu, X., Zhou, L., Lv, B.: On graph-based name disambiguation. J. Data Inf. Qual. **2**(2), 10 (2011)
3. Ferreira, A.A., Gonçalves, M.A., Laender, A.H.F.: A brief survey of automatic methods for author name disambiguation. ACM Sigmod Rec. **41**(2), 15–26 (2012)
4. Han, H., Zha, H., Giles, C.L.: Name disambiguation in author citations using a K-way spectral clustering method. In: Proceedings of the Joint Conference on Digital Libraries, JCDL 2005, Denver, CO, USA, pp. 334–343. ACM (2005)
5. Hoffmann, O., Reitz, F.: hdblp: historical data of the dblp collection, April 2018. Zenodo [dataset]. https://doi.org/10.5281/zenodo.1213051
6. Kang, I., Kim, P., Lee, S., Jung, H., You, B.: Construction of a large-scale test set for author disambiguation. Inf. Process. Manag. **47**(3), 452–465 (2011)
7. Levin, M., Krawczyk, S., Bethard, S., Jurafsky, D.: Citation-based bootstrapping for large-scale author disambiguation. JASIST **63**(5), 1030–1047 (2012)
8. Momeni, F., Mayr, P.: Evaluating co-authorship networks in author name disambiguation for common names. In: Fuhr, N., Kovács, L., Risse, T., Nejdl, W. (eds.) TPDL 2016. LNCS, vol. 9819, pp. 386–391. Springer, Cham (2016). https://doi.org/10.1007/978-3-319-43997-6_31
9. Müller, M., Reitz, F., Roy, N.: Data sets for author name disambiguation: an empirical analysis and a new resource. Scientometrics **111**(3), 1467–1500 (2017)
10. Reitz, F.: Two test collections for the author name disambiguation problem based on DBLP, April 2018. Zenodo [dataset]. https://doi.org/10.5281/zenodo.1215650
11. Reitz, F., Hoffmann, O.: Did they notice? – a case-study on the community contribution to data quality in DBLP. In: Gradmann, S., Borri, F., Meghini, C., Schuldt, H. (eds.) TPDL 2011. LNCS, vol. 6966, pp. 204–215. Springer, Heidelberg (2011). https://doi.org/10.1007/978-3-642-24469-8_22
12. Reuther, P.: Namen sind wie Schall und Rauch: Ein semantisch orientierter Ansatz zum Personal Name Matching. Ph.D. thesis, University of Trier, Germany (2007)
13. Santana, A.F., Gonçalves, M.A., Laender, A.H.F., Ferreira, A.A.: On the combination of domain-specific heuristics for author name disambiguation: the nearest cluster method. Int. J. Dig. Libr. **16**(3–4), 229–246 (2015)
14. Shin, D., Kim, T., Choi, J., Kim, J.: Author name disambiguation using a graph model with node splitting and merging based on bibliographic information. Scientometrics **100**(1), 15–50 (2014)
15. Sun, C., Shen, D., Kou, Y., Nie, T., Yu, G.: Topological features based entity disambiguation. J. Comput. Sci. Technol. **31**(5), 1053–1068 (2016)

Homonym Detection in Curated Bibliographies: Learning from dblp's Experience

Marcel R. Ackermann$^{(\boxtimes)}$ (ORCID) and Florian Reitz (ORCID)

Schloss Dagstuhl LZI, dblp computer science bibliography, 66687 Wadern, Germany
{marcel.r.ackermann,florian.reitz}@dagstuhl.de

Abstract. Identifying (and fixing) homonymous and synonymous author profiles is one of the major tasks of curating personalized bibliographic metadata repositories like the *dblp computer science bibliography*. In this paper, we present a machine learning approach to identify homonymous profiles. We train our model on a novel gold-standard data set derived from the past years of active, manual curation at dblp.

Keywords: Machine learning · Artificial neural networks
Digital libraries · Homonym detection · Metadata curation
dblp

1 Introduction

Modern digital libraries are compelled to provide accurate and reliable author name disambiguation (AND). One such database is the *dblp computer science bibliography*, which collects, curates, and provides open bibliographic metadata of scholarly publications in computer science and related disciplines [1]. As of January 2018, the collection contains metadata for more than 4 million publications, which are listed on more than 2 million author profiles. As can be easily seen from those numbers, purely manual curation of author profiles is impracticable. Therefore, algorithmic methods for supporting AND tasks are necessary. The two most notorious problem categories that lead to incorrect attribution of authorship are: (1) cases when different persons share the same name (known as the *homonym* problem), and (2) cases when the name of a particular author is given in several different ways (known as the *synonym* problem).

We present and evaluate a machine learning approach to detect homonymous author profiles in large bibliographic databases. To this end, we train a standard multilayer perceptron to classify an author profile into either of the two classes "homonym" or "non-homonym". While the setup of our artificial neural network is pretty standard, we make use of two original components to build our classifier:

F. Reitz—Research funded by a grant of the Leibniz Competition, grant no. LZI-SAW-2015-2.

E. Méndez et al. (Eds.): TPDL 2018, LNCS 11057, pp. 59–65, 2018.
https://doi.org/10.1007/978-3-030-00066-0_5

(1) We use historic log data from the past years of active, manual curation at dblp to build a "golden" training and testing data set of more than 24,000 labeled author profiles for the homonym detection task. (2) We define a vectorization scheme that maps inhomogeneously sized and structured author profiles onto numerical vectors of fixed dimension. The design of these numerical features is based on the practical experience and domain knowledge obtained by the dblp team and uses only a minimal amount of core bibliographic metadata. Please note that our approach has been designed as an effort to improve dblp. Instead of trying to algorithmically resolve the defect, it intends to keep a human curator in the loop and just uncovers defective profiles. Reliable, fully automatic approaches are still an open problem.

Related Work. The vast majority of recent approaches [2] tackle AND as a batch task by re-clustering all existing publications at once. However, in the practice of a curated database, AND is performed rather incrementally as new metadata is added. Only recently, a number of approaches have considered this practice-driven constraint [3–7]. With the recent advances made in artificial intelligence, a number of (deep) artificial neural network methods have also been applied to AND problems [8,9]. However, previous approaches focus on learning semantic similarity of individual publications. It is still unclear how this can be used to assess whole profiles. There exist many data sets derived from dblp that are used to train or evaluate AND methods [10]. To the best of our knowledge, there is no data set that considers the evolution of curated author profiles beside the test collection of Reitz [11], which is the foundation of our contribution (see Sect. 2).

2   Learning Homonymous Author Bibliographies

One of dblp's characteristic features is the assignment of a publication to its individual author and the curation of bibliographies for all authors in computer science. In order to guarantee a high level of data quality, this assignment is a semi-automated process that keeps the human data curator in the loop and in charge of all decisions. In detail, for each incoming publication, the mentioned author names are automatically matched against existing author profiles in dblp using several specialized string similarity functions [12]. Then, a simple co-author network analysis is performed to rank the potential candidate profiles. If a match is found, the authorship record is assigned, but only after the ranked candidate lists have been manually checked by the human data curator. In cases that remain unclear even after a curator checked all candidates, a manual in-depth check is performed, often involving external sources. However, the amount of new publications processed each day makes exhaustive detailed checking impossible, which inevitably leads to incorrect assignments. Thus, while the initial checking of assignments ensures an elevated level of data quality, a significant number of defective author profiles still find their way into the database.

To further improve the quality of the database, another automated process checks all existing author profiles in dblp on a daily basis. This process is designed to uncover defects that become evident as a result of recently added or corrected data entries, and to present its findings to a human curator. By analyzing a profile and its linked coauthors for suspicious patterns, this process can detect probably synonymous profiles [13]. For the detection of probably homonymous profiles, no automated process has existed prior to the results presented here.

If a suspicious case of a synonymous or homonymous profile is validated by manual inspection, then it is corrected by either merging or splitting the profiles, or by reassigning a selection of publications. In 2017 alone, 9,731 profiles were merged and 3,254 profiles were split, while in 6,213 cases partial profiles were redistributed. This curation history of dblp forms a valuable set of "golden" training and testing data set for curating author profiles [11].

A Gold-Data Set for Homonym Detection. We use the historic dblp curation data from the embedded test collection as described by Reitz [11, Sect. 3.2] to build a "golden" data set for homonym detection. This collection compares dblp snapshots from different timestamps $t_1 < t_2$ and classifies the manual corrections made to the author profiles between t_1 and t_2. For this paper, we use historic data for the observation interval $[t_1, t_2]$ with $t_1 =$ "2014-01-01" and $t_2 =$ "2018-01-01". This test collection is available online as open data [11].

Within this collection, we selected all source profiles of defect type "Split" as our training and testing instances of label class "homonym". These are profiles at t_1 where a human curator at some point later between t_1 and t_2 decided to split the profile (i.e., the profile has actually been homonymous at t_1).

Additionally, from all other profiles in dblp at t_1, we selected the profiles which did either (a) contain non-trivial person information like a homepage URL or affiliation, or (b) at least one of the author's names in dblp ends by a "magic" 4-digit number (i.e., the profile had been manually disambiguated [1] prior to t_1) as instances of label class "non-homonym". Those profiles had all been checked by a human curator at some point prior to t_1, and the profiles were not split between t_1 and t_2. While this is not necessarily a proof of non-homonymity, such profiles are generally more reliable than an average, random profile from dblp.

To further rule out trivial cases for both labels, we dropped profiles that at t_1 did list either less than two publications or less than two coauthors. We ended up with 2,802 profiles labeled as "homonym" and 21,576 profiles labeled as "non-homonym" (i.e., a total of 24,378 profiles) from dblp at t_1.

Vectorization of Author Profiles. In order to train an artificial neural network using our labeled profiles, we need to represent the non-uniformly sized author profiles at timestamp t_1 as numerical vectors of fixed dimension. Our vectorization makes use of two precomputed auxiliary structures: (1) For each profile, we use a very simple connected component approach to cluster its set of coauthors as follows. First, we consider the local (undirected) subgraph of the dblp coauthor network containing only the current person and all direct coauthors as nodes. We call this the local coauthor network. Then, we remove the current person and all incident edges from the local coauthor network. The remaining connected components form the *local coauthor clustering* of the current person. (2) We train a vector representation of all title words in the dblp corpus using the word2vec algorithm [14]. In the vectorization below, we use this *title word embedding* model as basis to compute paragraph vectors (also known as doc2vec [15]) of whole publication titles, or even collections of titles. For the concrete model hyperparameters, see the full version of this paper [16].

The design of the feature components of our vectorization is based on the experience and domain knowledge obtained by the dblp team during the years of actively curating dblp's author profiles. That is, we identified the following feature groups that are implicitly and explicitly taken into consideration whenever a human curator is assessing the validity of a profile. A detailed listing of all features is given in the full version of this paper [16]. In Sect. 3, we will study the impact of each feature group on the classifier's performance.

- *group B (3 dims):* Basic, easy-to-compute facts of the author's profile, i.e., the number of publications, coauthors, and coauthor relations on that profile.
- *group C (7 dims):* Features of the local coauthor clustering, like the number of clusters and features of their size distribution. The aim of this feature set is to uncover the incoherence of local coauthor communities, which is symptomatic of homonymous profiles, as experience shows.
- *group T (12 dims):* Geometric features (in terms of cosine distance) of the embedded paragraph vectors for all publication titles listed on that profile. This feature set aims to uncover inhomogeneous topics of the listed publications, which might be a sign of a homonymous profile.
- *group V (13 dims):* Geometric features (in terms of cosine distance) of the embedded paragraph vectors for all venues (i.e., journals and conference series) listed on that profile, where each venue is represented by the complete collection of all titles published in that venue. This feature set also aims to uncover inhomogeneous topics by using the aggregated topical features of its venue as a proxy for the actual publication.
- *group Y (4 dims):* Features of the publication years listed on that profile. The aim of this feature group is to uncover profiles that mix up researchers with different years of activity.

3   Evaluation

We implemented and trained a standard multilayer perceptron with three hidden layers using the open-source Java library DeepLearning4J [17]. For the concrete model and evaluation setup see the full version of this paper [16].

Before running our experiments, we randomly split our gold-data profiles into fixed sets of 80% training and 20% testing profiles. Since neural networks work best when data is normalized, we rescaled all profile features to have an empirical mean of 0.0 and a standard error of 1.0 on the training data. For each set of vectorization feature groups we studied, 25 models had been trained independently on the training data and evaluated on the testing data.

Since we find our case label classes to be unbalanced, measures like precision, recall, and F1-score are known to give misleading scores [18]. Hence, we rather use Matthews correlation coefficient (MCC) [19] or the area under the receiver operating characteristic (AUROC) [20] instead for evaluation. Both measures are known to yield reliable scores for diagnostic tests even if class labels are severely unbalanced [18]. However, in Table 1, we still give precision, recall, and F1-score in order to allow for our results to be compared with other studies.

Results. As can be seen from Table 1 – and probably not surprisingly – our classifier is most effective if all studied feature groups are taken into consideration (i.e., feature set "BCTVY"). Note that for this set of features, precision is much higher than recall. However, this is actually tolerable in our real-world application scenario of unbalanced label classes: We need to severely limit the number of false-positively diagnosed cases (i.e., we need a high precision) in order to have our classifier output to be practically helpful for a human curator, while at the same time in a big bibliographic database, the ability to manually curate defective profiles is more likely limited by the team size than by the number of diagnosed cases (i.e., recall does not necessarily need to be very high).

One interesting observation that can be made in Table 1 is that the geometric features of the publication titles alone do not seem to be all too helpful (see feature set "BT"), while the geometric features of the aggregated titles of the venues seem to be the single most helpful feature group (see feature set "BV"). We conjecture that this is due to mere title strings of individual publications not being expressive and characterful enough in our setting to uncover semantic similarities. One way to improve feature group T would be to additionally use keywords, abstracts, or even full texts to represent a single publication, provided that such information is available in the database. However, it should be noted that even in its limited form, feature group T is still able to slightly improve the classifier if combined with feature group V (see feature set "BTV" in Table 1).

Table 1. Result scores of 25 independently trained classifiers for the different vectorization feature groups we studied, given as "mean $\pm$ standard deviation".

Features	Precision	Recall	F1-score	MCC	AUROC
B	0.823 ± 0.311	0.024 ± 0.012	0.047 ± 0.022	**0.130 ± 0.055**	0.799 ± 0.013
BC	0.818 ± 0.173	0.057 ± 0.016	0.106 ± 0.028	**0.197 ± 0.045**	0.842 ± 0.005
BT	0.542 ± 0.177	0.051 ± 0.030	0.092 ± 0.052	**0.138 ± 0.060**	0.786 ± 0.009
BV	0.745 ± 0.040	0.232 ± 0.047	0.350 ± 0.068	**0.372 ± 0.055**	0.815 ± 0.006
BY	0.781 ± 0.022	0.153 ± 0.014	0.256 ± 0.020	**0.314 ± 0.016**	0.820 ± 0.004
BTV	0.709 ± 0.011	0.268 ± 0.013	0.389 ± 0.015	**0.393 ± 0.013**	0.832 ± 0.003
BCTVY	**0.793 ± 0.009**	**0.424 ± 0.011**	**0.552 ± 0.010**	**0.541 ± 0.008**	**0.890 ± 0.002**

In addition to our experiments, we implemented a first prototype of a continuous homonym detector to be used by the dblp team in order to curate the author profiles of the live dblp database. To this end, all dblp author profiles are vectorized and assessed by our classifier on a regular basis. This prototype does not just make use of the binary classification as in our analysis of Table 1, but rather ranks suspicious profiles according to the probability of label "homonym" as inferred by our classifier (i.e., the softmax score of label "homonym' in the output layer). As a small sample from practice, we computed the top 100 ranked profiles from the dblp XML dump of April 1, 2018 [21], and we checked those profiles manually. We found that in that practically relevant top list, 74 profiles where correctly uncovered as homonymous profiles, while 12 profiles where false positives, and for 14 profiles the true characteristic could not be determined even after manually researching the case.

References

1. Ley, M.: DBLP - some lessons learned. PVLDB **2**(2), 1493–1500 (2009)
2. Ferreira, A.A., Gonçalves, M.A., Laender, A.H.F.: A brief survey of automatic methods for author name disambiguation. SIGMOD Rec. **41**(2), 15–26 (2012)
3. de Carvalho, A.P., Ferreira, A.A., Laender, A.H.F., Gonçalves, M.A.: Incremental unsupervised name disambiguation in cleaned digital libraries. JIDM **2**(3), 289–304 (2011)
4. Esperidião, L.V.B., et al.: Reducing fragmentation in incremental author name disambiguation. JIDM **5**(3), 293–307 (2014)
5. Qian, Y., Zheng, Q., Sakai, T., Ye, J., Liu, J.: Dynamic author name disambiguation for growing digital libraries. Inf. Retr. J. **18**(5), 379–412 (2015)
6. Santana, A.F., Gonçalves, M.A., Laender, A.H.F., Ferreira, A.A.: Incremental author name disambiguation by exploiting domain-specific heuristics. JASIST **68**(4), 931–945 (2017)
7. Zhao, Z., Rollins, J., Bai, L., Rosen, G.: Incremental author name disambiguation for scientific citation data. In: DSAA 2017, pp. 175–183. IEEE (2017)
8. Tran, H.N., Huynh, T., Do, T.: Author name disambiguation by using deep neural network. In: Nguyen, N.T., Attachoo, B., Trawiński, B., Somboonviwat, K. (eds.) ACIIDS 2014. LNCS (LNAI), vol. 8397, pp. 123–132. Springer, Cham (2014). https://doi.org/10.1007/978-3-319-05476-6_13
9. Müller, M.-C.: Semantic author name disambiguation with word embeddings. In: Kamps, J., Tsakonas, G., Manolopoulos, Y., Iliadis, L., Karydis, I. (eds.) TPDL 2017. LNCS, vol. 10450, pp. 300–311. Springer, Cham (2017). https://doi.org/10.1007/978-3-319-67008-9_24
10. Müller, M., Reitz, F., Roy, N.: Data sets for author name disambiguation: an empirical analysis and a new resource. Scientometrics **111**(3), 1467–1500 (2017)
11. Reitz, F.: Two test collections for the author name disambiguation problem based on DBLP, March 2018. https://doi.org/10.5281/zenodo.1215650
12. Ley, M., Reuther, P.: Maintaining an online bibliographical database: The problem of data quality. In: EGC 2006. RNTI, vol. E-6, pp. 5–10. Éditions Cépaduès (2006)
13. Reuther, P.: Personal name matching: new test collections and a social network based approach. Technical report 06-1, University of Trier (2006)
14. Mikolov, T., Sutskever, I., Chen, K., Corrado, G.S., Dean, J.: Distributed representations of words and phrases and their compositionality. In: NIPS 26, pp. 3111–3119 (2013)

15. Le, Q.V., Mikolov, T.: Distributed representations of sentences and documents. In: ICML 2014. JMLR Proceedings, vol. 32, pp. 1188–1196. JMLR.org (2014)
16. Ackermann, M.R., Reitz, F.: Homonym detection in curated bibliographies: learning from dblp's experience (full version). arXiv:1806.06017 [cs.DL] (June 2018)
17. Gibson, A., Nicholson, C., Patterson, J.: Eclipse DeepLearning4J v0.9.1. https://deeplearning4j.org
18. Powers, D.M.W.: Evaluation: from precision, recall and F-measure to ROC, informedness, markedness and correlation. Technical report SIE-07-001, Flinders University (2007)
19. Matthews, B.W.: Comparison of the predicted and observed secondary structure of T4 phage lysozyme. BBA Protein Struct. **405**(2), 442–451 (1975)
20. Ling, C.X., Huang, J., Zhang, H.: AUC: a statistically consistent and more discriminating measure than accuracy. In: IJCAI 2003, pp. 519–526. Morgan Kaufmann (2003)
21. DBLP: XML of 1 April 2018. https://dblp.org/xml/release/dblp-2018-04-01.xml.gz

Data Management

Research Data Preservation Using Process Engines and Machine-Actionable Data Management Plans

Asztrik Bakos[ID], Tomasz Miksa[(✉)][ID], and Andreas Rauber[ID]

SBA Research and TU Wien, Favoritenstrasse 16, 1040 Wien, Austria
miksa@ifs.tuwien.ac.at

Abstract. Scientific experiments in various domains require nowadays collecting, processing, and reusing data. Researchers have to comply with funder policies that prescribe how data should be managed, shared and preserved. In most cases this has to be documented in data management plans. When data is selected and moved into a repository when project ends, it is often hard for researchers to identify which files need to be preserved and where they are located. For this reason, we need a mechanism that allows researchers to integrate preservation functionality into their daily workflows of data management to avoid situations in which scientific data is not properly preserved.

In this paper we demonstrate how systems used for managing data during research can be extended with preservation functions using process engines that run pre-defined preservation workflows. We also show a prototype of a machine-actionable data management plan that is automatically generated during this process to document actions performed. Thus, we break the traditional distinction between platforms for managing data during research and repositories used for preservation afterwards. Furthermore, we show how researchers can easier comply with funder requirements while reducing their effort.

Keywords: Data managament · Repositories
Machine-actionable DMPs · Data management plans
Digital preservation · BPMN

1 Introduction

In data driven research projects it is important to have access to previous experiments or results [7]. To enable citation, linking or referring to older results, it is vital to have a repository to store data [8]. Repositories guarantee long-term access, over several projects, researcher groups and years [11]. Thus, they support reproducibility and enable verification and validation of scientific findings [10]. Policy makers, such as the European Commission, require data to be open and available for a broader public [5]. Storing research data in a shared folder is not a viable solution, since uploading, handling versions, annotating files, citation

© Springer Nature Switzerland AG 2018
E. Méndez et al. (Eds.): TPDL 2018, LNCS 11057, pp. 69–80, 2018.
https://doi.org/10.1007/978-3-030-00066-0_6

and configuring access are beyond the functionality of a simple network drive and require technical skills.

Following recommendations on data management and preservation, such as FAIR [9], not only requires technical skills, but also requires real effort, meaning time to complete all necessary tasks. Thus, data management and preservation distracts researchers from their core interests and is a time consuming side activity. For example, submitting files into a repository is regarded as an administrative task. If it is not only pressing a button and answering few inputs, then researchers tend not to go through all the steps. Research projects often have hundreds of input or intermediate files: raw data, experiment configurations, stored on a local fileshare or disk. At the end of the project the files are preserved. If preserving one file takes only one or two minutes, preserving some hundred files would take half of a workday. This is a time consuming activity for researchers and/or repository operators who facilitate this process. Another problem is the complexity of the preservation workflow. Various file types require different handling, storing, access method or rights. If images are processed using one tool, and archives using another, the manual decision and tool selection again increases the time spent preserving each research file.

In this paper we propose a system that acts as a platform for data management during research projects and a repository for preserving data. Our solution is based on a content management system that can execute business processes and can process information from machine-actionable data management plans. Researchers use the system to organise their re-search data and select those files that must be kept for a long term. Then pre-defined preservation processes modelled as BPMN processes are executed by the workflow engine. The workflow engine uses information on data obtained directly from the platform, for example, by running profiling tools, and appends it to the machine-actionable DMP. As a result, researchers make sure their data is properly preserved, save effort and comply with funder policies. We created a working example of the proposed system using Alfresco. Describing the concrete preservation tasks is not in the scope of this paper. We propose an approach to automate research data preservation. The actual conditions for branching and tools depend on the given research project and the stakeholder's needs.

The paper is organized as follows. Related works are discussed in Sect. 2. Section 3 provides an overview of the proposed system. Section 4 introduces its architecture and discusses main building blocks. Section 5 describes the proof of concept implementation. Section 6 provides conclusions and future work.

2   Related Work

Data repositories store all sorts of research related data. Tasks of a repository are: handling up- and downloads, grant access to files. Apart from the basic features a repository can support a preservation process, for example Archivematica where a complete OAIS [15] workflow is run when a research file is uploaded [3]. Similar open platforms are DSpace and Islandora, with additional versioning. The trend shows that with every newer version there are less restrictions regarding accepted file types [14]. Cloud-based repositories like Zenodo or Digital Commons [3] are accessible for larger projects, but the physical location of the data is not known. This problem can be addressed by developing a local institutional repository [2].

Data Management principles are typically defined by research funders [1]. Usually the repository operator's task is to manage data following a data management plan conform to the policies [12]. This plan contains information about the identification of the data object and its preservation. To help writing data management plans, one can use the checklist provided by the DCC [6]. Even if a data management plan contains every information and a repository operator still has to manually execute it. To automate this process, we need to make data management plans *machine actionable*. This means that the plan is software readable and writable too: a preservation system can execute and control the plan [13]. However machine actionable data management plans [16] are still a work in progress of the Research Data Alliance[1]. In this paper we suggest a possible application for machine actionable data management plans and a specific data model too.

Storing research data can be realised with a variety of systems. The initial approach to store research files is a file share (e.g. ActiveDirectory or Samba), where a researcher can upload files without additional features. The Open Science Framework[2] is an online collaboration platform combining services to manage research projects. Only preservation tools are not provided as services, and configuring OSF to follow data management policies is not possible when requirements differ for file types. A similar approach is the CERN Analysis preservation [4] where the file preservation is included into the analysis workflow. In other words the preservation is done right when the research data is being processed. In this paper we use a storage-specialized content management system, Alfresco[3] that handles workflows and custom data types. By using Alfresco we can set up a content management system with the same functionality as OSF and extended it with workflow and user management, as well as preservation workflow.

Process Models are structured descriptions of processes that prescribe specific tasks which should be executed taking into account decision criterion and available inputs. In this paper, we use Business Process Model Notation (BPMN)[4]. BPMN has a wide range of flow elements: tasks, gateways, flows and most importantly a standard BPMN model is read and executed by process engines.

3   Proposed Solution - Big Picture

In this section we describe a data management system that reduces the researcher's workload during research data management and preservation. It is

---

[1] https://www.rd-alliance.org/groups/dmp-common-standards-wg.

[2] https://osf.io/.

[3] https://www.alfresco.com/.

[4] http://www.omg.org/bpmn/.

based on a typical content management system that allows researchers to upload files, edit and share them, etc. The system is also equipped with a process engine that allows executing pre-defined processes. Such systems exist and are popular in business settings. For example, IBM[5] offers a system allowing to share files among users, and define tasks that are parts of automatically executed business processes. Non-commercial equivalents exist and are discussed in Sect. 2. In the reminder of this section we use a running example to show how such a system can be customised to fit into the landscape of research data management, preservation, and funder requirements.

Assume, we are in a research team, where the research data is stored in various images, coming from many sources in many formats and sizes. Our research team has to preserve input and result files and make them FAIR. The situation is depicted in Fig. 1. We have the following actors with the following responsibilities:

Management: delivers and enforces data management policies.
Scientist: Uploads research data, and triggers the preservation process.
Repository administrator: creates a BPMN process based on policies.
System: executing the process, handling files, storage and access pages.

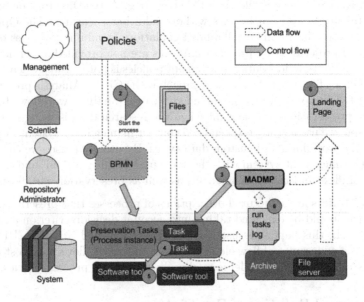

Fig. 1. The big picture showing essential components of the proposed system and actors involved.

The policy defines that *reusable and long-term accessible* format is a JPEG file with lossless compression. Due to the size and resolution of our images, all

---

[5] IBM Business Process Manager: https://www.ibm.com/us-en/marketplace/ business-process-manager.

of them must be resized to 70% of the original size before being preserved. Now if we upload an incoming PNG file into our system, the following events happen and are depcited in Fig. 1:

1. For preparation a basic instance of the data management plan is created, along with the process model for the whole process (see more on process models in Sect. 4.1). Here the repository operator designs the process, for example: if a PNG file is uploaded, then converti it and resize it.
2. A scientist starts the preservation process on the selected files.
3. The files get identified. One is a 2000×3000 pixel PNG. The identification tool fills the fields *file format* and *dimensions* automatically. If a field needs an expert's interaction (where one needs to understand what the image shows) then it is inputted via an auto-generated form.
4. The file's data management plan instance is appended with the file info (see more on machine-actionable data management plans in Sect. 4.2). Now the process has all the information required to proceed with the preservation. In the following steps the process engine will read the metadata fields written in this step.
5. The preservation is executed on the file as the process model describes: since the file was identified as a PNG, it needs to be converted. This means that the process engine will call a software to do the conversion, and later the resize too.
6. The user is notified by the system that the JPEG file has been preserved and is redirected to a landing page. The user can also download the data management plan as JSON-LD[6] or PDF.

As we can see in Fig. 1, most of the action in the process takes place below the level of the user interaction, i.e. the researcher only needs to start the process, fill some (and not all) of the metadata fields and the rest is done by the underlying system. Furthermore the files are stored in the content management system exactly as they should be preserved according to the policy. This mean there is no need for a preservation after the project is finished (apart from requirements stemming from preservation planning performed in the future), everything is on its place during the project. Thus the researcher's (or the repository operator's) work is reduced to the necessary minimum. In the next section we explain what special process definition and data management plan is necessary to make the design implementable.

The demonstrated workflow is simple, however it contains the core building elements of a complex preservation process. We showed that writing metadata and branching based on the read values to run specific external tools is possible. In a workflow one can combine conditions, and reference any program or script needed for the preservation.

---

[6] A JSON format supporting the use of ontologies: https://json-ld.org/.

4   Architecture for Automating Research Data Preservation

In this section we describe how content management systems can be extended with digital preservation capabilities. For this purpose we use process modeling engines available in a content management system. We also demonstrate how information from data management plans can become machine-actionable, that is, we describe a model we developed to organize the information in machine-actionable way and demonstrate how this information is used by the system to take actions and thus reduce the effort required from both researchers and repository operators.

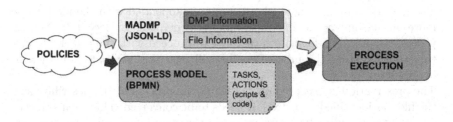

Fig. 2. Schema of the proposed architecture

The essential building blocks of the proposed solution are: (1) a content management system; (2) a model of a preservation process; (3) policy-based machine-actionable data management plans. Figure 2 illustrates the relation between the components. Policies provide a high level requirements for both repository and machine actionable data management plans. They prescribe actions and conditions applying to data and data management. Typically, updating the metadata for a research data object has to happen manually. This metadata is later compared with the values in the data management plan to decide further execution. In our solution, we propose to obtain data automatically from the data object and thus reduce the manual workload.

4.1   BPMN Workflows

Researchers use the platform to manage their data. A data management plan contains information describing processed files. This information is needed by the preservation process that is modelled as a BPMN process. When researchers upload a new data object to the platform, the system automatically generates a DMP. This initial DMP is *empty* - it has only fields that need to be filled, but not yet the values. System analyses uploaded files and adds information like file type, size, resolution (in case of images) to the data management plan. All this information will be later used by a pre-defined BPMN workflows to make decisions when users trigger the corresponding preservation process.

Each property of the data management plan is regarded as a decision point, since it influences how the data object is handled later on. The exact method how the data object is handled, is called an execution branch. The execution branch consists of tasks, transformations, tools. There might be other branches for different object types. For example an image will not be handled the same way as a text file would be: an image can be resized, but not a text file.

If all the decision points are defined, and all the branches are known, repository operators can draw the process model using the BPMN-notation. The process model is created once only for a given preservation process. An example of a preservation process is depicted in Fig. 3.

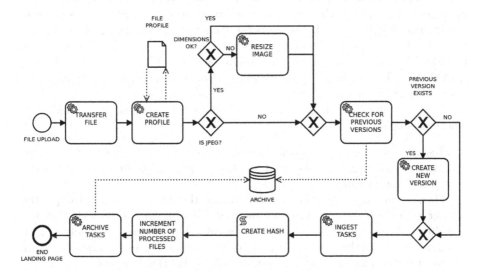

Fig. 3. Preservation workflow as a BPMN diagram

In the diagram depicted in Fig. 3 the *Create profile* machine task refers to a tool execution that outputs a profile. The profile contains for example the file format. In this example the system decides whether it is a JPEG image. If so, then it refers again to the profile to see the image dimensions. If the data management plan does not allow images wider than 2000 pixels, and the file being processed happens to be wider than 2000 pixels, then it will be automatically re-sized in the task *Resize image*. The next machine task, *Check for Previous Versions* looks into the system to see whether this data object was already processed. If yes, then the system needs to create a new version of the data object. After that follows the *Ingest Tasks* to create identifiers, descriptive info, quality assurance data. For example if a data management plan requires a hash to be calculated, then it will be done in the *Create hash* script task. Finally, a file counter is increased and the rest of the archive tasks will run.

We see that the workflow depicted in Fig. 3 is not complex, since the source data management plan contains only three decision points. This gives us only three exclusive gateways in the BPMN. In more complex scenarios, where the data management plan contains more regulations for file handling, differentiating we would need to specify exclusive gateways and task going point-by-point through the data management plan. If we translated the data management plan into a flow of decisions and branches, which we can easily rewrite into one of the standard process definition markups, like BPMN. The key to the automation is using a process engine, driven by the data management plan based BPMN.

4.2   Machine Actionable Data Management Plans

A BPMN process model is only one half of the automation: understanding a data management plan on such a level that we know the properties requires project-specific knowledge. Also a BPMN process cannot express every theme and point from the data management plan. Especially those that contain provenance or metadata information about the data object and which have to be stored along with the file. Therefore, we need to create a data management plan that can be read and written by the process.

The creation of the machine actionable data management plan (MADMP) has three steps:

1. First its data model is defined - which fields are necessary and which standards and ontologies can be used for that purpose. This step happens at the preservation planning stage.
2. When submitting a dataset into the system, the upload process prompts the user with a simple form to enter the metadata. It is important that all the field are fixed inputs, i.e. there is no free-text. This way we can safely rely on ontologies; since the checkboxes and dropdown contents are generated directly out of the ontologies. Then an instance of the MADMP is created.
3. The system processes the dataset by running some basic file identification tools. Based on the results now the data management plan can be completed, the rest of the file information (like size, type, format, encoding) will be inserted.

The resulting MADMP - as seen in Listing 1.1 - has a header part with the identifiers and specific metadata to the actual DMP, and one or more data object part ("hasDataObject"). In this second part each object is characterized using known ontologies listed in the header. The header also identifies the data management plan, and its author. For our task the part with the data objects is the most important. The key to machine-actionability is the fact that all the fields are identified by ontologies, even the domain specific ones. The preservation system can not know in advance all the possible metadata needed for biological data. So it will load the domain specific metadata fields from the ontology ("hasDomainSpecificExtension"). In Listing 1.1 on line 31 we load the hypothetical "PROV" extension, which has a field named "hasSpecialStandard", and a value set to "UHDTV" (line 33). And since the fields are clearly identifiable, we can fill those fields using an identification tool run by the preservation process automatically. Then in some later step an other tool can read this data.

Listing 1.1. Sample MADMP

```
 1  {
 2    "header": {
 3        "@context": {
 4                "dmp": "http://purl.org/madmps#",
 5                "dc": "http://purl.org/dc/elements/1.1/",
 6                "prov": "http://www.w3.org/ns/prov#",
 7        },
 8        "@id": "http://example.org/dmp/madmp-proto",
 9        "@type": "dmp:DataManagementPlan",
10        "dcterms:title": "Sample Machine Actionable DMP",
11        "dcterms:description": "Sample MADMP",
12        "dc:creator": [{
13                "@id": "orcid.org/0000-1111-2222-3333",
14                "foaf:name": "Asztrik Bakos",
15                "foaf:mbox": "mailto:author@uni.ac.at"
16        }],
17        "dcterms:hasVersion": "v1.0.2",
18        "dc:date": "2017-07-31"
19    },
20    "dmp:hasDataObject": [{
21        "@id": "123",
22        "@type": "dmp:File",
23        "dc:date": "2017-07-11",
24        "dcterms:hasVersion": "1.0",
25        "prov:wasDerivedFrom": {
26                "dmp:DataObject": {
27                        "@id": "122"
28                }
29        },
30        "dmp:hasMetadata": {
31                "dmp:hasDomainSpecificExtension": "PROV",
32                "dcterms:description": "File 1",
33                "prov:hasSpecialStandard": "dbo:UHDTV",
34        },
35        "premis:hasRestriction": "no",
36        "dmp:hasLicense": {
37                "premis:hasLicenseTerms": "CC-BY-3.0"
38        }
39    }]
40  }
```

The MADMP after archiving of the files will contain a list of the processed data objects. These data objects will include the file states before and after the preservation. Here we will use the *Basis Object* field of the MADMP; the after-preservation file will have its before-preservation counterpart as basis. It is important to note that the after and before preservation data object's field values might differ significantly. For example a file which was uploaded for archiving might be a personal document, without any licenses attached. But after being submitted to the repository, it might get a license as seen on line 37 of Listing 1.1.

5   Proof of Concept

In this section we describe the implementation of the proposed system using the Alfresco content management system which we extended with a process model and custom data types to support MADMPs. This evaluation also proves that the proposed content management system design is feasible.

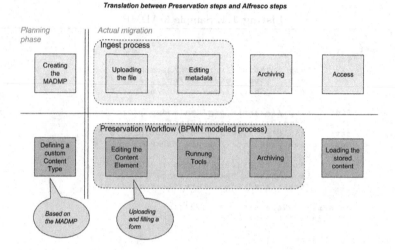

Fig. 4. Comparison of the design and the Alfresco setup

Figure 4 depicts how we configured and extended Alfresco to implement the proposed solution. The upper row contains the elements of the design, the lower row their Alfresco counterparts. To create the MADMP in Alfresco, we used a custom content type as an extension to Alfresco in which we defined the fields with ontologies. Then we created a BPMN process model. We used an online BPMN drawing tool[7] and modeled the preservation workflow we had. We entered manually the delegate expressions Alfresco uses, so that the process model instances could communicate with the system. Then we uploaded the BPMN file to the Alfresco server.

Running the setup requires a front end login into Alfresco to select and start our process model on one or more data files in the system. It is important to have the file as a part of our custom data type. The process engine then puts through the file, identifies it, runs the specified tools, transforms the file and writes the obtained metadata into the custom data type's fields. As a result, we can view the data object: the uploaded and transformed file with the new metadata.

Apart from making the preservation easier, writing the process model and configuring the repository takes more time than setting up a file server. The reason is that it needs a deeper understanding of the preservation process and so the initial setup is more complex, but brings long term benefits, because repetitive tasks performed by researchers are automated. We plan to evaluate the system further, by modelling typical preservation workflows known from systems like Archivematica using Alfresco and BPMN. Thus, we will be able to identify the necessary services to be added, as well as measure the complexity of workflows. Since Archivematica has a flat workflow without branches its workflow can easily be modelled in BPMN and included in the system proposed by us.

---

[7] http://bpmn.io/.

We will also experiment with more complex workflows preserving heterogeneous research data, such as NetCDF, HDF5.

6   Conclusions and Future Work

In this paper we discussed the problem of managing and preserving scientific data. We proposed a new system that acts both as a platform for data management during research project and a repository for preserving files. Thus, we showed how the effort and complexity of moving files between systems can be reduced.

Our solution is based on a content management system that executes business processes and processes information from machine-actionable data management plans. Researchers use the system to organise research data and select those files that must be kept for long term. Then pre-defined preservation processes modelled as BPMN processes are executed by the workflow engine. The workflow engine uses information on data obtained directly from the platform, for example, by running profiling tools, and appends it to the machine-actionable DMP. As a result, researchers make sure their data is properly preserved, save effort and comply with funder policies. We created a working example of the proposed system using Alfresco.

Future work will focus on modelling further preservation workflows and modelling more complex BPMN diagrams to include elements such as: user tasks, parallel gateways, messages. We will also investigate in what way other data management systems can be extended with preservation functions.

Acknowledgments. This research was carried out in the context of the Austrian COMET K1 program and publicly funded by the Austrian Research Promotion Agency (FFG) and the Vienna Business Agency (WAW).

References

1. H2020 programme guidelines on fair data management in horizon 2020. EC Directorate General for Research and Innovation (2016)
2. Bankier, J.G.: Institutional repository software comparison. UNESCO (2014)
3. Castagne, M.: Institutional repository software comparison: DSpace, EPrints, Digital Commons, Islandora and Hydra. University of British Columbia (2013)
4. Chen, X., et al.: CERN analysis preservation: a novel digital library service to enable reusable and reproducible research. In: Fuhr, N., Kovács, L., Risse, T., Nejdl, W. (eds.) TPDL 2016. LNCS, vol. 9819, pp. 347–356. Springer, Cham (2016). https://doi.org/10.1007/978-3-319-43997-6_27
5. European Commission: European Open Science Cloud Declaration (2017)
6. DCC: Checklist for a Data Management Plan. v. 4.0. Edinburgh: Digital Curation Centre (2013). http://www.dcc.ac.uk/resources/data-management-plans. Accessed 29 Mar 2018

7. Darema, F.: Dynamic data driven applications systems: a new paradigm for application simulations and measurements. In: Bubak, M., van Albada, G.D., Sloot, P.M.A., Dongarra, J. (eds.) ICCS 2004. LNCS, vol. 3038, pp. 662–669. Springer, Heidelberg (2004). https://doi.org/10.1007/978-3-540-24688-6_86

8. Hey, T., Tansley, S., Tolle, K.: The Fourth Paradigm: Data-Intensive Scientific Discovery. Microsoft Research, Redmond (2009)

9. Wilkinson, M.D., et al.: The FAIR guiding principles for scientific data management and stewardship. Nature Sci. Data **3** (2016). Article no. 160018

10. Miksa, T., Rauber, A., Mina, E.: Identifying impact of software dependencies on replicability of biomedical workflows. J. Biomed. Inform. **64**(C), 232–254 (2016)

11. Otto, B.: Data governance. Microsoft Res. **4**, 241–246 (2011)

12. Proell, S., Meixner, K., Rauber, A.: Precise data identification services for long tail research data. In: iPRES 2016 (2016)

13. Rauber, A., Miksa, T., Ganguly, R., Budroni, P.: Information integration for machine actionable data management plans. In: IDCC 2017 (2017)

14. Rosa, C.A., Craveiro, O., Domingues, P.: Open source software for digital preservation repositories: a survey. Int. J. Comput. Sci. Eng. Surv. (IJCSES) **8**(3), 21–39 (2017)

15. Schembera, B., Bönisch, T.: Challenges of research data management for high performance computing. In: Kamps, J., Tsakonas, G., Manolopoulos, Y., Iliadis, L., Karydis, I. (eds.) TPDL 2017. LNCS, vol. 10450, pp. 140–151. Springer, Cham (2017). https://doi.org/10.1007/978-3-319-67008-9_12

16. Simms, S., Jones, S., Mietchen, D., Miksa, T.: Machine-actionable data management plans. Res. Ideas Outcomes **3**, e13086 (2017)

Maturity Models for Data and Information Management

A State of the Art

Diogo Proença[1,2(✉)] and José Borbinha[1,2]

[1] Instituto Superior Técnico, Universidade de Lisboa, Lisbon, Portugal
{diogo.proenca, jlb}@tecnico.ulisboa.pt
[2] INESC-ID, Lisbon, Portugal

Abstract. A Maturity Model is a widely used technique that is proved to be valuable to assess business processes or certain aspects of organizations, as it represents a path towards an increasingly organized and systematic way of doing business. A maturity assessment can be used to measure the current maturity level of a certain aspect of an organization in a meaningful way, enabling stakeholders to clearly identify strengths and improvement points, and accordingly prioritize what to do in order to reach higher maturity levels. This paper collects and analyzes the current practice on maturity models from the data and information management domains, by analyzing a collection of maturity models from literature. It also clarifies available options for practitioners and opportunities for further research.

Keywords: Maturity · Maturity model · Maturity assessment
Data management · Information management

1 Introduction

A maturity model is a technique that proved valuable in measuring different aspects of a process or an organization. It represents a path towards increasingly organized and systematic way of doing business in organizations.

A maturity model consists of a number of "maturity levels", often five, from the lowest to the highest, initial, managed, defined, quantitatively managed and optimizing (however, the number of levels can vary, depending on the domain and the concerns motivating the model). This technique provides organizations: (1) a measuring for auditing and benchmarking; (2) a measuring of progress assessment against objectives; (3) an understanding of strengths, weaknesses and opportunities (which can support decision making concerning strategy and project portfolio management).

We can trace the subject of maturity models back to 1973 [1], and recognize as highlights the Software Engineering Institute (SEI) Capability Maturity Model Integration (CMMI) [2] that was first presented in 1991, and in 2004 the ISO/IEC 15504 [3]. Both the CMMI and ISO/IEC 15504 are key references, born in the Software Engineering domain, culminating decades of development and refinement of the corresponding models. Moreover, there is certification for these two references, which are

© Springer Nature Switzerland AG 2018
E. Méndez et al. (Eds.): TPDL 2018, LNCS 11057, pp. 81–93, 2018.
https://doi.org/10.1007/978-3-030-00066-0_7

the de facto assessment techniques to use when benchmarking organizations for their software engineering process implementation and maturity. As such, in order for the results to be comparable, there is a detailed maturity assessment method behind each of these maturity models. These methods define in detail how to plan, conduct, and determine the maturity levels of an assessment and how to present the results to the organization. These methods make each assessment repeatable and comparable with results from other organizations, allowing for benchmarking.

This paper is structured as follows. The first sections of this paper provide a background to the concepts of maturity and maturity model, the rationale for the selection of the maturity models examined, and a brief description of each maturity model. A more detailed analysis follows. The next section provides a discussion on the analysis of the selected maturity models. A concluding section highlights gaps in the current range of maturity models and identifies opportunities for further research.

2 Background

To evaluate maturity, organizational assessment models are used, which are also known as stages-of-growth models, stage models, or stage theories [4].

Maturity is a state in which, when optimized to a particular organizational context, is not advisable to proceed with any further action. It is not an end, because it is a "mobile and dynamic goal" [4]. It is rather a state in which, given certain conditions, it is agreed not to continue any further action. Several authors have defined maturity, however many of these definitions fit into the context in which each the maturity model was developed.

In [5], the authors define maturity as a specific process to explicitly define, manage, measure and control the evolutionary growth of an entity. In turn, in [6] the authors define maturity as a state in which an organization is perfectly able to achieve the goals it sets itself. In [7] it is suggested that maturity is associated with an evaluation criterion or the state of being complete, perfect and ready and in [8] as being a concept which progresses from an initial state to a final state (which is more advanced), that is, higher levels of maturity. Similarly, in [9] maturity is related with the evolutionary progress in demonstrating a particular capacity or the pursuit of a certain goal, from an initial state to a final desirable state. In [10] maturity is seen as the "extent to which an organization has explicitly and consistently deployed processes that are documented, managed, measured, controlled, and continually improved."

Most maturity model definitions found in literature clarify that maturity models are particularly important for identifying strengths and weaknesses of the organizational context to which they are applied, and the collection of information through methodologies associated with benchmarking.

In [4] the authors define maturity models as a series of sequential levels, which together form an anticipated or desired logical path from an initial state to a final state of maturity. Röglinger et al. [4] explain that a maturity model includes "a sequence of levels (or stages) that together form an anticipated, desired, or logical path from an initial state to maturity."

Some definitions involve common organizational concepts. For example, the definition in [11] defines a maturity model as "... a framework of evaluation that allows an organization to compare their projects and against the best practices or the practices of their competitors, while defining a structured path for improvement." This definition is deeply embedded in the concept of benchmarking. In other definitions, such as the in [12], we identify the concern of associating a maturity model to the concept of continuous improvement. In [13], it was concluded that the great advantage of maturity models is that they show that maturity must evolve through different dimensions and, once reached a maturity level, sometime is needed for it to be actually sustained.

The SEI explains that a maturity model "contains the essential elements of effective processes for one or more areas of interest and describes an evolutionary improvement path from ad hoc, immature processes to disciplined, mature processes with improved quality and effectiveness."

Currently, the lack of a generic and global standard for maturity models has been identified as the cause of poor dissemination of this concept.

3   Overview of Maturity Models

Many maturity models are referenced in discussion in the domains related with data and information management. We performed literature searches in Scopus and Google Scholar for variations of 'data management', 'information management', 'model', 'maturity', capability', 'assessment', 'improvement', 'measurement', as well sampling the citations that resulted from this initial set of literature sources. Community websites were also explored, such as the Research Data Alliance (RDA), the Digital Preservation Coalition, and the CMMI Institute.

We additionally included models based on our knowledge with existing publications. Next, we developed a set of inclusion and exclusion criteria. First, the maturity assessment must be explicitly focused on the domains that surround the data and information management topics. There are models that measure almost every perceivable aspect of an organization without focusing on the aspects relating with the management of data and information. One example is the ISACA's COBIT5 which focuses on measuring IT Governance. This model has several measures for aspects relating with data and information in the IT context. However, the model does not explicitly focus on data and information management aspects and for this reason was not considered for this work.

Further, we excluded models that their explicit outputs are not a set of maturity or capability levels. These models can be used to assess an organization however their output can be a percentage value or a grade without the underlying concept of maturity. Such models include, the ISO16363 [15] and the Data Seal of Approval[1] from the digital preservation domain, and MoRec[2] from the records management domain.

---

[1] https://www.datasealofapproval.org/en/.

[2] http://www.moreq.info/.

Table 1. Maturity models analyzed.

Name	Abbreviation	Year	Reference
Gartner Enterprise Information Management Maturity Model	EIMM	2008	[25]
ARMA Information Governance Maturity Model	IGMM	2010	[26]
Enterprise Content Management Maturity Model	ECMM	2010	[27]
Recordkeeping Maturity Model and Roadmap	RKMM	2010	[28]
Asset Management Maturity Model	AMMM	2011	[29]
Stanford Data Governance Maturity Model	DGMM	2011	[30]
Digital Preservation Capability Maturity Model	DPCMM	2012	[31]
Brown Digital Preservation Maturity Model	BDPMM	2013	[32]
JISC Records Management Maturity Model	RMMM	2013	[33]
CMMI Institute Data Management Maturity Model	DMMM	2014	[34]
SU Capability Maturity Model for Research Data Management	CMMRDM	2014	[35]
Preservica Digital Preservation Maturity Model	PDPMM	2015	[36]
Digital Asset Management Maturity Model	DAMM	2017	[37]
E-ARK Information Governance Maturity Model	A2MIGO	2017	[40, 41]

In total 14 maturity models met all the criteria, described briefly below in chronological order and listed together with the respective abbreviation and main reference in Table 1.

Gartner Enterprise Information Management Maturity Model (2008) – Enterprise Information Management (EIM) is defined by Gartner as the organizational commitment to "structure, secure and improve the accuracy and integrity of enterprise information; solve semantic inconsistencies across boundaries and; support the objectives of enterprise architecture and the business strategy" [25]. Gartner proposes six phases of maturity regarding EIM. Where in level 0 there are no EIM activities in place and in level 5 EIM is fully implemented in the organization. It provides examples on how to get questions from the maturity criteria. However, it does not provide a method or guidelines for assessment using this maturity model. This maturity model consists of five maturity levels with no attributes defined.

ARMA Information Governance Maturity Model (2010) – Builds on the generally accepted recordkeeping principles developed by ARMA[3]. The principles provide high-level guidelines of good practice for recordkeeping although they do not go into detail to the implementation of these principles and do not have further details on policies, procedures, technologies and roles [26]. There are a series of steps to assess the current maturity level and identify the desired level. These steps are not formal and consist of simple statements of what to do without defined guidance on how to perform the steps. This maturity model consists of five maturity levels with eight attributes defined which are called principles.

---

[3] www.arma.org/principles.

Enterprise Content Management (ECM) Maturity Model (2010) – Provides the tools to build a roadmap for ECM improvement by providing the current state of ECM implementation as well as a roadmap to reach the required maturity level [27]. Its aim is to build a roadmap for ECM improvement, in a step-by-step fashion ranging from basic information collection and simple control to refined management and integration. No assessment method is described, the way of getting the current level is done by the organization itself by checking if the organization possesses all the requirements for a given level regarding a specific dimension. This maturity model consists of five maturity levels with attributes at two levels, three attributes called categories which are further decomposed into 13 dimensions.

Recordkeeping Maturity Model and Roadmap (2010) – Developed to improve recordkeeping practice is Queensland public authorities. It builds on the premise that "good business practice incorporates regular review and assessment to ensure optimal performance o processes and systems" [28]. The maturity model identifies the minimum mandatory requirements for recordkeeping which represent maturity level 3. In other words, organizations should be at least at level 3 to be compliant with the minimum requirements. The assessment results can also be used to prioritize a strategic recordkeeping implementation plan. This maturity model consists of five maturity levels with attributes at two levels, nine attributes called principles which are further decomposed into 36 key areas.

Asset Management Maturity Model (2011) – Originated from an evaluation in the Netherlands to investigate how asset managers deal with long-term investment decisions [29]. This evaluation took into consideration organizations that control infrastructures, such as, networks, roads and waterways and focus on the strategy, tools, environment and resources. The maturity model consists of five maturity levels and is detailed through four dimensions. Its aim is to understand how asset managers deal with long-term investment decisions and provide an improvement path for organization to improve the long-term investment decisions. This maturity model consists of five maturity levels with four attributes defined which are called categories.

Stanford Data Governance Maturity Model (2011) – Based on the Data Governance Program from Stanford and is centered on the institution as it was developed having in mind the goals, priorities and competences of Stanford. It focuses on both the foundations and the project aspects of data governance and measures the core data governance capabilities and development of the program resources [30]. The name for each of the maturity levels is not described in this model. Its aim is to measure the foundational aspects and project components of the Stanford's Data Governance program. This maturity model consists of five maturity levels with three attributes defined which are called dimensions.

Digital Preservation Capability Maturity Model (2012) – A tool to chart the evolution from a disorganized and undisciplined electronic records management program, or one that does not exist, into increasingly mature stages of digital preservation capability [31]. The DPCMM is designed to help identify, protect and provide access to long-term and permanent digital assets. Consists of 75 statements where which has an integer value ranging from zero to four designated as an index value [31]. These index values are then mapped to a certain capability level. There is an on-line assessment tool available at http://www.digitalok.org. This maturity model consists of five maturity levels with attributes at two levels, three attributes called domains which are further decomposed into 15 components.

Brown Digital Preservation Maturity Model (2013) – Examines the notion of "trusted" digital repositories and proposes a maturity model for digital preservation. The author defined the digital preservation process perspectives, which are the set of processes that together establish the digital preservation capability. Then, for each of these processes there are a set of requirements that organizations must achieve to reach a certain maturity level for a certain process [32]. An assessment method is not defined. The organization should assess themselves against the requirements of the maturity model and position themselves among the maturity levels. However, there is no method or tool to facilitate this assessment. This maturity model consists of five maturity levels with ten attributes defined which are called process perspectives.

JISC Records Management Maturity Model (2013) – Created by JISC infoNet and stands as a self-assessment tool for higher education institution in England and Wales [33]. It is based on a code of practice and its aim is to help in the compliance with this code although it is independent from the code. Its aim is to help higher education institutions to assess their current approach on records management regarding recommendations issued by the United Kingdom government and bench-mark against other similar organizations. Self-assessment is conducted by using a spreadsheet, consisting of statements for each of the nine sections. Users should choose the level that best suits the organization for each statement. This maturity model consists of five maturity levels with nine attributes defined which are called sections.

CMMI Institute Data Management Maturity Model (2014) – A reference model for data management process improvement created by the CMMI Institute. Defines the fundamental business processes of data management and specific capabilities that constitute a path to maturity [34]. It allows organizations to evaluate themselves against documented best practices, determine gaps, and improve data management across functional, business, and geographic boundaries. Its aim is to facilitate an organization's understanding of data management as a critical infrastructure, through increasing capabilities and practices. This maturity model consists of five maturity levels with attributes at two levels, six attributes called categories which are further decomposed into 25 process areas.

Syracuse University Capability Maturity Model for Research Data Management (2014) – Developed by the school of information studies at the University of Syracuse in the USA [35]. It is based on the number and name of levels of CMMI, as well as, the principles of each level. RDM has become a treading topic in data management as increased importance from government agencies, such as, the US National Science Foundation. These funding agencies are raising the issue of maintaining good RDM practices for the projects that are funded by them. There is no assessment method specified. This maturity model consists of five maturity levels with five attributes defined which are called key process areas.

Preservica Digital Preservation Maturity Model (2015) – Created on the premise that organizations have realized that is critical for their business that information is retained over a long period of time [36]. Preservica defines three main sections for the maturity model. The first section is durable storage which comprehends levels 1 to 3, where raw bit storage increases in safety and security. The second section comprehends levels 4 to 5, where the raw bits in storage become preserved and organized. The third and last section is information preservation which comprehends level 6, where the information survives the lifetime of the application that created it. There is no assessment method specified. This maturity model consists of six maturity levels with no attributes defined.

Digital Asset Management (DAM) Maturity Model (2017) – Provides a description of where an organization is, where it needs to be so that it can perform gap analysis and comprehend what it needs to do to achieve the desired state of DAM implementation [37]. There is a description on how to do a self-assessment. It should begin by identifying the stakeholders who identified the need for DAM and can advocate in favor of it. Then, a set of questionnaires must be created and administered to each of the stakeholders identified. Then the levels can be determined using the answers to the questionnaires. This maturity model consists of five maturity levels with attributes at two levels, four attributes called categories which are further decomposed into 15 dimensions.

E-ARK Information Governance Maturity Model (2017) – Based on the OAIS/ISO 14721 [38], the TRAC/ISO 16363 [15] and PAIMAS/ISO 20652 [39]. A2MIGO uses the dimensions described in ISO9001 (Management, Processes and Infrastructure) and the maturity levels defined in SEI CMMI (Initial, Managed, Defined, Quantitatively Managed, Optimizing) [40]. The SEI CMMI levels were selected due to their broader scope making them suitable for wider fields such as that of information governance. This maturity model provides a self-assessment questionnaire, details how the results are analyzed, and clarifies the concepts being used [41]. This maturity model consists of five maturity levels with three attributes defined which are called dimensions.

4   Analysis

There is a growing body of knowledge studying maturity models in other domains, and we draw from this work in our analysis. Mettler et al. [16] note the variety of research that exists on maturity models, and we have attempted to cover a wide range to form our theoretical foundation here. First, the works by Maier et al. [17] and Proenca et al. [18] provide a similar survey of models in a variety of domains and that focus on the application of models. Second, to understand the models as artefacts, we have drawn on work in Design Science research, including the examples and approaches to define requirements for the process of developing a maturity model [14], as well as general design principles for maturity models [4].

We determined a set of attributes to analyze the existing options available for data and information management maturity assessment, the results of which are detailed in Table 2. We first determined the domain of the maturity model. For this work, we identified three domains that deal with data and information management. These are Information Management (IM), Digital Preservation (DP), and Records Management (RM).

We also examined the nature of the assessment process and expected outputs. Specific requirements are necessary for different types of intended audience of the

Table 2. Analyzed maturity models for assessment in data and information management.

Name	Domain	Audience	AM	PRA	CER	Origin	IOP	SWPI	ACC
EIMM	IM	External	TPA	GR	No	C	No	No	Free
IGMM	RM	Both	SA	SIA	No	C	No	Yes	Charged
ECMM	IM	Both	SA	GR	No	C	No	No	Free
RKMM	RM	Internal	SA	SIA	No	P	Yes	Yes	Free
AMMM	IM	Internal	TPA	GR	No	A	No	Yes	Free
DGMM	DM	Internal	SA	SIA	No	A	No	Yes	Free
DPCMM	DP	Internal	TPA	GR	No	P	No	Yes	Free
BDPMM	DP	Internal	SA	SIA	No	C	No	Yes	Charged
RMMM	RM	Both	SA	GR	No	P	No	No	Free
DMMM	DM	Both	CP	SIA	Yes	C	Yes	Yes	Charged
CMMRDM	DM	Internal	N/A	GR	No	A	No	Yes	Free
PDPMM	DP	Both	N/A	GR	No	C	No	No	Free
DAMM	DM	Both	SA, TPA	GR	No	C	No	No	Charged
A2MIGO	DP	Internal	SA, TPA	SIA	No	A	Yes	Yes	Free

Legend: *Columns* - AM = Assessment Method; PRA = Practicality;
CER = Certification; IOP = Improvement Opportunities Prioritization;
SWPI = Strong/Weak Point Identification; ACC = Accessibility. ***Column Domain* -**
IM = Information Management; DP = Digital Preservation; RM = Records Management.
***Column Assessment Method* -** SA = Self-Assessment; TPA = Third-party Assisted;
CP = Certified Professionals. ***Column Practicality* -** GR = General recommendations;
SIA = Specific improvement activities. ***Column Origin* -** C = Commercial;
A = Academic; P = Practitioner.

model, e.g. to be shared internally in the organization, with external stakeholders, or both. As well, we considered the assessment method, e.g. whether it is performed as a self-assessment, third-party assisted, or by a certified practitioner.

Next, we examined the practicality of the maturity model, which details if the practicality of the recommendations is problem-specific or general in nature. For this aspect, we examined if the maturity model provided just general recommendations or if it detailed specific improvement activities. A maturity model that just provides general recommendations is categorized as a descriptive model. One that details specific improvement activities is considered a prescriptive model according Röglinger et al. [4].

Another aspect relevant when analyzing a maturity model is whether it provides a certification or not. A certification can be used as a universal proof of conformance with a specific maturity model which intent is to recognize an organization as excellence in a given area of interest. There are cases when certifications are taken into consideration when organizations apply for research grants or projects.

Additionally, we examined the origin of the model, whether it originated in the academia, from practitioners, or with a commercial intent. This aspect was useful to understand that many of the maturity models that have a commercial intent recommend the acquisition of a certain service or product from the organization that developed the maturity model. This means that the purpose of the maturity model is to help organizations identify which products are the most relevant for their scenario the vendor organization's portfolio. On the other hand, practitioner and academic maturity models draw their criteria from the respective body of knowledge in their respective domains, with the intent of aiding organizations become efficient and effective in their respective domains. A distinction between the maturity model that originates from practitioners and the academic community, is that the academic maturity models are well-founded, with exhaustive documentation, rationale for the assessment criteria, empirical evidence, and detailed assessment method. The maturity models developed by practitioners often don't have an assessment method or a well-founded definition for the "maturity" concept and the maturity levels.

Moreover, we examined whether each maturity model allows for the prioritization of the improvement opportunities. The intention is to verify whether a maturity model defines a generic set of requirements to reach higher maturity levels or whether it provides a prioritized set of requirements relevant for the organization being assessed to improve their maturity. This aspect is closely related with the following aspect, whether the maturity model identifies strong and weak points of the organization. Without an identification of the weaknesses prioritizing the improvement opportunities is next to impossible.

Finally, the final aspect took into consideration in our analysis was the accessibility of the model. This means whether the maturity model documentation and assessment mechanisms are available for free or not. This is one of the most important aspects organizations consider when opting for a maturity model to be used in their organization. As one can expect, all of the maturity models that originated from academics and practitioners are freely available. On the other hand, the models that have a commercial intent can be charged. From our analysis maturity models which purpose is to identify products or services relevant for an organization to acquire are freely available. However, for commercial maturity models that provide a certification, as is the case with the DMMM, or were developed by an organization focused on custom development of products and solutions, the access is charged.

5   Discussion

This analysis of maturity models has generated a number of insights into both the domains that deal with data and information management and the maturity models themselves, as well as shedding light on the limitations around assessments performed using maturity models.

One significant trend to emerge from this comparison is a noticeable increase in the number and complexity of maturity models in recent years. This increasing interest in assessment models mirrors the increasing number of legislation surrounding the management of information. Increased interest and development of maturity models can indicate a domain's transition from a state of lack of definition and standardization towards optimization and improvement, although this shift is not always valuable or desired. Maturity models come with assumptions that sometimes conflict with the reality in organization scenarios. Improvement is often oriented towards quality control and consistency, minimizing unpredictability of outcomes over time and reducing individual efforts to a minimum. However, the culture built around the work of skilled individuals can contrast abruptly with these assumptions, resulting in resistance to the transition to a streamlined approach.

The assumptions of the most recognized maturity models such as those compliant with ISO33000 [19], a family of standards for process assessment, include a process orientation which considers the availability of multiple instances of the assessed processes across the organization. Just as the CMMI was not universally praised in the software industry [20], current highly detailed standards prescribing functional requirements are not necessarily fit for all purposes. Additional limitations that result from using maturity models include the tendency to oversimplify reality and obscuring alternate paths to maturity [21].

Some of the maturity models examined in this paper declare adherence to a model such as the SEI CMMI, they often do not demonstrate awareness of the concepts and assumptions. One exception to this reality is A2MIGO where the authors clearly show the relevance and extent of use of the CMMI in their work. Despite this fact and in general, greater clarity about underlying concepts and a stronger adherence to design principles for maturity models is needed to introduce trust in these maturity models.

6 Conclusion and Outlook

This work presented a state of the art on the subject of maturity models. This work is built upon several works, and it intends to provide a detailed overview on the topic, with a detailed coverage of the different approaches to this domain throughout the years. In addition, by doing this literature review, and by raising the various issues still faced regarding the subject, it is hoped that new research is raised from the points raised throughout this paper.

Indeed, there are still lots of interesting future research questions that can be derived from the analysis and conclusions provided by this paper that can help both Maturity Models developers and users. In this paper, we also described the concepts which form the foundation of maturity models. A description of the different aspects of current maturity models was presented, combining knowledge from the different domains analyzed.

As future work resulting from this paper, we concluded that current maturity assessment methods focus on highly complex and specialized tasks being performed by competent assessors in an organizational context [19]. These tasks mainly focus on manually collecting evidence to substantiate the maturity level determination. Because of the complexity of these methods, maturity assessment becomes an expensive and burdensome activity for organizations.

As such, this work motivated us to develop methods and techniques to automate maturity assessment. The wide spread of modeling practices of business domains, assisted by modeling tools, makes it possible to have access, for processing, to the data created and managed by these tools. There are several examples of models used to represent an organization architecture, such as, Archimate [23], BPMN [22] or UML [24]. These models are descriptive and can be detailed enough to allow to perform, to some extent, maturity assessment. For example, the collected evidence from an organization can be synthetized into a set of model representations that can then be used when analyzing and calculating the maturity levels.

Building on the knowledge of ontologies from the computer science and information science domains, these were used to represent maturity models and model representations. This was achieved by developing ontologies that express all the core concepts and relationships among them, as also the rules for a maturity assessment accordingly Then, by representing maturity models and models representations of concrete organizational scenarios using ontologies we can verify if an organization models representations matches the requirements to reach a certain maturity level using ontology query and reasoning techniques, such as Description Logics inference.

Acknowledgements. This work was supported by national funds through Fundação para a Ciência e a Tecnologia (FCT) with reference UID/CEC/50021/2013.

References

1. Nolan, R.L.: Managing the computer resource: a stage hypothesis. Commun. ACM **16**, 399–405 (1973)
2. Ahern, D.M., Clouse, A., Turner, R.: CMMI Distilled: A Practical Introduction to Integrated Process Improvement, 3rd edn. Addison Wesley Professional, Boston (2008)
3. ISO/IEC 15504:2004: Information technology - Process assessment. International Organization for Standardization and International Electrotechnical Commission (2004)
4. Röglinger, M., Pöppelbuß, J.: What makes a useful maturity model? A framework for general design principles for maturity models and its demonstration in business process management. In: Proceedings of the 19th European Conference on Information Systems, Helsinki, Finland, June 2011
5. Paulk, M., Curtis, B., Chrissis, M., Weber, C.: Capability Maturity Model for software, Version 1.1 CMU/SEI-93-TR-24. Carnegie Melon University, Pittsburgh (1993)
6. Anderson, E., Jessen, S.: Project maturity in organizations. Int. J. Proj. Manag. Account. **21**, 457–461 (2003)
7. Fitterer, R., Rohner, P.: Towards assessing the networkability of health care providers: a maturity model approach. Inform. Syst. E-bus. Manag. **8**, 309–333 (2010)
8. Sen, A., Ramammurthy, K., Sinha, A.: A model of data warehousing process maturity. IEEE Trans. Softw. Eng. **38**, 336–353 (2011)
9. Mettler, T.: A design science research perspective on maturity models in information systems. Institute of Information Management, University of St. Gallen, St. Gallen (2009)
10. CMMI Product Team: CMMI for development - version 1.3, Software Engineering Institute - Carnegie Mellon University, Technical report CMU/SEI-2010-TR-033 (2010)
11. Korbel, A., Benedict, R.: Application of the project management maturity model to drive organizational improvement in a state-owned corporation. In: Proceedings of 2007 AIPM Conference, Tasmania, Australia, pp. 7–10, October 2007
12. Jia, G., Chen, Y., Xue, X., Chen, J., Cao, J., Tang, K.: Program management organization maturity integrated model for mega construction programs in China. Int. J. Proj. Manag. **29**, 834–845 (2011)
13. Jamaluddin, R., Chin, C., Lee, C.: Understanding the requirements for project management maturity models: awareness of the ICT industry in Malaysia. In: Proceedings of the 2010 IEEE IEEM, pp. 1573–1577 (2010)

14. Becker, J., Knackstedt, R., Pöppelbuβ, J.: Developing maturity models for IT management: a procedure model and its application. Bus. Inf. Syst. Eng. **3**, 213–222 (2009)
15. ISO 16363:2012: Space data and information transfer systems – Audit and certification of trustworthy digital repositories (2012)
16. Mettler, T., Rohner, P., Winter, R.: Towards a classification of maturity models in information systems. In: D'Atri, A., De Marco, M., Braccini, A.M., Cabiddu, F. (eds.) Management of the Interconnected World, pp. 333–340. Physica-Verlag, Heidelberg (2010). https://doi.org/10.1007/978-3-7908-2404-9_39
17. Maier, A., Moultrie, J., Clarkson, P.: Assessing organizational capabilities: reviewing and guiding the development of maturity grids. IEEE Trans. Eng. Manag. **59**(1), 138–159 (2012)
18. Proenca, D., Borbinha, J.: Maturity models for information systems. In: Proceedings of the Conference on ENTERprise Information Systems 2016 (CENTERIS 2016) (2016)
19. ISO/IEC 33000:2015: Information technology - Process assessment. International Organization for Standardization and International Electrotechnical Commission (2015)
20. Fayad, M., Laitnen, M.: Process assessment considered wasteful. Commun. ACM **40**(11), 125–128 (1997)
21. Roglinger, M., Poppelbuß, J., Becker, J.: Maturity models in business process management. BPMJ **18**(2), 328–346 (2012)
22. Object Management Group: Business Process Model and Notation (BPMN), Version 2.0, OMG Standard, formal/2011-01-03 (2011)
23. The Open Group: ArchiMate 3.1 Specification. Van Haren Publishing (2017)
24. Object Management Group: OMG Unified Modelling Language (OMG UML), Version 2.5. (2015)
25. Newman, D., Logan, D.: Gartner Introduces the EIM Maturity Model. Gartner (2008)
26. ARMA International: Generally Accepted Recordkeeping Principles - Information Governance Maturity Model. http://www.arma.org/principles
27. Pelz-Sharpe, A., Durga, A., Smigiel, D., Hartmen, E., Byrne, T., Gingras, J.: ECM Maturity Model - Version 2.0. Wipro - Real Story Group - Hartman (2010)
28. Queensland State Archives: Recordkeeping maturity model and road map: improving recordkeeping in Queensland public authorities. The State of Queensland, Department of Public Works (2010)
29. Lei, T., Ligtvoet, A., Volker, L., Herder, P.: Evaluating asset management maturity in the Netherlands: a compact benchmark of eight different asset management organizations. In: Proceedings of the 6th World Congress of Engineering Asset Management (2011)
30. Stanford University: Data Governance Maturity Model. http://web.stanford.edu/dept/pres-provost/cgi-bin/dg/wordpress/
31. Dollar, C.M., Ashley, L.J.: Assessing digital preservation capability using a maturity model process improvement approach. Technical report, February 2013
32. Brown, A.: Practical Digital Preservation: A How-to Guide for Organizations of Any Size. Facet Publishing, London (2013)
33. JISC InfoNet: Records Management Maturity Model. http://www.jiscinfonet.ac.uk/tools/maturity-model/
34. CMMI Institute: Data Management Maturity (DMM) Model. Version 1.0, August 2014
35. Syracuse University: CMM for RDM – A Capability Maturity Model for Research Data Management. http://rdm.ischool.syr.edu/
36. Preservica: Digital Preservation Maturity Model. White Paper (2014)
37. Real Story Group, DAM Foundation: The DAM Maturity Model. http://dammaturitymodel.org/
38. ISO 14721:2010: Space data and information transfer systems – Open archival information system – Reference model (2010)

39. ISO 20652:2006: Space data and information transfer systems – Producer-archive interface – Methodology abstract standard (2006)
40. E-ARK Project: D7.5 – A Maturity Model for Information Governance - Final Version. E-ARK Project Consortium (2017)
41. E-ARK Project: D7.6 – Final Assessment and Evaluation. E-ARK Project Consortium (2017)

An Operationalized DDI Infrastructure to Document, Publish, Preserve and Search Social Science Research Data

Claus-Peter Klas(✉)🆔 and Oliver Hopt🆔

GESIS - Leibniz Institute for the Social Sciences, Cologne, Germany
{Claus-Peter.Klas,Oliver.Hopt}@gesis.org

Abstract. The social sciences are here in a very privileged position, as there is already an existing meta-data standard defined by the Data Documentation Initiative (DDI) to document research data such as empirical surveys. But even so the DDI standard already exists since the year 2000, it is not widely used because there are almost no (open source) tools available. In this article we present our technical infrastructure to operationalize DDI, to use DDI as living standard for documentation and preservation and to support the publishing process and search functions to foster re-use and research. The main contribution of this paper is to present our DDI architecture, to showcase how to operationalize DDI and to show the efficient and effective handling and usage of complex meta-data. The infrastructure can be adopted and used as blueprint for other domains.

Keywords: Research data · Research data management
Standards and interoperability · DDI · SKOS

1 Introduction

The Data Documentation Initiative (DDI) meta-data standard[1] is the best known and widely used standard for describing the social sciences survey data lifecycle from the conceptual start of a *study* to the support of reusing data found in data archives. DDI is organized as an alliance of several social science data archives and several other research institutions like universities. The alliance defined the first standard format for structuring codebook data (DDI 1.0) based on XML in 2000 and renewed it in 2003 (DDI 2.0). The recent version, DDI 3.2 published in 2008, contains 487 complex type definitions for 1154 element definitions [1] and follows the research lifecycle [5].

Making use of DDI in software systems is a high investment on the implementation side. For several software architectural approaches this investment will last only until the next version of the standard and has to be re-implemented. This

---

[1] http://www.ddialliance.org/.

© Springer Nature Switzerland AG 2018
E. Méndez et al. (Eds.): TPDL 2018, LNCS 11057, pp. 94–99, 2018.
https://doi.org/10.1007/978-3-030-00066-0_8

results in high costs of development with respect to time and money. To overcome these high investments, we originate a way to model the binding of DDI to applications independent of most (even upcoming) DDI version changes and handle interpretative differences in DDI without continuous reimplementation. Based on our DDI-FlatDB architecture, shown first at the community conferences EDDI 2015 [2] & 2016 [3] & 2017 [6] and IASSIST 2017 [7], we present our technical infrastructure to operationalize DDI, to use DDI as standard for on the one hand deep documentation and preservation of social science survey data and on the other hand for publishing and searching survey data to foster re-use and research.

This paper is structured as following: The main ideas behind the DDI-FlatDB are presented in Sect. 2. In Sect. 3 we show use cases and existing portals based on our DDI infrastructure. A summary concludes the paper.

2 DDI-FlatDB

There are two main ideas to behind the DDI-FlatDB architecture. The first idea is to store the DDI XML file as is and split it into identifiable entities for efficient access. The second idea is to provide a configurable access to the different DDI versions and dialects, see Sect. 2.1.

2.1 Splitting DDI XML

The first idea behind the DDI-FlatDB architecture is first to not create an entity-relationship model for each existing and upcoming DDI version, but to store the DDI XML files on the hard disk and keep them managed in a version control system, such as GIT.

To enable efficient access to the DDI elements (e.g. studyunit, question, variable) within the DDI files, a *customizable* split mechanism is provided, which splits each DDI XML file into identifiable elements, according to the functional requirements of the developed application. This split is necessary, because the process of loading larges XML files is costly. To access one element the whole XML document needs to be parsed, either with DOM or SAX parser. By splitting the XML, we get smaller XML snippets to be parsed and are able to incorporate caching into the process to offer the users fast access. The splitting process also validates the DDI XML and is currently built on VTD-XML library[2].

These split elements are stored in a simple (*flat*) one table database model, which holds the main keys and ids of the elements along with its raw XML content. This database is intended as *proxy only* to provide efficient read and write access, including caching mechanisms and the entities should be split strictly to functional and application specific requirements. That means, only entities, which need fast access according to the application use case, should be split and made accessible.

---

[2] https://vtd-xml.sourceforge.io/.

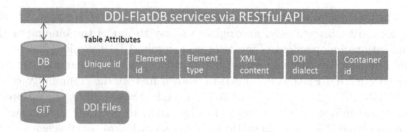

Fig. 1. DDI-FlatDB architecture

The XML elements within the database, respective in the file storage, are accessible via DB API directly, but we introduced a RESTful API including a client to provide secure and homogenous access, depicted in Fig. 1.

The overall technology stack is based on the Java Spring Framework[3]. The database currently selected is MySQL including the auditing addition based on Hibernate Envers[4]. Auditing, e.g. keeping track of changes in questionnaires, is an important use case for social science research.

Upload, Split, Validate and Versioning of DDI XML Documents. The upload and split process is implemented as depicted in Fig. 2. The initial step is to upload the DDI XML document incl. the application specific split configuration via REST POST (or place it in a specific directory directly). A validation of the XML is processed. The next step is to store the original document in the DDI-FlatDB. This is an important step wrt. versioning information. The final step is to apply the split configuration, inserting the elements from the split XML document into the DDI-FlatDB. Through a scheduled poll or event driven service the document is finally stored in the file system and version controlled via GIT.

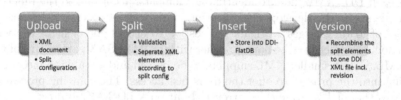

Fig. 2. Upload, split and versioning process

Connecting the Elements from DDI-FlatDB to Any User Interface. Handling large and complex DDI meta-data by splitting it up into smaller snippets was one part of the solution to operationalize DDI. With respect to different

---

[3] https://spring.io/.

[4] https://docs.jboss.org/envers/docs/.

DDI versions, dialects and interpretations we needed also a flexible and efficient way to map the split entities for programming. Therefore we developed a model to map the XML by a configuration file. At GESIS we have a technology stack based on Java (see Sect. 3 for applications), but the model can be used with any programming language. We are aware, that there exists APIs to map XML to Java objects, e.g. by annotation, but they are fixed to one XML model. Beside general configuration information about the entity (e.g. multiple languages or single vs. multiple entries), the XPath to fetch data is the main configuration entry. For each DDI dialect and version the XPath is specific. For example the configuration to access the `title` element of a studyunit for DDI 3.2 the following line expresses the XPath expression:

ddi32 . StudyUnit . Title . path=/DDIInstance/Group/SubGroup
/StudyUnit/Citation/Title/String [@xml : lang='%lang%']

Listing 1.1. Example XPath mapping for title field in DDI 3.2

Other lines can be added to reflect other DDI versions such as DDI Codebook. In our case., using Java, the XPath expression fills the title attribute of the Java bean via Java reflection using the setter and getter of the bean. The main advantage is now, that no Java code change to our applications is necessary, we only need to change a configuration file, if a new DDI dialect or version needs to be processed.

Summary. The described DDI-FlatDB architecture allows a flexible and efficient handling of any XML represented meta-data. It provides a fast and efficient access to any element in the meta-data element through the *splitting mechanism* with a state of the art database and caching technology. Through the *configurable mapping*, the developer can focus on the application and user needs, rather than developing the next DDI database model.

In the following section we will present several use cases and applications developed on the bases of the DDI-FlatDB.

3 Use Cases and Applications Based on DDI-FlatDB

3.1 Portals Based on DDI Meta-Data

Migration and Conversion. Within GESIS the survey research data is documented on several levels with several tools resulting in several databases, each having their own database modeling. The study description is documented using the *Data Catalog Editor* (DBK) connected to a MySQL database. The basic variable description is documented using the *Dataset Documentation Manager* (DSDM) based on Microsoft Access databases. Further variable documentation, e.g. classification of variables provided through the *Codebook Explorer*, also based on MS-Access databases. All the underlying database models are created keeping a DDI schema in mind, but were developed to support each use case of documenting each part of the research data. This heterogeneous situation is common not only at GESIS.

The objective at GESIS is to create a central DDI-FlatDB storage for editing and publishing research data and to provide flexible access for further use cases. Within GESIS the standard is DDI Lifecycle 3.2. For migration we exported the MySQL and Access databases as close as possible to DDI 3.2. This export resulted in a study unit, variable and question and variable categorization description in separate DDI XML exports. Finally, after an optional manual conversion, all exports are imported and integrated in the DDI-FlatDB.

We currently setup the above described process for the approx. 6000 archived studies at GESIS. For the study description from the DBK the process is finalized. For the variable description the process is setup for part of the data, but we already have two services in place, using the first integrated variable documentation.

Portals. Two new developed portals make use of the migrated data. The first portal is the *GESIS Search Portal*[5]. The GESIS Search Portal combines currently six different entities – research data, publications, variables and questions, webpages and instruments and tools – into one vertical search portal for the social sciences. The entities are all connected via a link database to support search and browse between entities.

The second portal *ExploreData* provides a sophisticated search for research data and variables based on categories and other facets. In addition it will be possible to compare variables and download subsets of research data.

The entity information for both portals is provided by the DDI-FlatDB infrastructure. To enable the search functionality an Elastic Search index is generated based on the centralized and harmonized study and variable information.

As third portal we develop a *Questionnaire editor for social science empirical surveys* to support several national and international surveys programs like the German General Social Survey (ALLBUS), the German Longitudinal Election Study (GLES) or the European Values Study (EVS).

The objective is to provide a web-based and collaborative questionnaire editor to create questionnaires storing all information directly in the DDI-FlatDB infrastructure. The editor should be able to create, edit, and document a questionnaire using DDI meta-data. Furthermore a generic export to PDF and in statistical file formats should support the hand-over to the survey institute to avoid any overhead in conversion. Further information about the questionnaire editor use cases and the underlying workflow can be found in [4]. The editor currently contains more than 5000 questions in 29 questionnaires. It is planned to open the questionnaire editor for public use.

Finally we can show the transfer of the DDI-FlatDB architecture using DDI meta-data to a completely different meta-data format. Controlled vocabularies (CV), e.g. from the DDI Alliance are used to categorized social survey studies and variables. A common format for CVs is the Simple Knowledge Organization System (SKOS)[6]. The adaptation of the DDI-FlatDB allowed us to focus on the

---

[5] http://gesis.org.

[6] https://www.w3.org/2004/02/skos/.

user interface requirements right from the start of the development. The creation, editing and storing of a CV worked directly on the SKOS data model. We did not need to develop any new data model, but could rely directly on SKOS. The current development version is accessible on the CESSDA development servers[7] and is planned to be finalized by end of this year.

4 Summary and Future Work

In this paper we presented the DDI-FlatDB architecture to operationalize a complex meta-data schema such as DDI for social survey research data. Besides the developed portals, our approach can support conversion and migration of DDI meta-data. Furthermore version and structural changes of the meta-data standard can be adopted. In addition we showed transferability given other meta-data schemata such as SKOS.

For future work we currently aim to integrate a link database, realizing all internal and external links within a meta-data standard to be directly accessible for more efficient access. In addition, we want to provide an editor tool to create the configuration file described in Sect. 2.1, as this can become quite complex.

The DDI-FlatDB is published as open source at the GESIS GIT repository: https://git.gesis.org/stardat/stardat-ddiflatdb. We also provide a Docker container available for direct usage.

References

1. DDI Alliance: DDI 3.2 XML schema documentation (2014). http://www.ddialliance. org/Specification/DDI-Lifecycle/3.2/XMLSchema/FieldLevelDocumentation/
2. Klas, C.-P., Hopt, O., Zenk-Möltgen, W., Mühlbauer, A.: DDI-FlatDB: efficient access to DDI (2016)
3. Klas, C.-P., Hopt, O., Zenk-Möltgen, W., Mühlbauer, A.: DDI-Flat-DB: a lightweight framework for heterogeneous DDI sources (2016)
4. Blumenberg, M., Zenk-Möltgen, W., Klas, C.-P.: Implementing DDI-lifecycle for data collection within the German GLES project (2015)
5. Vardigan, M., Heus, P., Thomas, W.: Data documentation initiative: toward a standard for the social sciences. Int. J. Digit. Curation **3**(1), 107–113 (2008). https://doi.org/10.2218/ijdc.v3i1.45. http://www.ijdc.net/index.php/ijdc/article/view/66
6. Hopt, O., Klas, C.-P., Mühlbauer., A.: DDI-FlatDB: next steps (2017). www.eddi-conferences.eu/ocs/index.php/eddi/eddi17/paper/view/301
7. Hopt, O., Klas, C.-P., Zenk-Möltgen, W., Mühlbauer, A.: Efficient and flexible DDI handling for the development of multiple applications (2017)

---

[7] http://cv-dev.cessda.eu/cvmanager.

Scholarly Communication

Unveiling Scholarly Communities over Knowledge Graphs

Sahar Vahdati[1] , Guillermo Palma[2(✉)] , Rahul Jyoti Nath[1],
Christoph Lange[1,4] , Sören Auer[2,3] , and Maria-Esther Vidal[2,3]

[1] University of Bonn, Bonn, Germany
{vahdati,langec}@cs.uni-bonn.de, s6ranath@uni-bonn.de
[2] L3S Research Center, Hannover, Germany
{palma,auer,vidal}@L3S.de
[3] TIB Leibniz Information Centre for Science and Technology, Hannover, Germany
Maria.Vidal@tib.eu
[4] Fraunhofer IAIS, Sankt Augustin, Germany

Abstract. Knowledge graphs represent the meaning of properties of real-world entities and relationships among them in a natural way. Exploiting semantics encoded in knowledge graphs enables the implementation of knowledge-driven tasks such as semantic retrieval, query processing, and question answering, as well as solutions to knowledge discovery tasks including pattern discovery and link prediction. In this paper, we tackle the problem of knowledge discovery in scholarly knowledge graphs, i.e., graphs that integrate scholarly data, and present KORONA, a knowledge-driven framework able to unveil scholarly communities for the prediction of scholarly networks. KORONA implements a graph partition approach and relies on semantic similarity measures to determine relatedness between scholarly entities. As a proof of concept, we built a scholarly knowledge graph with data from researchers, conferences, and papers of the Semantic Web area, and apply KORONA to uncover co-authorship networks. Results observed from our empirical evaluation suggest that exploiting semantics in scholarly knowledge graphs enables the identification of previously unknown relations between researchers. By extending the ontology, these observations can be generalized to other scholarly entities, e.g., articles or institutions, for the prediction of other scholarly patterns, e.g., co-citations or academic collaboration.

1 Introduction

Knowledge semantically represented in knowledge graphs can be exploited to solve a broad range of problems in the respective domain. For example, in scientific domains, such as bio-medicine, scholarly communication, or even in industries, knowledge graphs enable not only the description of the meaning of data, but the integration of data from heterogeneous sources and the discovery of previously unknown patterns. With the rapid growth in the number of publications, scientific groups, and research topics, the availability of scholarly datasets

© Springer Nature Switzerland AG 2018
E. Méndez et al. (Eds.): TPDL 2018, LNCS 11057, pp. 103–115, 2018.
https://doi.org/10.1007/978-3-030-00066-0_9

has considerably increased. This generates a great challenge for researchers, particularly, to keep track of new published scientific results and potential future co-authors. To alleviate the impact of the explosion of scholarly data, knowledge graphs provide a formal framework where scholarly datasets can be integrated and diverse knowledge-driven tasks can be addressed. Nevertheless, to exploit the semantics encoded in such knowledge graphs, a deep analysis of the graph structure as well as the semantics of the represented relations, is required. There have been several attempts considering both of these aspects. However, the majority of previous approaches rely on the topology of the graphs and usually omit the encoded meaning of the data. Most of such approaches are also mainly applied on special graph topologies, e.g., ego networks rather than general knowledge graphs. To provide an effective solution to the problem of representing scholarly data in knowledge graphs, and exploiting them to effectively support knowledge-driven tasks such as pattern discovery, we propose KORONA, a knowledge-driven framework for scholarly knowledge graphs. KORONA enables both the creation of scholarly knowledge graphs and knowledge discovery. Specifically, KORONA resorts to community detection methods and semantic similarity measures to discover hidden relations in scholarly knowledge graphs. We have empirically evaluated the performance of KORONA in a knowledge graph of publications and researchers from the Semantic Web area. As a proof of concept, we studied the accuracy of identifying co-author networks. Further, the predictive capacity of KORONA has been analyzed by members of the Semantic Web area. Experimental outcomes suggest the next conclusions: (i) KORONA identifies co-author networks that include researchers that both work on similar topics, and attend and publish in the same scientific venues. (ii) KORONA allows for uncovering scientific relations among researchers of the Semantic Web area. The contributions of this paper are as follows:

- A scholarly knowledge graph integrating data from DBLP datasets;
- The KORONA knowledge-driven framework, which has been implemented on top of two graph partitioning tools, semEP [8] and METIS [3], and relies on semantic similarity to identify patterns in a scholarly knowledge graph;
- Collaboration suggestions based on co-author networks; and
- An empirical evaluation of the quality of KORONA using semEP and METIS.

This paper includes five additional sections. Section 2 motivates our work with an example. The KORONA approach is presented in Sect. 3. Related work is analyzed in Sect. 4. Section 5 reports on experimental results. Finally, Sect. 6 concludes and presents ideas for future work.

2   Motivating Example

In this section, we motivate the problem of knowledge discovery tackled in this paper. We present an example of co-authorship relation discovery between researchers working on data-centric problems in the Semantic Web area. We checked the Google Scholar profiles of three researchers between 2015 and

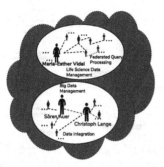
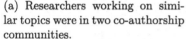
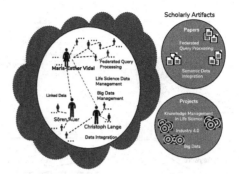

(a) Researchers working on similar topics were in two co-authorship communities.

(b) Researchers working on similar topics constitute a co-authorship community and produce a large number of scholarly artifacts.

Fig. 1. Motivating Example. Co-authorship communities from the Semantic Web area working on data-centric problems. Researchers were in different co-authorship communities (2016) (a) started a successful scientific collaboration in 2016 (b), and as a result, produced a large number of scholarly artifacts.

2017, and compared their networks of co-authorship. By 2016, Sören Auer and Christoph Lange were part of the same research group and wrote a large number of joint publications. Similarly, Maria-Esther Vidal, also working on data management topics, was part of a co-authorship community. Figure 1b illustrates the two co-authorship communities, which were confirmed by the three researchers. After 2016, these three researchers started to work in the same research lab, and a large number of scientific results, e.g., papers and projects, was produced. An approach able to discover such potential collaborations automatically would allow for the identification of the best collaborators and, thus, for maximizing the success chances of scholars and researchers working on similar scientific problems. In this paper, we rely on the natural intuition that successful researchers working on similar problems and producing similar solutions can collaborate successfully, and propose KORONA, a framework able to discover unknown relations between scholarly entities in a knowledge graph. KORONA implements graph partitioning methods able to exploit semantics encoded in a scholarly knowledge graph and to identify communities of scholarly entities that should be connected or related.

3 Our Approach: Korona

3.1 Preliminaries

The definitions required to understand our approach are presented in this section. First, we define a scholarly knowledge graph as a knowledge graph where nodes represent scholarly entities of different types, e.g., publications, researchers, publication venues, or scientific institutions, and edges correspond to an association between these entities, e.g., co-authors or citations.

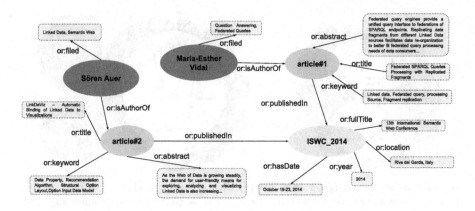

Fig. 2. Korona Knowledge Graph. Scholarly entities and relations.

Definition 1. *Scholarly Knowledge Graph. Let U be a set of RDF URI references and L a set of RDF literals. Given sets V_e and V_t of scholarly entities and types, respectively, and given a set P of properties representing scholarly relations, a scholarly knowledge graph is defined as $SKG = (V_e \cup V_t, E, P)$, where:*

- *Scholarly entities and types are represented as RDF URIs, i.e., $V_e \cup V_t \subseteq U$;*
- *Relations between scholarly entities and types are represented as RDF properties, i.e., $P \subseteq U$ and $E \subseteq (V_e \cup V_t \times P \times V_e \cup V_t \cup L)$*

Figure 2 shows a portion of a scholarly knowledge graph describing scholarly entities, e.g., papers, publication venues, researchers, and different relations among them, e.g., co-authorship, citation, and collaboration.

Definition 2. *Co-author Network. A co-author network $\mathcal{CAN} = (V_a, E_a, P_a)$ corresponds to a subgraph of $SKG = (V_e \cup V_t, E, P)$, where*

- *Nodes are scholarly entities of type* researcher*,*

$$V_a = \{a \mid (a\ rdf{:}type\ {:}Researcher) \in E\}$$

- *Researchers are related according to co-authorship of scientific publications,*

$$E_a = \{(a_i\ {:}co\text{-}author\ a_j) \mid \exists p . a_i, a_j \in V_a \wedge (a_i\ {:}author\ p) \in E \wedge$$
$$(a_j\ {:}author\ p) \in E \wedge (p\ rdf{:}type\ {:}Publication) \in E\}$$

Figure 3 shows scholarly networks that can be generated by KORONA. Some of these networks are among the recommended applications for scholarly data analytics in [14]. However, the focus on this work is on co-author networks.

3.2   Problem Statement

Let $SKG' = (V_e \cup V_t, E', P)$ and $SKG = (V_e \cup V_t, E, P)$ be two scholarly knowledge graphs, such that SKG' is an *ideal* scholarly knowledge graph that contains

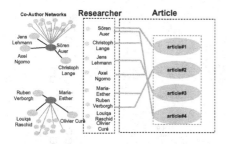 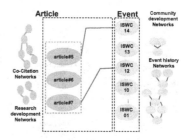

(a) Network of Researchers and Articles.   (b) Networks of Events and Articles.

Fig. 3. Scholarly networks. (a) Co-authors networks from researchers and articles.(b) Co-citation networks from discovered from events and articles.

all the *existing and successful relations* between scholarly entities in V_e, i.e., an oracle that knows whether two scholarly entities should be related or not. $\mathcal{SKG} = (V_e \cup V_t, E, P)$ is the *actual* scholarly knowledge graph, which only contains a portion of the relations represented in $\mathcal{SKG}'$, i.e., $E \subseteq E'$; it represents those relations that are known and is not necessarily complete. Let $\Delta(E', E) = E' - E$ be the set of relations existing in the ideal scholarly knowledge graph $\mathcal{SKG}'$ that are not represented in the actual scholarly knowledge graph $\mathcal{SKG}$. Let $\mathcal{SKG}_{\text{comp}} = (V_e \cup V_t, E_{\text{comp}}, P)$ be a *complete* knowledge graph, which includes a relation for each possible combination of scholarly entities in V_e and properties in P, i.e., $E \subseteq E' \subseteq E_{\text{comp}}$. Given a relation $e \in \Delta(E_{\text{comp}}, E)$, the problem of discovering scholarly relations consists in determining whether $e \in E'$, i.e., whether a relation $r = (e_i\ p\ e_j)$ corresponds to an existing relation in the ideal scholarly knowledge graph $\mathcal{SKG}'$.

In this paper, we specifically focus on the problem of discovering *successful co-authorship relations* between researchers in scholarly knowledge graph $\mathcal{SKG} = (V_e \cup V_t, E, P)$. Thus, we are interested in finding the co-author network $\mathcal{CAN} = (V_a, E_a, P_a)$ composed of the maximal set of relationships or edges that belong to the ideal scholarly knowledge graph, i.e., the set E_a in $\mathcal{CAN}$ that corresponds to a solution of the following optimization problem:

$$\underset{E_a \subseteq E_{comp}}{\text{argmax}} |E_a \cap E'| \tag{1}$$

3.3   Proposed Solution

We propose KORONA to solve the problem of discovering meaningful co-authorship relations between researchers in scholarly knowledge graphs. KORONA relies on information about relatedness between researchers to identify communities composed of researchers that work on similar problems and publish in similar scientific events. KORONA is implemented as an unsupervised machine learning method able to partition a scholarly knowledge graph into subgraphs or

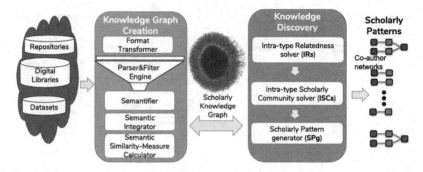

Fig. 4. The Korona Architecture. KORONA receives scholarly datasets and outputs scholarly patterns, e.g., co-author networks. First, a scholarly knowledge graph is created. Then, community detection methods and similarity measures are used to compute communities of scholarly entities and scholarly patterns.

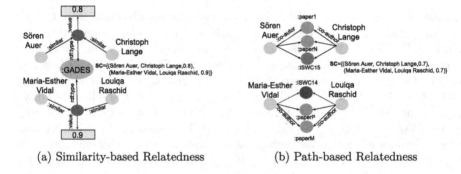

   (a) Similarity-based Relatedness   (b) Path-based Relatedness

Fig. 5. Intra-type Relatedness solver (IRs). Relatedness across scholarly entities. (a) Relatedness is computed according to the values of a semantic similarity metrics, e.g., GADES. (b) Relatedness is determined based on the number of paths between two scholarly entities.

communities of co-author networks. Moreover, KORONA applies the *homophily* prediction principle over the communities of co-author networks to identify successful co-author relations between researchers in the knowledge graph. The *homophily* prediction principle states that similar entities tend to be related to similar entities [6]. Intuitively, the application of the *homophily* prediction principle enables KORONA to relate two researchers r_i and r_j whenever they work on similar research topics or publish in similar scientific venues. The relatedness or similarity between two scholarly entities, e.g., researchers, research topics, or scientific venues, is represented as RDF properties in the scholarly knowledge graph. Semantic similarly measures, e.g., GADES [10] or Doc2Vec [5], are utilized to quantify the degree of relatedness between two scholarly entities. The identified degree shows the relevance of entities and returns the most related ones.

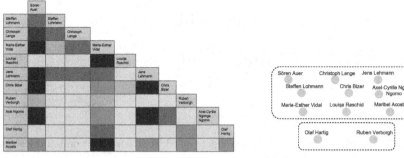

(a) Relatedness Across Researchers   (b) Communities of Researchers

Fig. 6. Intra-type Relatedness solver (IRs). Communities of similar researchers are computed. (a) The tabular representation of SC; lower and higher values of similarity are represented by lighter and darker colors, respectively. (b) Two communities of researchers; each one includes highly similar researchers.

Figure 4 depicts the KORONA architecture; it implements a knowledge-driven approach able to transform scholarly data ingested from publicly available data sources into patterns that represent discovered relationships between researchers. Thus, KORONA receives scholarly data sources and outputs co-author networks; it works in two stages: *(a)* Knowledge graph creation and *(b)* Knowledge graph discovery. During the knowledge graph creation stage, a semantic integration pipeline is followed in order to create a scholarly knowledge graph from data ingested from heterogeneous scholarly data sources. It utilizes mapping rules between the KORONA ontology and the input data sources to create the scholarly knowledge graph. Additionally, semantic similarity measures are used to compute the relatedness between scholarly entities; the results are explicitly represented in the knowledge graph as scores in the range of 0.0 and 1.0. The knowledge graph creation stage is executed offline and enables the integration of new entities in the knowledge graph whenever the input data sources change. On the other hand, the knowledge graph discovery step is executed *on the fly* over an existing scholarly knowledge graph. During this stage, KORONA executes three main tasks: *(i)* Intra-type Relatedness solver (**IRs**); *(ii)* Intra-type Scholarly Community solver (**IRSCs**); and *(iii)* Scholarly Pattern generator (**SPg**).

Intra-type Relatedness solver (IRs). This module quantifies relatedness between the scholarly entities of the same type in a scholarly knowledge graph $SKG = (V_e \cup V_t, E, P)$. **IRs** receives as input $SKG = (V_e \cup V_t, E, P)$ and a scholarly type V_a in V_t; it outputs a set SC of triples $(e_i, e_j, score)$, where e_i and e_j belong to V_a and $score$ quantifies the relatedness between e_i and e_j. The relatedness can be just computed in terms of the values of similarity represented in the knowledge graph, e.g., according to the values of the semantic similarity according to GADES or Doc2Vec. Alternatively, the values of relatedness can be computed based on the number of paths in the scholarly knowledge graph that connect the scholarly entities e_i and e_j. Figure 5

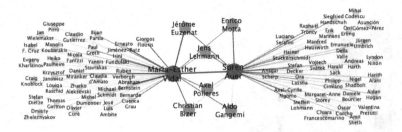

Fig. 7. Co-author network. A network generated from scholarly communities.

depicts two representations of the relatedness of scholarly entities. As shown in Fig. 5a, **IRs** generates a set $\mathcal{SC}$ according to the GADES values of semantic similarity; thus, **IRs** includes two triples (Sören Auer, Christoph Lange, 0.8), (Maria-Esther Vidal, Louiqa Raschid, 0.9) in $\mathcal{SC}$. On the other hand, if paths between scholarly entities are considered (Fig. 5b), the values of relatedness can different, e.g., in this case, Sören Auer and Christoph Lange are equally similar as Maria-Esther Vidal and Louiqa Raschid.

Intra-type Scholarly Community solver (IRSCs). Once the relatedness between the scholarly entities has been computed, communities of highly related scholarly entities are determined. **IRSCs** resorts to unsupervised methods such as METIS or semEP, and to relatedness values stored in $\mathcal{SC}$, to compute the scholarly communities. Figure 6 depicts scholarly communities computed by **IRSCs** based on similarity values; as observed, each community includes researchers that are highly related; for readability, $\mathcal{SC}$ is shown as a heatmap where lower and higher values of similarity are represented by lighter and darker colors, respectively. For example, in Fig. 6a, Sören Auer, Christoph Lange, and Maria-Esther Vidal are quite similar, and they are in the same community.

Scholarly Pattern generator (SPg). **SPg** receives communities of scholarly entities and produces a network, e.g., a co-author network. **SPg** applies the *homophily* prediction principle on the input communities, and connects the scholarly entities in one community in a network. Figure 7 shows a co-author network computed based on a scholarly knowledge graph created from DBLP; as observed, Sören Auer, Christoph Lange, and Maria-Esther Vidal are included in the same co-author network. In addition to computing the scholarly networks, **SPg** scores the relations in a network and computes the *weight of connectivity* of a relation between two entities. For example, in Fig. 7, thicker lines represent strongly connected researchers in the network. **SPg** can also filter from a network the relations labeled with higher values of weight of connectivity. All the relations in a network correspond to solutions to the problem of discovering *successful co-authorship relations* defined in Eq. 1. To compute the weights of connectivity, **SPg** considers the values of similarity of the scholarly entities in a community C; weights are computed as aggregated values using an aggregation function $f(.)$, e.g., average or triangular norm. For each pair (e_i, e_j) of scholarly entities in C, the weight of connectivity between e_i and e_j, $\phi(e_i, e_j \mid C)$, is defined as: $\phi(e_i, e_j \mid C) = \{f(score) \mid e_z, e_q \in C \wedge (e_z, e_q, score) \in \mathcal{SC}\}$.

4   Empirical Evaluation

4.1   Knowledge Graph Creation

A scholarly knowledge graph has been crafted using the DBLP collection (7.83 GB in April 2017[1]); it includes researchers, papers, and publication year from the International Semantic Web Conference (ISWC) 2001–2016. The knowledge graph also includes similarity values between researchers who have published at ISWC (2001–2017). Let PC_{e_i} and PC_{e_j} be the number of papers published by researchers e_i and e_j together (as co-authors), respectively at ISWC (2001–2017). Let TP_{e_i} and TP_{e_j} be the total number of papers that e_i and e_j have in all conferences of the scholarly knowledge graph, respectively. The similarity measure is defined as: $SimR(e_i, e_j) = \frac{PC_{e_i} \cap PC_{e_j}}{TP_{e_i} \cup TP_{e_j}}$. The similarities between ISWC (2002–2016) are represented as well. Let RC_i and RC_j the number of the authors with papers published in conferences c_i and c_j respectively. The similarity measure corresponds to $SimC(c_i, c_j) = \frac{RC_i \cap RC_j}{RC_i \cup RC_j}$. Thus, the scholarly knowledge graph includes both scholarly entities enriched with their values of similarity.

4.2   Experimental Study

The effectiveness of KORONA has been evaluated in terms of the quality of both the generated communities of researchers and the predicted co-author networks.

Research Questions: We assess the following research questions: **(RQ1)** Does the semantics encoded in scholarly knowledge graphs impact the quality of scholarly patterns? **(RQ2)** Does the semantics encoded in scholarly knowledge graph allow for improving the quality of the predicting co-author relations?

Implementation: KORONA is implemented in Python 2.7. The experiments were executed on a macOS High Sierra 10.13 (64 bits) Apple MacBook Air machine with an Intel Core i5 1.6 GHz CPU and 8 GB RAM. METIS 5.1[2] and SemEP[3] are part of KORONA and used to obtain the scholarly patterns.

Evaluation Metrics: Let $Q = \{C_1, \ldots C_n\}$ be the set of communities obtained by KORONA: *Conductance*: measures relatedness of entities in a community, and how different they are to entities outside the community [2]. The inverse of the conductance $1 - Conductance(S)$ is reported. *Coverage*: compares the fraction of intra-community similarities among entities to the sum of all similarities among entities [2]. *Modularity*: is the value of the intra-community similarities among

---

[1] http://dblp2.uni-trier.de/e55477e3eda3bfd402faefd37c7a8d62/.

[2] http://glaros.dtc.umn.edu/gkhome/metis/metis/download.

[3] https://github.com/gpalma/semEP.

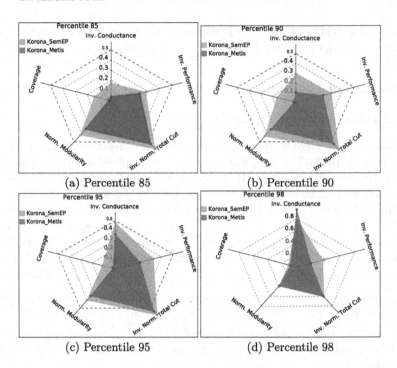

(a) Percentile 85   (b) Percentile 90

(c) Percentile 95   (d) Percentile 98

Fig. 8. Quality of Korona. Communities evaluated in terms of prediction metrics (higher values are better); percentiles 85, 90, 95, and 98 are reported. KORONA exhibits the best performance at percentile 95 and groups similar researchers according to research topics and events where they publish.

the entities divided by the sum of all the similarities among the entities, minus the sum of the similarities among the entities in different communities, in the case they were randomly distributed in the communities [7]. The value of the modularity lies in the range $[-0.5, 1]$, which can be scaled to $[0, 1]$ by computing $\frac{Modularity(Q)+0.5}{1.5}$. *Performance*: sums the number of intra-community relationships, plus the number of non-existent relationships between communities [2]. *Total Cut*: sums all similarities among entities in different communities [1]. Values of total cut are normalized by dividing by the sum of the similarities among the entities; inverse values are reported, i.e., $1 - NormTotalCut(Q)$.

Experiment 1: Evaluation of the Quality of Collaboration Patterns. Prediction metrics are used to evaluate the quality of the communities generated by KORONA using METIS and semEP; relatedness of the researchers is measured in terms of $SimR$ and $SimC$. Communities are built according to different similarity criteria; percentiles of 85, 90, 95, and 98 of the values of similarity are analyzed. For example, in percentile 85 only 85% of all similarity values among entities have scores lower than the similarity value in the percentile 85. Figure 8 presents the results of the studied metrics. In general, in all percentiles, the communities include closely related researchers. However, both implementations of

Table 1. Survey. Questions to validate the recommended collaborations.

Q1. Do you know this person? Have you co-authored before? To avoid confusion, the meaning of "knowing" was kept simple and general. The participants were asked to only consider if they were aware of the existence of the recommended person in their research community

---

Q2. Have you co-authored "before" with this person at any event of the ISWC series? With the same intent of keeping the survey simple, all types of collaboration on papers in any edition of this event series were considered as "having co-authored before"

Q3. Have you co-authored with this person after May 2016? Our study considered scholarly metadata of publications until May 2016. The objective of this question was to find out whether a prediction had actually come true, and the researchers had collaborated

---

Q4. Have you ever planned to write a paper with the recommended person and you never made it and why? The aim is to know whether two researchers who had been predicted to work together actually wanted to but then did not and the reason, e.g., geographical distance

---

Q5. On a scale from 1–5, (5 being most likely), how do you score the relevance of your research with this person? The aim is to discover how close and relevant are the collaboration recommendations to the survey participant

---

KORONA exhibit quite good performance at percentile 95, and allow for grouping together researchers that are highly related in terms of the research topics on which they work, and the events where their papers are published. On the contrary, KORONA creates many communities of no related authors for percentiles 85 and 90, thus exposing low values of coverage and conductance.

Experiment 2: Survey of the Quality of the Prediction of Collaborations among Researchers. Results of an online survey[4] among 10 researchers are reported; half of the researchers are from the same research area, while the other half was chosen randomly. Knowledge subgraphs of each of the participants are part of the KORONA research knowledge graph; predictions are computed from these subgraphs. The predictions for each were laid out in an online spreadsheet along with 5 questions and a comment section. Table 1 lists the five questions that the survey participants were asked to validate the answers, while Table 2 reports on the results of the study. The analysis of results suggests that KORONA predictions represent potentially *successful co-authorship relations*; thus, they provide a solution to the problem tackled in this paper.

5   Related Work

Xia et al. [14] provides a comprehensive survey of tools and technologies for scholarly data management, as well as a review of data analysis techniques, e.g.,

---

[4] https://bit.ly/2ENEg2G.

Table 2. Survey results. Aggregated normalized values of negative answers provided by the study participants during the validation of the recommended collaborations (Q.1(a), Q.1(b), Q.2, Q.3, and Q.4); average (lower is better) and standard deviation (lower is better) are reported. For Q.5, average and standard deviation of the scale from 1–5 are presented; higher average values are better.

Korona	%	Q.1(a)	Q.1(b)	Q.2	Q.3	Q.4	Q.5
Korona-METIS	85	0.26±0.25	0.72±0.29	0.99±0.04	0.86±0.13	0.86±0.20	3.10±0.59
Korona-semEP	85	0.24±0.21	0.80±0.34	1.00±0.00	0.97±0.07	0.93±0.16	3.35±0.85
Korona-METIS	90	0.39±0.24	0.91±0.19	1.00±0.00	1.00±0.00	0.98±0.04	3.03±0.79
Korona-semEP	90	0.13±0.18	0.89±0.18	1.00±0.00	1.00±0.00	0.85±0.23	3.12±1.06
Korona-METIS	95	0.40±0.34	0.93±0.08	1.00±0.00	0.80±0.45	0.95±0.10	3.20±0.81
Korona-semEP	95	0.14±0.30	0.81±0.40	0.67±0.58	0.60±0.55	0.69±0.47	3.83±0.76

social networks and statistical analysis. However, all the proposals have been made over raw data and knowledge-driven methods were not considered. Wang et al. [13] present a comprehensive survey of link prediction in social networks, while Paulheim [9] presents a survey of methodologies used for knowledge graph refinement; both works show the importance of the problem of knowledge discovery. Traverso-Ribón et al. [12] introduces a relation discovery approach, $\mathcal{KOI}$, able to identify hidden links in TED talks; it relies on heterogeneous bipartite graphs and on the link discovery approach proposed in [8]. In this work, Palma et al. present semEP, a semantic-based graph partitioning approach, which was used in the implementation of KORONA-semEP. Graph partitioning of semEP is similar to $\mathcal{KOI}$ with the difference of only considering isolated entities, whereas $\mathcal{KOI}$ is desired for ego networks. However, it is only applied to ego networks, whereas KORONA is mainly designed for knowledge graphs. Sachan and Ichise [11] propose a syntactic approach considering dense subgraphs of a co-author network created from the DBLP. They discover relations between authors and propose pairs of researchers belonging to the same community. A link discovery tool is developed for the biomedical domain by Kastrin et al. [4]. Albeit effective, these approaches focus on the graph structure and ignore the meaning of the data.

6   Conclusions and Future Work

KORONA is presented for unveiling unknown relations; it relies on semantic similarity measures to discover hidden relations in scholarly knowledge graphs. Reported and validated experimental results show that KORONA retrieves valuable information that can impact the research direction of a researcher. In the future, we plan to extend KORONA to detect other networks, e.g., affiliation networks, co-citation networks and research development networks. We plan to extend our evaluation over big scholarly datasets and study the scalability of KORONA; further, the impact of several semantic similarity measures will be included in the study. Finally, KORONA will be offered as an online service that will enable researchers to explore and analyze the underlying scholarly knowledge graph.

Acknowledgement. This work has been partially funded by the EU H2020 programme for the project iASiS (grant agreement No. 727658).

References

1. Buluç, A., Meyerhenke, H., Safro, I., Sanders, P., Schulz, C.: Recent advances in graph partitioning. In: Kliemann, L., Sanders, P. (eds.) Algorithm Engineering. LNCS, vol. 9220, pp. 117–158. Springer, Cham (2016). https://doi.org/10.1007/978-3-319-49487-6_4
2. Gaertler, M.: Clustering. In: Brandes, U., Erlebach, T. (eds.) Network Analysis: Methodological Foundations. LNCS, vol. 3418, pp. 178–215. Springer, Heidelberg (2005). https://doi.org/10.1007/978-3-540-31955-9_8
3. Karypis, G., Kumar, V.: A fast and high quality multilevel scheme for partitioning irregular graphs. Sci. Comput. **20**, 359–392 (1998)
4. Kastrin, A., Rindflesch, T.C., Hristovski, D.: Link prediction on the semantic MEDLINE network. In: Džeroski, S., Panov, P., Kocev, D., Todorovski, L. (eds.) DS 2014. LNCS (LNAI), vol. 8777, pp. 135–143. Springer, Cham (2014). https://doi.org/10.1007/978-3-319-11812-3_12
5. Le, Q.V., Mikolov, T.: Distributed representations of sentences and documents. CoRR abs/1405.4053 (2014)
6. Liben-Nowell, D., Kleinberg, J.: The link-prediction problem for social networks. JASIST **58**(7), 1019–1031 (2007)
7. Newman, M.E.: Modularity and community structure in networks. Proc. Natl. Acad. Sci. **103**(23), 8577–8582 (2006)
8. Palma, G., Vidal, M.-E., Raschid, L.: Drug-target interaction prediction using semantic similarity and edge partitioning. In: Mika, P., et al. (eds.) ISWC 2014. LNCS, vol. 8796, pp. 131–146. Springer, Cham (2014). https://doi.org/10.1007/978-3-319-11964-9_9
9. Paulheim, H.: Knowledge graph refinement: a survey of approaches and evaluation methods. Semant. Web J. **8**(3), 489–508 (2017)
10. Ribón, I.T., Vidal, M., Kämpgen, B., Sure-Vetter, Y.: GADES: a graph-based semantic similarity measure. In: SEMANTICS (2016)
11. Sachan, M., Ichise, R.: Using semantic information to improve link prediction results in network datasets. IJET **2**(4), 334–339 (2010)
12. Traverso-Ribón, I., Palma, G., Flores, A., Vidal, M.-E.: Considering semantics on the discovery of relations in knowledge graphs. In: Blomqvist, E., Ciancarini, P., Poggi, F., Vitali, F. (eds.) EKAW 2016. LNCS (LNAI), vol. 10024, pp. 666–680. Springer, Cham (2016). https://doi.org/10.1007/978-3-319-49004-5_43
13. Wang, P., Xu, B., Wu, Y., Zhou, X.: Link prediction in social networks: the state-of-the-art. Sci. China Inf. Sci. **58**(1), 1–38 (2015)
14. Xia, F., Wang, W., Bekele, T.M., Liu, H.: Big scholarly data: a survey. In: IEEE Big Data (2017)

Metadata Analysis of Scholarly Events of Computer Science, Physics, Engineering, and Mathematics

Said Fathalla[1,3](✉) iD, Sahar Vahdati[1] iD, Sören Auer[4,5] iD,
and Christoph Lange[1,2] iD

[1] Smart Data Analytics (SDA), University of Bonn, Bonn, Germany
{fathalla,vahdati,langec}@cs.uni-bonn.de
[2] Fraunhofer IAIS, Sankt Augustin, Germany
[3] Faculty of Science, Alexandria University, Alexandria, Egypt
[4] Computer Science, Leibniz University of Hannover, Hannover, Germany
[5] TIB Leibniz Information Center for Science and Technology, Hannover, Germany
soeren.auer@tib.eu

Abstract. Although digitization has significantly eased publishing, finding a relevant and a suitable channel of publishing remains challenging. Scientific events such as conferences, workshops or symposia are among the most popular channels, especially in computer science, natural sciences, and technology. To obtain a better understanding of scholarly communication in different fields and the role of scientific events, metadata of scientific events of four research communities have analyzed: Computer Science, Physics, Engineering, and Mathematics. Our transferable analysis methodology is based on descriptive statistics as well as exploratory data analysis. Metadata used in this work have been collected from the OpenResearch.org community platform and SCImago as the main resources containing metadata of scientific events in a semantically structured way. There is no comprehensive information about submission numbers and acceptance rates in fields other than Computer Science. The evaluation uses metrics such as continuity, geographical and time-wise distribution, field popularity and productivity as well as event progress ratio and rankings based on the SJR indicator and h5-indices. Recommendations for different stakeholders involved in the life cycle of events, such as chairs, potential authors, and sponsors, are given.

Keywords: Scientific events · Metadata analysis
Scholarly communication · Citation count · OpenResearch.org

1 Introduction

Information emanating from scientific events as well as journals have become increasingly available through online sources. However, because of the rapidly

© Springer Nature Switzerland AG 2018
E. Méndez et al. (Eds.): TPDL 2018, LNCS 11057, pp. 116–128, 2018.
https://doi.org/10.1007/978-3-030-00066-0_10

growing amount of scholarly metadata, exploration of information still stays challenging for researchers, scholarly metadata managers, social scientists, and librarians. Aggregation of metadata from several data repositories, digital libraries, and scholarly metadata warehouses enables comprehensive analysis and services to the users of such services. An earlier version of this work focused on an analysis of CS sub-communities in terms of continuity, geographical and time distribution, field popularity, and productivity [7]. In this work, we aim to analyze the development of scholarly knowledge dissemination in the Computer Science (CS), Physics (PHY), Engineering (ENG) and Mathematics (MATH) domains. Furthermore, four additional metrics are computed: SJR indicator, h5-index, citation count, and Progress Ratio (PR), the latter being defined by ourselves. An empirical study has been conducted to gain more insights on the significance of the mentioned challenges in order to ultimately devise novel means for scholarly communication. Researchers of different communities use different channels for publishing. The balance between these channels is based on the grown culture of the respective community and community-defined criteria for the quality of these channels. In some fields, such as medical science, publishing in journals is the main and most valuable channel, whereas in other fields, such as computer science, events are highly important. Beside general criteria such as the impact factor of journals and the acceptance rate of events, there are community-defined criteria for the ranking of journals and events. Such criteria are not standardized nor maintained by a central instance, but are transferred from seniors to juniors. To shed light on these criteria across disciplines, our research aims at answering the following research questions: (RQ1) *How important are events for scholarly communication in the respective communities?* and (RQ2) *What makes an event a high-ranked target in a community?* Statistical analysis of metadata of events, such as title, acronym, start date, end date, number of submissions, number of accepted papers, city, state, country, event type, field, and homepage, can give answers to such questions. For instance, the existence of long and continual events depicts the importance of scholarly events (RQ1), and high-ranked events have a high h-index and a high continuity, exceeding 90% (RQ2). To do so, a list of common criteria has been defined to analyze the importance of events for publishing in different research fields. This study attempts to close an important gap in analyzing the progress of a CS community in terms of submissions and publications, which are not available for MATH, PHY, and ENG, over time and provides overviews for the stakeholders of scholarly communication, particularly to: (1) *event organizers* – to trace their events' progress/impact, (2) *authors* – to identify prestigious events to submit their research results, and (3) *proceedings publishers* – to know the impact of the events whose proceedings they are publishing.

The rest of this article is organized as follows: Sect. 2 gives an overview of related work. Section 3 presents the research workflow. Section 4 presents an exploratory analysis of the metadata of selected events. Section 5 discusses the results of this study. Section 6 concludes and outlines future work.

2   Related Work

The next decade is likely to witness a considerable rise in metadata analyses of scientific events because of the mega-trend of digitization and the preparation of manuscripts, as well as the organization of scientific events, have become considerably easier. Several approaches for tracking the evolution of a specific scientific community by analyzing the metadata of scholarly artifacts have been developed [1,2,10,12]. Aumuller and Rahm [1] have analyzed the author affiliations of database publications appeared in SIGMOD and VLDB conferences in the period 2000–2009. Barbosa et al. [2] have analyzed full papers published in the IHC conference series in the period of 1998–2015, while Hiemstra et al. [10] have analyzed publications of the SIGIR information retrieval conference in the period of 1978–2007. Guilera et al. [9] presented an overview of meta-analytic research activity and show the evolution of publications in psychology over time. El-Din et al. [4] presented a descriptive analysis of Egyptian publications using several indicators such as the total number of citations, authors, and their affiliations. Different computer science communities have been analyzed with respect to the history of publications of event series [3,7]. Fathalla and Lange [6] published EVENTSKG, a knowledge graph representing a comprehensive semantic description of 40 top-prestigious event series from six computer science communities since 1969. The EVENTSKG dataset is a new release of their previously presented dataset called EVENTS [5] with 60% additional event series. Notably, EVENTSKG uses the Scientific Events Ontology (SEO)[1] as a reference ontology for event metadata modeling and connects related data that was not previously connected in EVENTS. Biryukov and Dong [3] addressed collaboration patterns within a research community using information about authors, publications, and conferences. The analysis in this work is based on a comprehensive list of metrics considering quality in four fields of science.

3   Method

Metadata of scholarly events in the four research fields has been studied involving conferences, workshops, meetings, and symposia. The method of choice for this study is a meta-analysis that refers to the statistical methods used in research synthesis for drawing conclusions and providing recommendations from the results obtained from multiple individual studies. The overall workflow of this study comprises five steps: (1) data gathering, (2) identification of prestigious events, (3) data preprocessing, and (4) data analysis and visualization, and finally, (5) recording observations and drawing conclusions (see Fig. 1). The following subsections provide further details about each step.

3.1   Data Identification and Gathering

Target Research Communities Identification: In order to identify target research communities, all metrics used by well-known services have been surveyed. These

---

[1] http://sda.tech/SEOontology/Documentation/SEO.html.

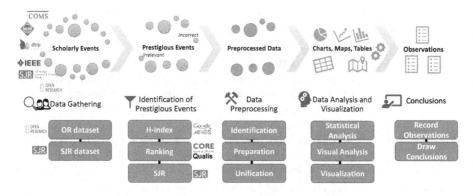

Fig. 1. Research Workflow. The input of this research is a set of unstructured scholarly metadata that turned to analysis and recommendations after defining and applying quality measures on a cleansed and structured subset.

communities and their sub-fields involved in this study are listed in Table 1. The following metrics have been used because they are available and implemented in a reusable way: *Google Scholar Metrics (GSM)* ranks events and journals by scientific field based on a 5-year impact analysis over the Google Scholar citation data[2]. *SCImago Journal Rank (SJR indicator)*[3] is a measure of the scientific influence of scholarly journals and events based both on the number of citations received by a journal or event proceedings and the prestige of the journals/events where such citations come from [8]. This rank is calculated based on information contained in the Scopus[4] database from 1996 till 2018.

Data Harvesting Sources: Two major datasets are used in this study: (1) OpenResearch dataset (ORDS) (5,500+ events) and (2) SCImago dataset (SCIDS) (about 2,200 events). The reason for using two datasets is that the CS community archives more information about past events than others, such as the numbers of submissions and accepted papers. OR supports researchers in collecting, organizing, sharing and disseminating information about scientific events, tools, projects, people and organizations in a structured way [13].

3.2   Data Preprocessing

Several manual preprocessing tasks have been carried out to prepare the data for the analysis, including: (1) *Data cleansing and preparation*: inadequate, incorrect, inaccurate or irrelevant data have been identified. Then, fill in missing data, delete dirty data, and resolve inconsistencies, (2) *Data integration:* involves combining data from multiple sources into meaningful and valuable information, (3) *Data structure transformation:* involves converting cleaned data values from unstructured formats into a structured one, and (3) *Name Unification:* involves

---

[2] https://scholar.google.com/intl/en/scholar/metrics.html.

[3] http://www.scimagojr.com.

[4] https://www.scopus.com/.

Table 1. Research fields and corresponding sub-fields

Research fields	Sub-communities
Computer Science	World Wide Web (Web), Computer Vision (CV), Software Engineering (SE), Data Management (DM), Security and privacy (SEC), Knowledge Representation and Reasoning (KR), Computer Architecture (ARCH), Machine Learning (LRN)
Physics	Astronomy, High energy physics, Particle accelerators, Applied physics and mathematics, Nuclear Science, Nanomaterials, Neutrino detectors, Geophysics
Engineering	Civil engineering, Mechanical engineering, Chemical engineering, Electrical engineering
Mathematics	Algebra, Mathematical logic, Applied mathematics, General mathematics, Discrete Mathematics

integrating all events of a series with multiple names under its most recent name. Usually, events change their names because of changing the event to a bigger scale, e.g., from a symposium to a conference, e.g., ISWC and ISMAR. Also, the change sometimes happens because of adding a new scope or topic, e.g., SPLASH. This led us to perform a unification process before beginning the analysis.

4   Data Analysis

Generally, the data analysis process is divided into three categories: descriptive statistics analysis (DSA), exploratory data analysis (EDA), and confirmatory data analysis (CDA) [11]. The analysis in this study is based on DSA and EDA. A list of 10 metrics is defined over numeric values, as well as metrics having other complex datatypes, focusing on high impact events on research communities. Compared to previous work, four additional metrics are computed: SJR indicator, h5-index, citation count, and Progress Ratio (PR), the latter being defined by ourselves.

(i) *Acceptance rate:* is defined as the ratio between the number of submitted articles and the number of accepted ones.

(ii) *Continuity:* refers to how continuously an event has been held over its history. Using the formula $c = min\{100\%, (e * r)/a\}$, where c is the continuity, e is the number of editions of the event so far, r is the regularity of the event editions (e.g., 1 for 'every year' , 2 for 'every two years'), and a is the age. Year is the granularity for this metric.

(iii) *Geographical distribution:* refers to the *location* of the event from which the number of distinct locations visited by an event is derived.

(iv) *Time distribution:* refers to the period of time each year in which the event takes place. It is important for a researcher interested in a particular event to know when this event will be held to know when to prepare and, if accepted, present their work.

(v) *Community popularity:* reveals how popular an event is in a research community, in terms of a high number of submissions.

(vi) *Field productivity (FP):* reveals how productive, in terms of the number of publications, a field is in a given year within a fixed-time interval. FP_y^f for a field f in a year y, where $C_{i,y}^f$ is the number of publications for an event i in year y and n is the number of events belonging to sub-field f, and m is the number of years in the time span of the study, as defined in Eq. 1.

(vii) *Progress ratio (PR):* is the ratio of the number of publications of an event in a given year to the total number of publications in a given period of time. PR_e^y for an event (e) in a year (y), where P_y is the number of publications of that event in that year and n is the number of years in the time span of the study, is defined in Eq. 1.

$$FP_y^f = \frac{\sum_{i=1}^{n} C_{i,y}^f}{\sum_{k=1}^{m} \sum_{i=1}^{n} C_{i,y_k}^f}, \quad PR_e^y = \frac{P_e}{\sum_{i=1}^{n} P_i} \tag{1}$$

(viii) *SJR indicator:* is the average number of weighted citations received in the selected year by the papers published by the event in the three previous years [8].

(ix) *h5-index:* is the largest number h such that h articles published in the last 5 complete years have at least h citations each.

(x) *Citation count:* is the number of citations papers receive, according to SCIDS.

5 Results and Discussion

This section discusses the analysis results for the two datasets about events in the last 50 years, according to the metrics defined in Sect. 4. One prominent observation is that there is no comprehensive information about submission numbers and acceptance rates in other fields than CS. Therefore, metrics such as acceptance rate, FP, community popularity and PR cannot be applied to non-CS events in practice.

5.1 Scientific Communities Analysis

The results of comparing events from all scientific communities involved have been presented in terms of Time and Geographical distribution, h5-index, continuity, SJR indicator and citation count.

Time Distribution (TD): It is observed that most editions of top events are held around the same month of each year.

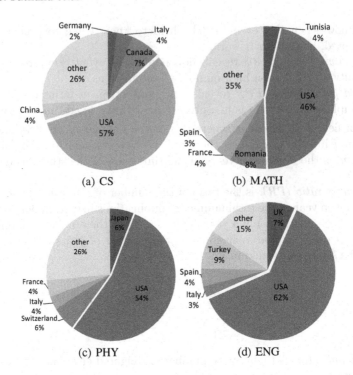

Fig. 2. Geographical distribution of CS, MATH, PHY and ENG events in 2017–1995.

Geographical Distribution: As shown in Fig. 2, venues in the USA hosted about 50% of the scientific events in all communities during the whole period. Other countries have almost similar, relatively low percentages. For instance, Italy hosted 4% of CS and ENG events, while France hosted 4% and 6% of MATH and PHY events respectively.

H5-index: Fig. 3 shows the frequency distribution of events by categorizing the h-index of the events into four bins. The slices of each pie chart compare the frequency distribution of events in each community. CS community has the largest number of events (92) with $h > 30$, while ENG community has the smallest. The number of MATH events with $h > 30$ is as large as that of PHY, while each of them is twice as large as ENG. Also, the number of ENG events with h-index between 21 and 30 is as large as that of PHY. Overall, CS has the largest number of high-impact events, while ENG has the lowest. This can be, for example, attributed to the size of the community and its sub-communities and their fragmentation degree, since a larger community of course results in higher citation numbers.

Continuity: As shown in Fig. 4, all events in all communities have a continuity greater than 90% except for NNN and ICE-TAM, which have a continuity of 88% and 86% respectively. NNN was not held in 2003 and 2004, and ICE-TAM was not held only in 2013. In CS, the continuity of USENIX is 93% because it was

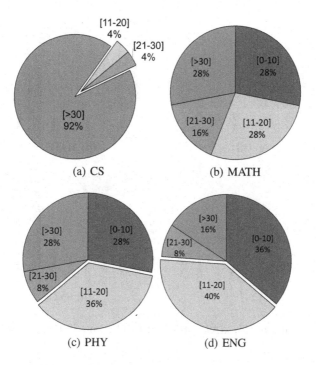

Fig. 3. H-index frequency of top-25 events in CS, MATH, PHY and ENG.

held every year from 1990 except for two years (1994 and 1997). Overall, a very high continuity among the most renowned events has been observed, which indicates stability and attractiveness of hosting and organizing such events in the community.

SJR Indicator: The average SJR of all events in each research field in 2016 have been computed[5]. As shown in Table 2, CS communities have an average SJR (avg. SJR) of 0.23, which is almost double the SJR indicator of PHY and ENG each; MATH comes next. As the SJR indicator is calculated based on the number of citations, therefore, it could be inferred that CS and MATH communities are more prolific or interconnected in terms of citations in 2016 compared to PHY and ENG. Since PHY has the largest number of papers published in 2016 this can rather be attributed to the number of citations per article. On average, a CS paper contains about 20 references (refs/paper), while a PHY paper contains only 15 references. In terms of number of references included in papers published in 2016 (total refs.), CS has the largest number of references, while ENG has the lowest.

Citation Count: The number of citations by all proceedings papers of events that took place in, e.g., Germany, have been collected for CS, ENG, MATH and PHY communities over the last 10 years. Figure 5 illustrates how the number of

---

[5] Since the data collection of event metadata was in 2017, SJR of 2016 is considered.

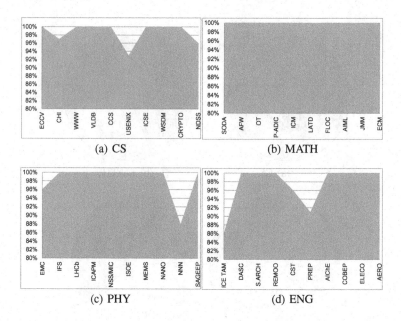

(a) CS   (b) MATH

(c) PHY   (d) ENG

Fig. 4. Continuity of CS, MAT, PHY and ENG events in the last two decades.

citations for each community has developed during that time. While the number of citations has increased for all communities during that time, the strongest increases can be observed in CS and ENG. The citations for PHY and MATH are relatively low and are almost similar. Overall, an upward trend in the number of citations is visible.

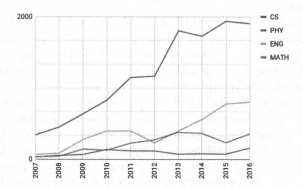

Fig. 5. Citation count by different communities in Germany.

Table 2. Scientometric profile of the top CS, PHY, ENG and MATH events.

Metrics	Avg (h)	h>10	Avg. SJR	Papers (2016)	Papers (2013–15)	Total refs (2016)	Refs/paper
CS	6.58	151	0.23	13,234	163,556	262,548	20
PHY	6.65	28	0.14	16,795	90,245	248,216	15
ENG	4.09	21	0.14	1,675	46,790	27,137	16
MATH	6.79	25	0.21	16,585	68,814	258,275	16

5.2 CS Sub-communities Analysis

This section focuses on analyzing events of CS sub-communities based on the number of submissions and accepted papers. *Community popularity:* The CV community has the largest number of submissions and accepted papers during the three 5-year time window. The lead of CV in submissions and accepted papers has continuously increased over the whole period. *Field productivity:* The slices of the pie chart in Fig. 6 compare the cumulative field productivity of eight CS communities over the last 10 years. It is observed that the CV community is the most productive community over the other communities with an FP of 22%; then the DM community comes, whereas the LRN community has a significantly lower ratio of 4%. *Progress ratio:* Fig. 7 shows the PR of the top events in each CS sub-community in the period 1997–2016. The PR of the five events had a slight rise in the period 1997–2005; then, they all rose noticeably in the last decade.

6 Discussion and Future Work

The existence of long and continual events depicts the importance of scholarly events of different scholarly fields (RQ1). Researchers consider scholarly event as a serious gate to disseminate their research results. At the time of submitting, researchers consider certain characteristics to select the target venue. Generally, the acceptance rate is considered one of the most important characteristics. However, the findings of this study indicate that the *success* of events depends on several characteristics such as continuity, broad coverage of topics, high reputation of organizers, participants and speakers etc. As a result of domain conceptualization to provide the base-camps of this study, a comprehensive list of event-related properties is presented (RQ2). The contributions of this research are as follows:

- Creation of a dataset with a potential of being imported to a scholarly event knowledge graph (OpenResearch.org a crowd sourcing platform for events),
- The conceptualization of the scholarly communication domain and the development of an event quality framework are based on a list of defined quality metrics,

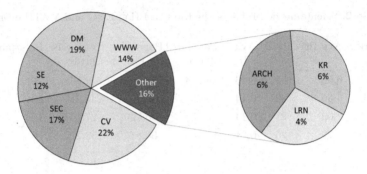

Fig. 6. FP of CS sub-communities in the period 1997–2016.

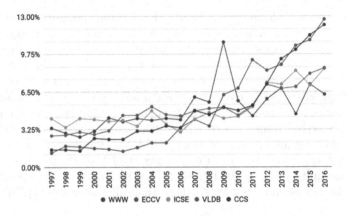

Fig. 7. PR of top events in each CS sub-community in the period 1997–2016.

- Four additional metrics have been applied for scholarly events impact assessment: Continuity, Community popularity, Field Productivity, and Progress Ratio (first defined in this work),
- An empirical evaluation of the quality of scholarly event metadata of CS, PHY, ENG and MATH research communities involving different event types such as conferences, workshops, meetings and symposiums, and
- Supporting communities by giving recommendations for different stakeholders of events.

The most remarkable findings indicated by this study are:

- There is not much information about publications of non-CS communities,
- Most editions of top events have been held around the same time each year,
- A wide geographical distribution increases the awareness of researchers about the existence of the event,
- High-ranked events have a high h-index and a long continuity (i.e., exceeding 90%),
- Among all countries hosting events, the USA has been the host of about 50% of the scientific events in all communities in the last two decades,

- CS and MATH communities are more prolific, and their publications have more citations among each other in 2016, compared to PHY and ENG,
- CV community had the largest number of submissions and accepted papers during the three 5-year time windows,
- CS community has the largest number of events with h-index exceeding 30 compared to other communities, which can be attributed to the key role of conferences in CS,
- Most of the research findings of non-CS communities were published as abstracts or posters, while research findings of CS were published as full papers.

Based on our observations, a number of recommendations is concluded to different stakeholders involved in the life cycle of publishing through events. The lessons learned are based on the three major defined characteristics: community productivity, community popularity and progress ratio of events. In the scope of this work, we addressed several lessons learned for three target groups of researchers: *To organizers*: The organizers adjusting the topics covered by their event to the most productive and popular ones, increase the *success* rate of their event. Possibility to have a progress ratio overview of other events enables organizers to compare their event with competing events and identify the organizational problems, e.g., publicity issues, the reputation of the members, location dynamicity. Therefore, in order to provide a high-profile event to the community, following certain strategies to fulfill the characteristics of high ranked events is necessary, e.g., keeping event topic coverage up to date with new research trends, involving high profile people and sponsors, continuity of the event and geographic distribution of event venues etc. *Researcher*: Community productivity and popularity change the research direction of individual scientists. Submitting to events with a broad range of up to date topics keeps the research productivity and publication profile of researchers aligned with growing communities. While searching for a venue to submit research results, consideration of multiple characteristics of the host conference or the journal changes the future visibility and impact of the submission. *Sponsors*: Progress ratio of prestigious events and considered characteristics gives insights about small size or preliminary events. Sponsoring such small size, but strong and valuable events supports their rapid growth as well as research topics popularity and directions. The analysis sheds light on the evolving and different publishing practices in different communities and help to identify novel ways for scholarly communication, as, for example, the blurring of journals and conferences or open-access overlay-journals as they already started to emerge.

In further research, we plan to extend the analysis to other fields of science and apply more metrics such as author and paper affiliation analysis. It is also interesting to assess the impact of digitization with regard to journals (which receive more attention than events in fields other than computer science) as well as awards. Although large parts of our analysis methodology have already been automated, we plan to further optimize the process so that analysis results can be almost instantly generated from the OpenResearch.org data basis. This work provides the foundations of discovery, recommendation and ranking services for scholarly events with transparent measures.

Acknowledgments. Said Fathalla would like to acknowledge the Ministry of Higher Education (MoHE) of Egypt for providing a scholarship to conduct this study. I would like to offer my special thanks to Heba Mohamed for her support in data gathering process.

References

1. Aumüller, D., Rahm, E.: Affiliation analysis of database publications. SIGMOD Rec. **40**(1), 26–31 (2011)
2. Barbosa, S., Silveira, M., Gasparini, I.: What publications metadata tell us about the evolution of a scientific community: the case of the Brazilian human-computer interaction conference series. Scientometrics **110**(1), 275–300 (2017)
3. Biryukov, M., Dong, C.: Analysis of computer science communities based on DBLP. In: Lalmas, M., Jose, J., Rauber, A., Sebastiani, F., Frommholz, I. (eds.) ECDL 2010. LNCS, vol. 6273, pp. 228–235. Springer, Heidelberg (2010). https://doi.org/ 10.1007/978-3-642-15464-5_24
4. El-Din, H., Eldin, A., Hanora, A.: Bibliometric analysis of Egyptian publications on Hepatitis C virus from PubMed using data mining of an in-house developed database. Scientometrics **108**(2), 895–915 (2016)
5. Fathalla, S., Lange, C.: EVENTS: a dataset on the history of top-prestigious events in five computer science communities. In: Workshop on Semantics, Analytics, Visualization: Enhancing Scholarly Dissemination. Springer, Heidelberg (2018, in press)
6. Fathalla, S., Lange, C.: EVENTSKG: a knowledge graph representation for top-prestigious computer science events metadata. In: Conference on Computational Collective Intelligence Technologies and Applications. Springer, Heidelberg (2018, in press)
7. Fathalla, S., Vahdati, S., Lange, C., Auer, S.: Analysing scholarly communication metadata of computer science events. In: Kamps, J., Tsakonas, G., Manolopoulos, Y., Iliadis, L., Karydis, I. (eds.) TPDL 2017. LNCS, vol. 10450, pp. 342–354. Springer, Cham (2017). https://doi.org/10.1007/978-3-319-67008-9_27
8. González-Pereira, B., Guerrero-Bote, V.P., Moya-Anegón, F.: A new approach to the metric of journals' scientific prestige: the SJR indicator. J. Inf. **4**(3), 379–391 (2010)
9. Guilera, G., Barrios, M., Gómez-Benito, J.: Meta-analysis in psychology: a bibliometric study. Scientometrics **94**(3), 943–954 (2013)
10. Hiemstra, D., Hauff, C., De Jong, F., Kraaij, W.: SIGIR's 30th anniversary: an analysis of trends in IR research and the topology of its community. In: ACM SIGIR Forum, vol. 41, no. 2, pp. 18–24. ACM (2007)
11. Martinez, W., Martinez, A., Martinez, A., Solka, J.: Exploratory Data Analysis with MATLAB. CRC Press, Boca Raton (2010)
12. Nascimento, M.A., Sander, J., Pound, J.: Analysis of SIGMOD's co-authorship graph. ACM SIGMOD Rec. **32**(3), 8–10 (2003)
13. Vahdati, S., Arndt, N., Auer, S., Lange, C.: OpenResearch: collaborative management of scholarly communication metadata. In: Blomqvist, E., Ciancarini, P., Poggi, F., Vitali, F. (eds.) EKAW 2016. LNCS (LNAI), vol. 10024, pp. 778–793. Springer, Cham (2016). https://doi.org/10.1007/978-3-319-49004-5_50

Venue Classification of Research Papers in Scholarly Digital Libraries

Cornelia Caragea[1]([⊠]) and Corina Florescu[2]

[1] Computer Science, Kansas State University, Manhattan, KS 66502, USA
ccaragea@ksu.edu
[2] Computer Science and Engineering, University of North Texas,
Denton, TX 76207, USA
CorinaFlorescu@my.unt.edu

Abstract. Open-access scholarly digital libraries crawl periodically a
list of URLs in order to obtain appropriate collections of freely-available
research papers. The metadata of the crawled papers, e.g., title, authors,
and references, are automatically extracted before the papers are indexed
in a digital library. The venue of publication is another important aspect
about a scientific paper, which reflects its authoritativeness. However,
the venue is not always readily available for a paper. Instead, it needs
to be extracted from the references lists of other papers that cite the
target paper. We explore a supervised learning approach to automatically classifying the venue of a research paper using information solely
available from the content of the paper and show experimentally on a
dataset of approximately 44,000 papers that this approach outperforms
several baselines and prior work.

Keywords: Text classification · Digital libraries · Venue classification

1 Introduction

Scholarly digital libraries such as CiteSeer[x] and ArnetMiner are powerful
resources for many applications that analyze scientific documents on a Web-wide
scale, including topic classification [2,12] and automatic keyphrase extraction
[1,5,7]. These digital libraries periodically crawl a list of URLs, e.g., URLs point-
ing to authors' homepages, in order to obtain freely-available research papers.
Metadata extraction tools [4,9] are then used to automatically extract the metadata of the crawled papers, e.g., the title, abstract, authors, and the references
section. The venue of publication is another important aspect about a paper,
which reflects the *authoritativeness* of a paper. The availability of paper venues
can help improve data organization and the search and retrieval of information
in digital libraries, e.g., when a user is interested in papers published in certain
venues. However, the venue of a paper is not always readily available, but it
needs to be extracted from other sources. For example, CiteSeer[x] [6] currently
extracts the venue of a target paper from the references lists of other papers

© Springer Nature Switzerland AG 2018
E. Méndez et al. (Eds.): TPDL 2018, LNCS 11057, pp. 129–136, 2018.
https://doi.org/10.1007/978-3-030-00066-0_11

that cite the target paper. Unfortunately, this approach is challenging when the venue of a newly published paper is sought because recent papers often accumulate citations over a period of time, and not immediately. Moreover, despite that the metadata extraction approaches implemented in CiteSeer^x are fairly accurate, they still result in noisy metadata extraction [3].

In this paper, we explore a text-based supervised learning approach to venue classification of research papers, using information solely available from the content of each paper. Our approach is based on two observations. First, in a scholarly domain, the topical influence of one paper on another is captured by means of the citation relation [10,15]. Second, scientific paper titles comprise a large fraction of the keywords of a paper [1,11]. Thus, we propose to take into account the topical influence of one paper on another by incorporating keywords from the titles of cited papers, in addition to the title and abstract of the target paper.

Our Contributions. We present an approach to venue classification of research papers that uses only information contained in each paper itself. The result of this classification task will aid indexing of papers in digital libraries, and hence, will lead to improved search and retrieval results. For example, accurate classification of papers' venues is highly needed in retrieval systems where one might be interested in searching for papers published in specific venues, rather than performing a generic search.

Our contributions are summarized as follows:

- We propose a supervised learning approach to venue classification that combines information available in the references list of a paper with its title and abstract. This idea is similar to *query expansion from information retrieval.* The analogy is that words/keywords from the titles in a references list act as "expanded" additional evidence for identifying the venue of a paper.
- We show experimentally on a dataset of $\approx 44,000$ papers that our approach yields classifiers that are more accurate compared with those trained using either the title and abstract of a paper or the references list. We also show experimentally that our approach outperforms several baselines and prior work on venue classification.

2   Related Work

The task of venue classification has similarities with topic classification [2,12] since the topic of a paper contributes majorly to its publication in a specific venue. Yet, significant differences exist between venue and topic classification, with venue classification being more challenging since generally there are overlapping topics between multiple venues, and precisely, there is no one-to-one mapping between topics and venues.

Yang and Davison [16] proposed a venue recommendation method based on a collaborative filtering technique. Context-free stylometric features, which are represented by lexical, syntactic and structural aspects are considered in addition to the content of the paper to better measure the similarity between documents.

The target paper is then assigned to the venue that occurs most frequently among the most similar documents. HaCohen-Kerner *et al.* [8] studied stylistic characteristics of research papers to differentiate the styles of writing of three Computer Science conferences, SIGIR, ACL and AAMAS, and found that a paper's vocabulary richness and the use of certain parts of speech can accurately classify the type of writing of a paper. An approach to venue recommendation that explores the structure of the co-authorship network is also proposed by Luong *et al.* [13]. The method first extracts the name of the authors of a paper, and then, collects information about each author's publications and co-authors using the Microsoft Academic. To recommend venues for a target paper, each candidate venue receives a score based on the number of papers published in that conference by authors and co-authors. Each co-author is weighted by the number of papers they have co-authored with the a main author to capture the strength of the connection between authors in the network. Medvet *et al.* [14] address venue recommendation by matching topics of a target paper with topics of possible venues for that paper.

Many of the previous works discussed above focus on venue recommendation and return a ranked list of venues to be recommended for a paper. In contrast, we explore the venue classification of papers, i.e., classifying a paper into one of the available venues. To our knowledge, this is the first work that uses information from the references list together with the title and abstract of a target paper to predict its venue of publication.

3   Approach

Problem Statement: Given a paper d and a set of venues $\mathcal{V}$, the task is to classify the paper d into one of the available venues from $\mathcal{V}$.

Leveraging two lines of research, one on keyword extraction and another on topic evolution and influence in citation networks within a *query expansion* like framework, we propose an approach to venue classification that combines information from both the title + abstract and the references list of a paper. This idea is similar to *query expansion* from information retrieval, where a user's query is expanded by adding new words (e.g., words from the relevant documents or synonyms and semantically related words from a thesaurus) to the terms used in the query in order to increase the quality of the search results. By analogy, we regard the papers that are cited by the target paper as the "relevant documents" and add terms from the titles of the cited papers as they appear in the references list of the target paper. In doing so, we incorporate: (1) stronger evidence (higher frequency) about the words/keywords in the paper - the more frequent the words are, the more indicative they are about the topic of the paper; and (2) word relatedness - terms that are related to the words used in the target paper, which may be indicative of broader topics. Specifically, instead of using a thesaurus or controlled vocabulary to choose the additional words, we leverage the words/keywords from the references list of the target paper, which represent a fusion of topics related to the paper in question. Thus, we posit that this mixture of topics can help predict the venue of a paper.

In our approach, we jointly model the content (title + abstract) and the references list in a supervised approach. We represented both the content and the references list using the "bag of words" (BoW) representation. For BoW, we first construct a vocabulary from the terms in the title and abstract (or from the references section) of the training set, and then, represent each paper as normalized *tf-idf* vectors. In order to jointly model the title + abstract and the references list of a paper, we employ two models (early and late fusion). In early fusion, the BoW from title + abstract and BoW from references list are concatenated, creating a single vector representation of the data. In contrast, late fusion trains two separate classifiers (one on each set of BoW features) and then combines their class probability distributions (using multiplication in our case) to obtain the class probability distribution of the aggregated model that produces the final output.

We preprocessed the papers in our collection to remove punctuation, numbers, and stop words, and performed stemming. In addition, we kept only words with document frequency (df) greater than 10. We experimented with different df values, and found $df \geq 10$ to give the best results for both content and references list.

4   Dataset

In order to evaluate our approach, we created our own dataset starting with a subcollection of the *CiteSeer^x* digital library [3].

We selected 22 conference venues focused on broad topics such as: machine learning (*NIPS, ECML, ICMl, WWW*), data mining (*KDD, ICDM, CIKM, SDM, SAC*), information retrieval (*SIGIR*), artificial intelligence (*IJCAI, AAAI, ICRA*), human computer interaction (*CHI*), natural language processing (*ACL, EMNLP, EACL, HLT-NAACL, LREC, COLING*),

Table 1. Dataset summary.

Venue	#papers	Venue	#papers
NIPS	5312	CIKM	1454
IJCAI	4722	WWW	1452
ICML	3888	COLING	1412
ICRA	2998	SDM	1191
ACL	2979	EMNLP	1128
VLDB	2972	ICDM	1111
CHI	2375	HLT-NAACL	920
AAAI	2201	LREC	891
CVPR	2039	EACL	760
KDD	1938	ECML	692
SIGIR	1601	SAC	585

computer vision (*CVPR*) and databases (*VLBD*), as shown in Table 1. A venue was included in our dataset if it had at least 500 papers. Our dataset, called *CiteSeer44K*, consists of *44,522* papers.

5   Experiments and Results

5.1   Features and Models Comparisons for Venue Classification

How does the performance of classifiers trained on features extracted from the references section compare with that of classifiers train on features extracted from the title and abstract for venue classification?

To answer this question, we study the performance of normalized *tf-idf* features from title + abstract and references list separately, using three classification algorithms: Naïve Bayes Multinomial (NBM), Random Forests (RF), and Support Vector Machines (SVM), using *10-fold cross-validation*.

Table 2. Comparison of features extracted from title + abstract and references list.

Features	Classifier	Precision	Recall	F1-score
Title + abstract	RF	0.705	0.666	0.660
	SVM	0.717	0.717	0.717
	NBM	0.646	0.634	0.634
References list	RF	0.741	0.721	0.713
	SVM	**0.759**	**0.736**	**0.742**
	NBM	0.696	0.684	0.683

Table 2 shows the performance achieved by these classifiers (RF, SVM and NBM) when they are trained separately on each set of features (the title + abstract and the references list). As can be seen from the table, regardless of the machine learning algorithm used (RF, SVM, NBM), all classifiers trained on features from the references list achieve better results compared with the classifiers trained on content-based (title + abstract) features. For example, the SVM trained on normalized *tf-idf* from the references section obtains a relative improvement of 3.48% in F1-score over the SVM trained on normalized *tf-idf* from the title + abstract. The results of this experiment show that the terms found in the references section of a research paper contain significant hints that have the potential to improve the venue classification of a research paper. Since SVM performs best among all classifiers, we use it in the next experiments.

Can we further improve the performance of venue classification of research papers by jointly modeling the references section of a paper and its title + abstract? In this experiment, we compare the classifiers trained on features extracted from title + abstract and references list (independently) with the early and late fusion classifiers, that jointly model the title and abstract with the references list.

Table 3. Results of early and late fusion classifiers on *CiteSeer44K*.

Features	Classifier	Precision	Recall	F1-score
Title + abstract	SVM	0.717	0.717	0.717
References list	SVM	0.759	0.736	0.742
Early fusion	SVM	**0.793**	**0.776**	**0.781**
Late fusion	Ensemble of two SVMs	0.793	0.765	0.771

Table 3 shows the performance of the four models on the *CiteSeer44K* dataset: SVM trained on title + abstract, SVM train on references list, the early fusion SVM trained on all information units (by concatenating title + abstract and references list), and the late fusion model (i.e., an ensemble of two SVMs, which combines the SVM from the title + abstract with the SVM from the references list). As can be seen from the table, the information in the references section substantially improves venue classification when it is jointly modeled with the information contained in the title + abstract, under both scenarios, early and late fusion. The late fusion model performs slightly worse than the early fusion model, but still outperforms the individual SVMs trained on either the title + abstract or the references list. These results illustrate that jointly modeling the information from the references list with that from the title and abstract of a paper yields improvements in performance over models that use only one type of information.

5.2   Baseline Comparisons

We compared the early fusion model with three baselines and a prior work as follows: (1) *venue heuristic*, i.e., assign the target paper to the most frequent venue found in its references list; (2) *a bag of venues*, i.e., use the 22 selected venues as the vocabulary, and encode each paper as venue frequency; (3) *a bag of words from title and abstract*; and (4) *Luong et al.'s approach* [13] that consists of building a social network for each author, capturing information about her co-authors and publication history. In this experiment, we kept only the papers that have at least one reference (in the references list) that is published in one of our 22 venues, for a fair comparison among all models. Thus, the dataset was reduced to *28,735* papers (called *CiteSeer29K*).

Table 4. Comparison with previous work and baselines on *CiteSeer29K*.

Model	Used classifier	Precision	Recall	**F1-score**
Venue heuristic	-	0.478	0.484	0.464
Bag of venues	SVM	0.558	0.566	0.546
Bag of words from title + abstract	SVM	0.720	0.709	0.713
Author's network [13]	-	0.646	0.642	0.630
Early fusion	SVM	**0.800**	**0.771**	**0.779**

Table 4 shows the results of the comparison of the early fusion SVM with the baselines and prior work by Luong et al. [13]. We can observe from this table that the early fusion model that uses information from both title + abstract and references list substantially outperforms all three baselines and the prior work.

6   Conclusion and Future Work

In this paper, we presented a supervised approach to venue classification, which combines information from the references section of a research paper with its title and abstract, within a *query expansion-like framework*. Our experimental findings on a dataset of $\approx 44,000$ papers are as follows: (1) We found that the references section of a paper contains hints that have the potential to improve the venue classification task; (2) We showed that the words/keywords from the references list act as "expanded" additional evidence for identifying the venue of a paper when added to its title + abstract; (3) Finally, we showed that our early fusion model performs better than several baselines and a prior work. In the future, it would be interesting to extend the classification models to venue recommendation to help authors in the decision making for an appropriate venue.

Acknowledgments. This research was supported by the NSF awards #1813571 and #1802358 to Cornelia Caragea. Any opinions, findings, and conclusions expressed here are those of the author and do not necessarily reflect the views of NSF.

References

1. Caragea, C., Bulgarov, F., Godea, A., Das Gollapalli, S.: Citation-enhanced keyphrase extraction from research papers: a supervised approach. In: Proceedings of EMNLP (2014)
2. Caragea, C., Bulgarov, F., Mihalcea, R.: Co-training for topic classification of scholarly data. In: Proceedings of EMNLP, pp. 2357–2366 (2015)
3. Caragea, C., et al.: CiteSeerx: a scholarly big dataset. In: Proceedings of ECIR (2014)
4. Councill, I.G., Giles, C.L., Kan, M.Y.: ParsCit: an open-source CRF reference string parsing package. In: International Language Resources and Evaluation (2008)
5. Florescu, C., Caragea, C.: PositionRank: an unsupervised approach to keyphrase extraction from scholarly documents. In: Proceedings of ACL (2017)
6. Giles, C.L., Bollacker, K., Lawrence, S.: CiteSeer: an automatic citation indexing system. In: Digital Libraries, pp. 89–98 (1998)
7. Gollapalli, S.D., Caragea, C.: Extracting keyphrases from research papers using citation networks. In: Proceedings of AAAI, pp. 1629–1635 (2014)
8. HaCohen-Kerner, Y., Rosenfeld, A., Tzidkani, M., Cohen, D.N.: Classifying papers from different computer science conferences. In: Motoda, H., Wu, Z., Cao, L., Zaiane, O., Yao, M., Wang, W. (eds.) ADMA 2013. LNCS (LNAI), vol. 8346, pp. 529–541. Springer, Heidelberg (2013). https://doi.org/10.1007/978-3-642-53914-5_45
9. Han, H., Giles, C.L., Manavoglu, E., Zha, H., Zhang, Z., Fox, E.A.: Automatic document metadata extraction using support vector machines. In: Proceedings of JCDL (2003)
10. Jo, Y., Lagoze, C., Giles, C.L.: Detecting research topics via the correlation between graphs and texts. In: Proceedings of KDD, pp. 370–379 (2007)
11. Litvak, M., Last, M.: Graph-based keyword extraction for single-document summarization. In: Multi-source Multilingual Information Extraction and Summarization (2008)
12. Lu, Q., Getoor, L.: Link-based classification. In: Proceedings of the Twentieth International Conference on Machine Learning (ICML) (2003)

13. Luong, H., Huynh, T., Gauch, S., Do, L., Hoang, K.: Publication venue recommendation using author network's publication history. In: Pan, J.-S., Chen, S.-M., Nguyen, N.T. (eds.) ACIIDS 2012. LNCS (LNAI), vol. 7198, pp. 426–435. Springer, Heidelberg (2012). https://doi.org/10.1007/978-3-642-28493-9_45
14. Medvet, E., Bartoli, A., Piccinin, G.: Publication venue recommendation based on paper abstract. In: IEEE International Conference on Tools with Artificial Intelligence (ICTAI) (2014)
15. Wang, X., Zhai, C., Roth, D.: Understanding evolution of research themes: a probabilistic generative model for citations. In: Proceedings of KDD (2013)
16. Yang, Z., Davison, B.D.: Writing with style: venue classification. In: International Conference of Machine Learning and Applications (2012)

Digital Humanities

Towards Better Understanding Researcher Strategies in Cross-Lingual Event Analytics

Simon Gottschalk[1]([✉]), Viola Bernacchi[2], Richard Rogers[3],
and Elena Demidova[1]

[1] L3S Research Center, Leibniz Universität Hannover, Hannover, Germany
{gottschalk,demidova}@L3S.de
[2] DensityDesign Research Lab, Milano, Italy
viola.bernacchi@live.it
[3] University of Amsterdam, Amsterdam, The Netherlands
rogers@uva.nl

Abstract. With an increasing amount of information on globally important events, there is a growing demand for efficient analytics of multilingual event-centric information. Such analytics is particularly challenging due to the large amount of content, the event dynamics and the language barrier. Although memory institutions increasingly collect event-centric Web content in different languages, very little is known about the strategies of researchers who conduct analytics of such content. In this paper we present researchers' strategies for the content, method and feature selection in the context of cross-lingual event-centric analytics observed in two case studies on multilingual Wikipedia. We discuss the influence factors for these strategies, the findings enabled by the adopted methods along with the current limitations and provide recommendations for services supporting researchers in cross-lingual event-centric analytics.

1 Introduction

The world's community faces an unprecedented number of events impacting it as a whole across language and country borders. Recently, such unexpected events included political shake-ups such as Brexit and the US pullout of the Paris Agreement. Such events result in a vast amount of event-centric, multilingual information that differs across sources, languages and communities and can reflect community-specific aspects such as opinions, sentiments and bias [18]. In the context of events with global impact, cross-cultural studies gain in importance.

Memory institutions are increasingly interested in collecting multilingual event-centric information and making this information available to interested researchers. For example, the Internet Archive provides the Archive-It service that facilitates curated collections of Web content[1]. Several recent research

---

[1] archive-it.org.

© Springer Nature Switzerland AG 2018
E. Méndez et al. (Eds.): TPDL 2018, LNCS 11057, pp. 139–151, 2018.
https://doi.org/10.1007/978-3-030-00066-0_12

efforts target the automatic creation of event-centric collections from the Web and large-scale Web archives (e.g. iCrawl [6,7]) as well as creation of event-centric knowledge graphs such as EventKG [10]. In this context one of the key Web resources to analyze cross-cultural and cross-lingual differences in representations of current and historical events is the multilingual Wikipedia [16,18].

However, at present very little is known about the strategies and the requirements of researchers who analyze event-centric cross-lingual information. In this paper we take the first important step towards a better understanding of researcher strategies in the context of event-centric cross-lingual studies at the example of multilingual Wikipedia. The goals of this paper are to better understand: (1) How do researchers analyze current events in multilingual settings? In particular, we are interested in the content selection strategies, analysis methods and features adopted along with the influence factors for this adoption. (2) Which findings can be facilitated through existing cross-lingual analytics methods, what limitations do these methods have and how to overcome them?

To address these questions we conducted two qualitative case studies that concerned the Brexit referendum and the US pullout of the Paris Agreement. We observed interdisciplinary and multicultural research teams who performed analyses of the event representations in the multilingual Wikipedia dataset during a week's time. As a first step, we used in-depth pre-session questionnaires aimed at collecting the participants' background. Following that, the participants defined their own research questions and several working sessions took place. During these sessions we observed the methods adopted by the participants and the findings they obtained. Finally, we interviewed the participants.

The main findings of our analysis are as follows: (1) *The content selection strategy* mostly depends on the event characteristics and the collection properties. (2) *The adoption of analysis methods and features* is most prominently influenced by the researcher backgrounds, the information structure and the analysis tools. (3) *The features involved* in the adopted analysis methods mostly include metadata (e.g. tables of content), rather than the actual texts. (4) *The main insights facilitated by the adopted analysis methods* include a variety of findings e.g. with respect to the shared and language-specific aspects, the interlingual dependencies, the event dynamics, as well as the originality and role of language editions. (5) *The limitations of the adopted methods* mostly concern the relatively low content and temporal resolution, as well as the lack of detailed insights into the communities and discussions behind the content. (6) *The recommendations to overcome these limitations* include the development of tools that better support cross-lingual overview, facilitate fact alignment, provide high temporal resolution, as well as community and discussion insights.

2   Study Context and Objectives

We focus on two political events with global impact that constitute important cases for the cross-cultural analysis:

- Brexit ($\mathcal{B}$): On 23 June 2016, a referendum took place to decide on the withdrawal of the United Kingdom (UK) from the European Union (EU).

51.9% of the voters voted to leave the EU, which lead to the withdrawal process called "Brexit".

– US **P**aris Agreement pullout ($\mathcal{P}$): On June 1, 2017, the US President Donald Trump announced to pull out of the Paris agreement, which was previously signed by 195 countries at the Paris climate conference.

To better understand researcher strategies, we asked the participants to conduct an analysis of the event, in particular with the focus on the international event perception. Overall, three main objectives (*O1–O3*) are addressed:

O1 - Content selection. How do researchers select articles, languages and revisions to analyze, given an ongoing event? Which factors influence this selection? Wikipedia articles are generated in a dynamic process where each edit of an article results in a new version called *revision*. Given the high velocity of discussions in different Wikipedia language editions surrounding an ongoing global event, there is a large amount of potentially relevant information. Thus, there is a need to identify the most relevant articles, their revisions and language versions as entry points for the detailed analysis.

O2 - Method and feature selection. Which methods and features can researchers use efficiently to perform cross-lingual event-centric analytics? Which factors influence this selection? Wikipedia articles describing significant events tend to cover a large number of aspects. The large number of articles, their revisions and the variety of language editions make the analysis particularly difficult. The challenge here is to choose analysis methods and features that can provide an overview of cross-lingual and temporal differences across multilingual event representations efficiently.

O3 - Findings and limitations. Which findings can be efficiently obtained by researchers when conducting research over multilingual, event-centric articles using particular analysis methods? What are the current limitations and how can they be addressed? The size, dynamics as well as cross-lingual and cross-cultural nature of Wikipedia articles pose challenges on the interpretation of research results, especially in case such an interpretation requires close reading of multilingual content. Our goal here is to better understand which findings can be obtained efficiently, and derive recommendations for future assistance.

We do not aim at the completeness of the considered strategies, methods, features and interpretation results, but focus on the participants' approaches as a starting point to better understand which methods and features appear most efficient from the participant perspective, which factors influence their selection and which findings they can enable in practice.

3   Methodology

The case studies were conducted by performing the following steps:

1. Introduction of the event to the participants.
2. Individual questionnaires to be filled out by the participants.
3. Working sessions in teams, observed by the authors.
4. Individual semi-structured interviews with the participants.

Table 1. Setup and the participant background. CS: Computer scientist, ID: Information designer, S: Sociologist.

	$\mathcal{B}$	$\mathcal{P}$
Study setup		
Event date	June 23, 2016	June 1, 2017
Study dates	June 27 - July 1, 2016	July 3 - July 7, 2017
Overall study duration	14 h	14 h
Participant background		
Number of participants	5 (ID: 3, S: 2)	4 (CS: 1, ID: 2, S: 1)
Native languages	IT (3), NL (1), UK (1)	IT(2), DE (2)
Languages spoken	EN, IT, NL, DE	EN, IT, DE, FR, ES
Wikipedia experience		
Role	Reader (5)	Reader (4), editor (2)
Frequency of use	Daily (1), weekly (4)	Daily (3), weekly (1)
Multilingual Wikipedia experience	Yes (2)	Yes (4)

3.1   Pre-session Questionnaires

Table 1 provides an overview of the study setup and the participant interdisciplinary and multicultural background, collected using pre-session questionnaires. According to these questionnaires, the participants estimated the language barrier to be a major problem in both of the studies and raised the question whether it is possible not only to detect cultural differences or commonalities, but also to reason about their origins.

3.2   Task Definition and Working Sessions

We asked the participants to: (1) define their own research questions and analysis methods; (2) conduct an analysis of the event-related articles across different Wikipedia language editions with respect to these questions; and (3) present their findings. This way, we kept the task description of the study rather open, as we intended to facilitate an open-minded discussion among the participants, to enhance their motivation and to reduce possible bias.

To enable in-depth insights, the studies implied high expenditures of approximately 14 h per participant, which overall translates into 126 person hours. The participants worked together as a team over four days.

4   The Participant Approach

The interdisciplinary expertise of the participants enabled them to tackle several facets of interest in the context of the considered events. In this section, we describe and compare the participant course of action in both case studies, from the definition of the research questions to the presentation of results.

Table 2. Overview of the datasets resulting from the data collection.

	$\mathcal{B}.wd$	$\mathcal{B}.ref$	$\mathcal{P}.agr$	$\mathcal{P}.wd$
#Languages	59	48	34	4
#Words (EN)	4, 468	12, 122	5, 950	5, 787
#Categories (EN)	3	9	159	9
#Other Articles	-		99	

4.1   Research Questions

At the beginning of the case studies, the participants agreed on the following research questions building the basis for the analysis:

- **Q0,$\mathcal{B}$**: What can Wikipedia tell us about the UK's changing place in the world after Brexit?
- **Q0,$\mathcal{P}$**: Has the announcement of the US pullout of the Paris Agreement changed the depiction of and attention to the issue of climate change?

In order to approach these research questions, the participants analyzed the following aspects in the course of their studies:

- **Q1:** How was the event-centric information propagated across languages?
- **Q2:** How coherent are the articles regarding the event across languages?
- **Q3:** Which aspects of the event are controversial across languages?

4.2   Data Collection

First of all, in both case studies the participants selected a set of relevant articles to be analyzed, which resulted in the datasets shown in Table 2. In $\mathcal{B}$, the participants selected two English articles: the "United Kingdom European Union membership referendum, 2016" article ($\mathcal{B}.ref$) and the "United Kingdom withdrawal from the European Union" article ($\mathcal{B}.wd$). For $\mathcal{P}$, there is a "Paris Agreement" article ($\mathcal{P}.agr$) in several languages, but only four language editions provided a "United States withdrawal from the Paris Agreement" article ($\mathcal{P}.wd$). Thus, the participants searched for paragraphs in Wikipedia articles linking to the articles $\mathcal{P}.agr$ and "Donald Trump", $\mathcal{P}.agr$ and "United States", and those linking to $\mathcal{B}.wd$ using the Wikipedia API and manual annotation. To address **Q0,$\mathcal{P}$**, the articles "Global warming", "2015 United Nations Climate Change Conference", "Climate change" and "Climate change denial" were considered.

4.3   Analysis Methods and Feature Groups

Overall, the analysis methods employed by the participants in both studies can be categorized into content, temporal, network and controversy analysis:

Table 3. For each of the four analysis methods, this table lists whether the participants employed features from the specified feature group in $\mathcal{B}$ or $\mathcal{P}$.

Analysis	Feat. group				
	Text-based	Multi-media	Edit-based	Link-based	Categories
Content	$\mathcal{B}, \mathcal{P}$	$\mathcal{B}$			
Temporal	$\mathcal{B}$	$\mathcal{P}$	$\mathcal{B}$		
Network	$\mathcal{P}$	$\mathcal{P}$		$\mathcal{B}$	$\mathcal{B}$
Controversy			$\mathcal{B}$		

- **Content analysis** to get detailed insights of how the event was described across languages (**Q0**).
- **Temporal analysis** to analyze when sub-events were reported (**Q1**).
- **Network analysis** to estimate the coherence between the event-centric articles across languages (**Q2**).
- **Controversy analysis** to identify the controversial event aspects (**Q3**).

Table 3 provides an overview of the analysis methods and the corresponding features employed in both case studies. For clarity, we categorize the features into groups that were covered to a different extent in the case studies:

- **Text-based features:** Features based on the Wikipedia article texts such as the textual content, terms, selected paragraphs and the table of contents.
- **Multimedia features:** Features based on the multimedia content (such as images) directly embedded in the articles.
- **Edit-based features:** Features based on the editing process in Wikipedia, including the discussion pages and the editors.
- **Link-based features:** Features employing the different types of links such as cross-lingual links between Wikipedia articles and links to external sources.
- **Category-based features:** Features employing Wikipedia categories.

4.4   Observations

Content Analysis. Due to the language barrier, the participants of $\mathcal{B}$ focused on the text-based features involving less text: tables of contents (TOCs) and images. Similarly, in $\mathcal{P}$ the terms from the extracted paragraphs were utilized.

Text-based features in $\mathcal{B}$: The participants arranged the ToC entries by their frequency across languages as shown in Fig. 1. This ToC comparison indicates that the articles differ in many aspects, e.g. the German article focuses on the referendum results in different regions, and the English Wikipedia focuses on Brexit's economical and political implications.

Multimedia features in $\mathcal{B}$: Using the Wikipedia Cross-lingual Image Analysis tool[2], it became evident that images containing the UK map and the referendum ballot paper were shared most frequently across languages.

---

[2] wiki.digitalmethods.net/Dmi/ToolWikipediaCrosslingualImageAnalysis.

English	German	French	Italian
Notes	Notes	Notes	Notes
Results	Results	Results	Results
References	References	References	Further reading
Date	History	Date	See also
Further reading	After the referendum	Legislation	Explanations of vote
History	Attitude of british media	Background	Positions expressed in the political debate
Legislation	Attitude of elected MPs and ministers	Campaign	Results by country
See also	Campaign before the referendum	Campaign suspension	Results by region
Administration	East midlands	Consequences	Text of the question
Business	East of England	Counting and results	
Cabinet ministers	England	Electoral body	
...	...	...	

Fig. 1. A comparison of the ToCs of the Brexit referendum article on the 24th June 2016 in four languages. The ToC entries are ordered by frequency and alphabetically. Darker colors correspond to a higher number of recurrences across languages, including standard Wikipedia sections and a section about the referendum results. (Color figure online)

Text-based features in $\mathcal{P}$: From the paragraphs extracted during the dataset collection, the participants extracted frequent words used in the context of the US pullout per language. This analysis showed different emphasis on the topic: The English Wikipedia mentioned oil and gas, the French Wikipedia included climate-related terms, and the Dutch one was focused on resistance.

Temporal Analysis. The description of ongoing events may vary substantially over time. In $\mathcal{B}$, the participants tracked this evolution using text-based and edit-based features. In $\mathcal{P}$, the participants focused on multimedia features.

Text-based features in $\mathcal{B}$: The participants extracted the ToC three times per day, in the time from the 22nd to the 24th of June 2016. The French version did not have the referendum results on the 23rd of June as opposed to other languages. On the following day, the English ToC stayed nearly unchanged, whereas a number of new German sections were added.

Edit-based features in $\mathcal{B}$ enabled observations of the Wikipedia editing process. The participants created a timeline depicted in Fig. 2 which is based on the data from the Wikipedia Edits Scraper and IP Localizer tool[3]. It illustrates the development of the $\mathcal{B}.wd$ article including its creation in other language editions. Article editions directly translated from other languages were marked and important events related to Brexit were added.

Multimedia features in $\mathcal{P}$: Motivated by **Q0,$\mathcal{P}$**, the $\mathcal{P}$ participants compared the images added to the set of climate-related articles before and after the US pullout became apparent. The majority of newly added images reflect statistics (in contrast to a mixture of photos and statistics added before), and some of them depict the US influence on the world's climate.

---

[3] wiki.digitalmethods.net/Dmi/ToolWikipediaEditsScraperAndIPLocalizer.

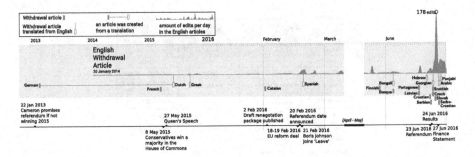

Fig. 2. A timeline of the $\mathcal{B}.wd$ article showing the English edit frequency over time and article editions in other languages. For example, the Dutch article was created on the 16th of August 2015 as a translation from German and English.

Network Analysis. The coherence of the Wikipedia language editions can provide useful insights. The participants of $\mathcal{B}$ focused on link-based and category-based features, while the participants of $\mathcal{P}$ focused on text-based features.

Category-based features in $\mathcal{B}$: The participants analyzed the categorization of the referendum and withdrawal articles in all available language editions by inserting the translated and aligned category names into the Table 2Net tool[4] and applying the ForceAtlas algorithm [12] to create the network shown in Fig. 3, which shows an isolated position of the English and Scottish articles.

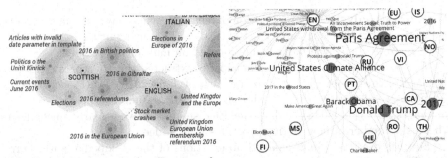

(a) Categorization of the Brexit referendum article across languages with language nodes and Wikipedia category nodes. The edges represent connections between categories and languages.

(b) Articles mentioning the US pullout of the Paris Agreement. Blue nodes represent languages and the others Wikipedia articles, where color and size denotes how many language editions link to the article.

Fig. 3. Network analysis in $\mathcal{B}$ (category-based) and $\mathcal{P}$ (text-based). (Color figure online)

Link-based features in $\mathcal{B}$: Links to external sources were extracted and compared using the MultiWiki tool [9]. For most of the language editions, the overlap

---

[4] http://tools.medialab.sciences-po.fr/table2net/.

of the linked web pages was rather low and reached higher values only in few cases, e.g. the English and German withdrawal articles shared 17.32% of links.

Text-based features in $\mathcal{P}$: The set of articles mentioning the US pullout of the Paris Agreement was put in a network shown in Fig. 3b. "Donald Trump", "Paris Agreement", "2017" and "United States Climate Alliance" are the articles mentioning the withdrawal in most languages, while some articles such as "Elon Musk" only mention it in very few languages. Another observation was the separation of political and science-related articles.

Multimedia features in $\mathcal{P}$: The participants retrieved a list of images used in the different language versions of the "Climate Change" article. A network where language nodes were connected to images revealed that some language editions (e.g. Dutch) and groups of languages (e.g. a group of Serbian, Slovakian, Serbo-Croatian and Faroese) differed from the others with respect to the image use.

Controversy Analysis: In $\mathcal{B}$, the participants observed controversies among the Wikipedia editors. In $\mathcal{P}$, no explicit controversy analysis was conducted due to the difficulties to resolve the origins of the extracted text paragraphs, the language barrier and the lack of extraction tools.

Edit-based features in $\mathcal{B}$: For each Wikipedia article, there is a discussion page, structured by a table of contents. The $\mathcal{B}$ participants reviewed the English discussion TOCs and identified an intense discussion among the Wikipedia editors on the question to which article the search term "Brexit"should link to.

5 Discussion

In this section we discuss our observations performed in the course of the case studies with respect to the objectives $O1$-$O3$ (Sect. 2).

5.1 *O1-O2*: Influence Factors for Content, Methods and Feature Selection

Overall, we observed that the adopted methods and their outcomes are influenced and also limited by a number of factors, including:

Characteristics of the Event. Relevant characteristics of the event include its topical breadth and global influence. For example, as the Brexit referendum was considered as an event of European importance, the participants of $\mathcal{B}$ focused on the European languages, while in $\mathcal{P}$ the US pullout of the Paris Agreement was studied in all languages due to its global impact. As the US pullout was considered to cover many aspects in politics and science, the participants focused on coverage, resulting in a larger set of articles to be analyzed compared to $\mathcal{B}$.

Participant Professional Background. The professional background of the participants influenced in particular the selection of analysis methods. Although the teams were interdisciplinary, the individual participants focused on the methods and features typical for their disciplines.

Table 4. Analysis methods: findings, limitations and tool recommendations.

Method	Facilitated findings	Current limitations	Recommendations
Content	- Shared article aspects - Interlingual dependencies - Overview of the context	- Lack of shared fact analysis - Lack of systematic content selection for close reading	- Cross-lingual fact alignment - Overview as an entry point to close reading
Temporal	- Event dynamics - Changes in public interest - Language version originality - Context shift	- Analysis is limited to specific revisions	- Higher temporal resolution
Network	- Cross-lingual similarity - Event coverage	- Roles of specific communities are hard to identify	- Editor community insights
Controversy	- Event perception	- Lack of cross-lingual discussion comparisons	- Discussion insights

Participant Language Knowledge. The participant language knowledge limited in particular their ability to apply analysis methods that require close reading. For example, the controversy analysis in $\mathcal{B}$ was only performed on the English discussion pages due to the language barrier. Nevertheless, the overall scope of the study was not limited by this factor. To cross the language barrier, the participants employed two techniques: (1) machine translation tools, and (2) content selection to reduce the amount of information in a foreign language to be analyzed (e.g. analyzing category names instead of full text).

Availability of the Analysis Tools. Existing analysis tools mostly support distant reading on larger collections using specific features, such as links, images, etc. and can be applied in the multilingual settings efficiently. Fewer tools, including for example MultiWiki [8], support close reading in the cross-lingual settings. The edit-based and text-based cross-lingual controversy analysis was not adequately supported by the existing tools.

Information Structure. One important factor of the event-centric cross-lingual analytics is the information structure. The features adopted in the case studies under consideration include rich text-based features such as hyperlinks and categories, as well as edit histories available in Wikipedia. Furthermore, the availability of comparable articles in different languages is an important feature of the Wikipedia structure in this context.

5.2   *O3*: Findings, Limitations and Recommendations

Table 4 provides an overview of the analysis methods.

Findings: The adopted analysis methods facilitated a range of findings. Content analysis using ToCs, images and word clouds enabled the identification of shared aspects, interlingual dependencies and provided a context overview. Temporal analysis involving ToCs, edit histories and images provided insights into the event dynamics, changes in the public interest within the language communities, originality of the language versions and the context shift. Network analysis resulted in an overview of the cross-lingual similarities, supported identification of the roles of the language editions, event coverage and the specific cross-lingual aspects. Controversy analysis conducted on the English Wikipedia (only) provided details on the event perception.

Limitations: The limitations of the adopted analysis methods mostly regard the relatively low content and temporal resolution, as well as the lack of detailed insights into the communities and discussions behind the content. With respect to the content analysis, the lack of close reading restricted the obtained insights to rather high-level comparative observations, such as shared aspects of the articles, rather than individual facts. In the temporal analysis, the information regarding content propagation was restricted to the origin of the first revision of the articles. The network analysis did not support insights in the specific communities behind the edits, such as the supporters and the opponents of Brexit. The controversy analysis based on discussion pages could not be applied in the cross-lingual settings, due to the lack of specific extraction tools and the language barrier. Overall, the limitations observed in our study are due to multilingual information overload, the language barrier and the lack of tools to systematically extract and align meaningful items (e.g., facts) across languages.

Recommendations: Our observations regarding the above limitations and the post-sessions interviews lead to the recommendations for future method and tool development summarized in Table 4. These recommendations include a zoom-out/in functionality to provide an overview, helping to select relevant content for close reading, extraction and cross-lingual alignment of information at a higher granularity level (e.g. facts), tracking article development over time including involved communities as well as a systematic analysis of discussion pages to better support controversy detection. In the future work we would also like to develop interactive cross-lingual search and exploration methods based on [4].

5.3 Limitations of the Study

In this qualitative study we limited our corpus to the multilingual Wikipedia, such that features and tools adopted by the participants are in some cases corpus-specific. Nevertheless, we believe that the results with respect to the participant strategies (e.g. the preferential usage of metadata to reduce close reading in a foreign language) are generalizable to other multilingual event-centric corpora, such as event-centric materials extracted from the Web corpora and Web archives.

6 Related Work

Multilingual and temporal analytics becomes an increasingly important topic in the research community (see e.g. [11,13] for recent studies on cross-lingual content propagation and editing activity of multilingual users). Till now only few studies focus on analyzing and effectively supporting the needs of researchers who conduct research on temporal content [5,15] and create event-centric temporal collections [6,7,17]. Whereas existing works on temporal collections focus on the monolingual case, working patterns and requirements of researchers analyzing multilingual temporal context remain largely non-investigated.

As Wikipedia language editions evolve independently and can thus reflect community-specific views and bias, multilingual Wikipedia became an important research target for different disciplines. One important area of interest in this context is the study of differences in the linguistic points of view in Wikipedia (e.g. [1,18]). Whereas several visual interfaces and interactive methods exist to support researchers in analysing Wikipedia articles across languages (e.g. MultiWiki [9], Contropedia [3], Manypedia [14] and Omnipedia [2]), our case study illustrates that substantial further developments are required to effectively support researchers in various aspects of cross-lingual event-centric analysis.

7 Conclusion

In this paper we presented two case studies in which we observed interdisciplinary research teams who conducted research on the event-centric information in the context of the Brexit referendum and the US pullout of the Paris Agreement. We summarized our observations regarding the content, method and feature selection, their influence factors as well as findings facilitated by the adopted methods and provided recommendations for services that can better support cross-lingual analytics of event-centric collections in the future.

Acknowledgements. This work was partially funded by the ERC ("ALEXAN-DRIA", 339233) and H2020-MSCA-ITN-2018-812997 "Cleopatra".

References

1. Al Khatib, K., Schütze, H., Kantner, C.: Automatic detection of point of view differences in wikipedia. In: COLING 2012, pp. 33–50 (2012)
2. Bao, P., Hecht, B., Carton, S., Quaderi, M., Horn, M., Gergle, D.: Omnipedia: bridging the wikipedia language gap. In: CHI 2012, pp. 1075–1084 (2012)
3. Borra, E., Laniado, D., et al.: A platform for visually exploring the development of wikipedia articles. In: ICWSM 2015 (2015)
4. Demidova, E., Zhou, X., Nejdl, W.: Efficient query construction for large scale data. In: Proceedings of SIGIR 2013, pp. 573–582 (2013)
5. Fernando, Z.T., Marenzi, I., Nejdl, W.: ArchiveWeb: collaboratively extending and exploring web archive collections - how would you like to work with your collections? Int. J. Digit. Libr. **19**(1), 39–55 (2018)
6. Gossen, G., Demidova, E., Risse, T.: iCrawl: improving the freshness of web collections by integrating social web and focused web crawling. In: JCDL (2015)

7. Gossen, G., Demidova, E., Risse, T.: Extracting event-centric document collections from large-scale web archives. In: Kamps, J., Tsakonas, G., Manolopoulos, Y., Iliadis, L., Karydis, I. (eds.) TPDL 2017. LNCS, vol. 10450, pp. 116–127. Springer, Cham (2017). https://doi.org/10.1007/978-3-319-67008-9_10

8. Gottschalk, S., Demidova, E.: Analysing temporal evolution of interlingual wikipedia article pairs. In: SIGIR 2016, pp. 1089–1092 (2016)

9. Gottschalk, S., Demidova, E.: MultiWiki: interlingual text passage alignment in wikipedia. TWEB 11(1), 6:1–6:30 (2017)

10. Gottschalk, S., Demidova, E.: EventKG: a multilingual event-centric temporal knowledge graph. In: Gangemi, A., et al. (eds.) ESWC 2018. LNCS, vol. 10843, pp. 272–287. Springer, Cham (2018). https://doi.org/10.1007/978-3-319-93417-4_18

11. Govind, Spaniol, M.: ELEVATE: a framework for entity-level event diffusion prediction into foreign language communities. In: WebSci 2017 (2017)

12. Jacomy, M., et al.: ForceAtlas2, a continuous graph layout algorithm for handy network visualization designed for the Gephi software. PloS One 9(6), e98679 (2014)

13. Kim, S., Park, S., Hale, S.A., Kim, S., Byun, J., Oh, A.H.: Understanding editing behaviors in multilingual wikipedia. PloS One 11(5), e0155305 (2016)

14. Massa, P., Scrinzi, F.: Manypedia: comparing language points of view of wikipedia communities. In: WikiSym 2012, p. 21 (2012)

15. Odijk, D., et al.: Supporting exploration of historical perspectives across collections. In: Kapidakis, S., Mazurek, C., Werla, M. (eds.) TPDL 2015. LNCS, vol. 9316, pp. 238–251. Springer, Cham (2015). https://doi.org/10.1007/978-3-319-24592-8_18

16. Pentzold, C., et al.: Digging wikipedia: the online encyclopedia as a digital cultural heritage gateway and site. J. Comput. Cult. Herit. 10(1), 5:1–5:19 (2017)

17. Risse, T., Demidova, E., Gossen, G.: What do you want to collect from the web? In: Proceedings of the Building Web Observatories Workshop (BWOW) 2014 (2014)

18. Rogers, R.: Digital Methods. MIT Press, Cambridge (2013)

Adding Words to Manuscripts: From PagesXML to TEITOK

Maarten Janssen[✉][iD]

CELGA-ILTEC, Universidade de Coimbra, Coimbra, Portugal
maartenjanssen@uc.pt

Abstract. This article describes a two-step method for transcribing historic manuscripts. In this method, the first step uses a page-based representation making it easy to transcribe the document page-by-page and line-by-line, while the second step converts this to the TEI/XML text-based format, in order to make sure the document becomes fully searchable.

Keywords: Manuscript transcription · TEI/XML
Linguistic annotation

1 Introduction

For the digitization of historic manuscripts and monographs located in libraries and archives around the world, the main focus typically is on two aspects related to the image: the quality of the image itself, and keeping track of the metadata to make sure it is clear which document the images pertain to. However, for properly disclosing the cultural heritage enclosed in those manuscripts ands monographs, it is necessary to take things one step further and also digitize the content of the documents, in order to allow searching through the text in order to get access to otherwise inaccessible information.

Therefore, there are various ongoing projects to digitize the content, in which manuscripts are manually transcribed or by handwriting recognition (HTR) where possible, and monographs are treated by optical character recognition (OCR). Almost without exception, such projects use a format for storing the content of the manuscript that describes the manuscript content page-by-page and line-by-line, corresponding to the structure of the physical object.

But for describing the text itself, a text-driven format, such as TEI/XML, in which the content of the manuscript is divided into paragraphs, sentences, and words is more appropriate. And although often the two types of formats are compatible, they conflict when the visual elements intersect with the textual elements: when paragraphs span across pages, or words across lines. Words split over two lines will not be found when searching the text, hence hampering the use of page-driven documents in queries.

In this article I will demonstrate how a two-stage work-flow can break this tension between the need to describe the page and the text. The idea behind

© Springer Nature Switzerland AG 2018
E. Méndez et al. (Eds.): TPDL 2018, LNCS 11057, pp. 152–157, 2018.
https://doi.org/10.1007/978-3-030-00066-0_13

it is simple: the initial stages of transcription are done in a page-driven format, which is converted to a text-driven format once the initial transcription is done. This leads to a fully searchable document in which all the page-driven content is kept. The last section briefly sketches how this approach is implement in the TEITOK corpus framework [1].

2   Pages vs. Paragraphs

The format typically used in digitization projects for historic manuscripts and monographs is page-driven. That is to say, the file (typically in XML), has a structure as exemplified in Fig. 1, where the document is subdivided into pages, which in turn are subdivided into manuscript lines (and optionally words). Page-driven formats include hOCR, used in OCR programs such as Tesseract (https://github.com/tesseract-ocr/), and ALTO and PageXML, which are formats designed by the academic community for the annotation of manually annotated documents, but also used as the output for HTR tools such as Transkribus (https://transkribus.eu/Transkribus).

```
<document>
    <page>
        <line>Manuscript line</line>
        <line>Manuscript line</line>
    </page>
</document>
```

Fig. 1. Structure of a page-driven XML format

Apart from the content itself, page-driven formats can also keep track of which part of the page corresponds to which line or word. The hOCR format does this by a bounding box - a set of four numbers defining the four corners of a rectangle containing the word or line. PageXML uses a more fine-grained system of a polygon tracing the outline of the line or word.

OCR and HTR tools always use a page-driven format, since they first need to recognize the lines on each page, and then break each line down into characters and words, determining the most likely string of characters based on an analysis of the image. This unavoidably leads to a structure made up of pages, lines, words, and characters. But also tools for guided manual transcription typically use a page-driven format, since the transcription has to be done page-by-page and line-by-line.

A good example of a tool using a page-driven XML format for editing is the TypeWright tool [2] (http://www.18thconnect.org/typewright). Typewright collects OCR documents from a number of different sources and then allows users to correct these automatically recognized texts on a line-by-line basis. The documents are kept in the ALTO/XML format, to which some optional mark-up from the TEI/XML standard has been added. Because pages and lines are kept as XML elements, it is easy to computationally get the content of a line out of the document, allow users to correct the OCR result, and save the result back to the XML node it came from.

But there is a significant problem with representing text in a page-based format: since pages and lines are XML nodes, nothing can cross a page or line. Therefore, any word crossing a line cannot be represented as a single unit, but has to be split into two graphical strings. And the same holds for paragraphs that crosses a page boundary. And therefore, search queries will not find any of the words that cross a line, which in historic manuscripts tend to be numerous. Even searching for sequences of words tends to be difficult if the two words are not on the same line. Probably in part because of that, very few databases with a page-based file format behind it allow searching for sequences of words, including major platforms such as Google Books.

Text-driven file formats do not have this problem, since they take the textual elements such as words, sentences, and paragraphs as primary units, meaning that any other mark-up that crosses with those elements has to be either split into parts, handled as stand-off annotation, or treated as empty elements. For page and line indications, the most common solution, for instance in TEI/XML, is to indicate only the beginning of lines and pages, as in Fig. 2.

```
<text>
   <p><pb/>
       <lb/>Manuscript line Manu
       <lb/>script line Manuscript line
   </p>
</text>
```

Fig. 2. Structure of a text-driven XML format

In and by itself the empty elements in Fig. 2 do not solve the problem of split words: the word *Manuscript* is still split into two strings. And marking on the line-break that it is not breaking a word, as is done with @break="false" in TEI, helps only in part. But since line breaks are empty elements, the line-break can simply be embedded in the word once tokenization is added, which in TEI would look like this: $\langle$w$\rangle$Manu-$\langle$lb/$\rangle$script$\langle$/w$\rangle$. So in a text-driven format, we can properly tokenize the text into words, while at the same time representing the segmentation into manuscript lines.

The downside of a text-driven file format is that it is difficult to extract a page or a line from the file - firstly because lines in this format are only implicitly represented as the segment between two subsequent $\langle$lb/$\rangle$ nodes, and secondly because those implicit segments are not guaranteed to be valid XML fragments. Hence editing a single page in a text-driven format is cumbersome and error-prone, although it is possible to reconstruct a full XML node for the fragment representing the page, as is for instance done in the Textual Communities tool (http://www.textualcommunities.usask.ca).

So there is a conflict between two different desires: on the one hand, the desire to represent manuscript pages and lines, and on the other hand, the desire to represent the text of the manuscript. And due to the fact that XML cannot have overlapping regions, these two desires are at times at odds. We can overcome this by using a two-stage approach, based on a simple idea: direct access to pages and lines is only really needed in the initial phases of the transcription process, when the actual lines of the document are transcribed either automatically or manually. Once the base transcription is finished, accessing pages and lines in a less direct manner is typically sufficient. So in the first stage, the manuscript is described in a page-based manner, making it easy to edit each page and line of the manuscript individually. But once the initial transcription is completed, the page-based format is converted to a text-based approach. And converting the page-driven format to a text-driven format does not need to loose any information: all the attributes on the page node can be kept on the empty node.

In the next section, I will briefly sketch how the TEITOK corpus platform implements exactly such a two-stage approach. More information about this implementation can be found via the website of the project: http://www.teitok. org.

3   TEITOK

TEITOK is an online corpus platform for visualizing, searching, and editing TEI/XML-based corpora. A corpus in TEITOK consists of a collection of individual files in a (tokenized) TEI/XML-compliant format, with small deviations from the P5 standard to make it easier to manipulate the files. For the alignment of an XML file with a manuscript, TEITOK uses an inline approach - the facsimile image of each page is kept on the ⟨pb/⟩ element, and to link an element in the transcription to a part of the image, it uses the structure of hOCR, in which each node on a page can be adorned with a *bounding box*, indicating the four corners of the element on the page. All nodes with such a bounding box attribute can be converted to a ⟨surface⟩ element in the ⟨sourceDoc⟩ to get a pure TEI/XML variant of the file when so desired.

3.1   Page-by-Page Transcription

For transcribing off a manuscript or monograph, TEITOK provides the two-stage approach mention before: in the first step, the transcription is created in a page-based representation. Rather than an existing framework, TEITOK uses a document format that is as close to TEI as possible, taking the full structure of a TEI/XML file, yet using instead of empty nodes ⟨pb/⟩ and ⟨lb/⟩ it uses full nodes ⟨page⟩ and ⟨line⟩.

The two-step method is built up of several stages: first, the facsimile images are uploaded as a PDF file. From this a page-based semi-TEI document is created that contains a page node for each page, linked to the corresponding facsimile image. In the image, the outer limits of the text are demarcated by hand, with an indication o how many lines the text block contains. With those indications, the system divides the image into lines of equal height, indicated as bounding boxes, which can later be corrected if necessary. Using these bounding boxes, it then creates an interface in which each line is displayed separately, with an edit box below it where the transcription can be typed in. The fact that the image is directly above the transcription makes it much easier to transcribe, and easier to keep track of the progress, where the interface can keep track of which lines have already been transcribed. Due to the fact that each line has to be self-contained, typesetting mark-up cannot cross a line-break, and has to be broken into parts. Semantic units that should not be split, such as paragraphs, have to entered as empty start and end tags, which will later be converted to their TEI counterparts.

Once the full transcription is done, the page-base file can be converted to its text-based counterpart, which is done by simply removing all closing tags for pages and lines, and turn the opening tags into $\langle pb/\rangle$ and $\langle lb/\rangle$ respectively, keeping the bounding boxes as they were. Since XML nodes can be added that cross lines and pages, either during the conversion or afterwards in the tokenization process, the conversion is irreversible.

Instead of using the TEITOK-internal transcription system, it is also possible to use a similar method with files created externally and saved in either hOCR or PageXML format. For both of these formats there is a conversion script to convert the page-based format into TEITOK-style TEI/XML, making it possible to use external transcription tools, and only use TEITOK after the initial transcription is done to create a searchable corpus, or to use OCR or HTR tools such as Tesseract or Transkribus, which can save to hOCR and PageXML respectively. The files converted from hOCR or PageXML will be more closely aligned with the images, since contrary to the manually transcribed files, also words will have bounding boxes.

3.2   Annotation and Searching

TEITOK offers the option to add multiple orthographic realizations to each word in the text. The words in the TEITOK document are XML nodes, and the inner content of those nodes corresponds to the semi-paleographic transcription of the word. But alternative spellings can be added as attributes over the node, representing the critical transcription, the expanded form, or the normalized form of the word. One of the attributes on the word node is always the searchable string corresponding to the word, which is to say, the word without the XML code, and elements that should be ignored such as hyphens preceding a line-break, or deleted characters. So the searchable string for the split word example in Fig. 2 would simply be *Manuscript*.

For all TEI/XML files in the corpus, TEITOK can export all words in the corpus to a searchable corpus in the Corpus Workbench (CWB) [3]. CWB is one of the most frequently used corpus query systems in linguistics, with an intuitive and powerful query language. In TEITOK, the CWB files are created directly from the XML files, and can contain any word-based attributes, as well as text-based attributes and intermediate level attributes such as paragraphs or verse lines.

In the CWB corpus, it is possible to define complex queries, combining all attributes on all levels. For instance the query in (1) will search for any word in the text that has a lemma ending in *-ion*, followed by a verb, in any text in the corpus written in 1885. Given that line and page breaks are kept in the XML files, but not in the searchable corpus, this query will find all occurrences in the corpus, independently of whether the two words occur on the same page, or whether or not there are line-breaks inside the words.

(1) [lemma=".*ion"] [pos="VERB"] :: match.text_year = "1885";

4   Conclusion

Despite the fact that there is a tension between the needs for a page-driven manuscript transcription and a text-driven variant, using a two-stage approach leads to a corpus that contains all the relevant information of a page-driven approach, while allowing text-specific treatment. A page-driven approach is only required in an initial stage, and can be converted to a text-based format such as TEI/XML without any loss of information - all information concerning pages or manuscript lines can be kept on self-closing XML nodes, that are implicitly closed by the start of a new page of line.

Once converted to a text-driven format, a full manuscript transcription can be tokenized, annotated, and made searchable in CWB, making it possible to formulate complex queries, which are not hampered by linebreaks in the middle of a word or page.

Since the two-stage approach in TEITOK leads to a heavily annotated file containing both the type of page-driven information desired by librarians and the type of text-driven information craved by linguists, the hope is that TEITOK will open up the way to collaborative projects, in which libraries take care of the digitalization process of the images, after which the transcription is handled either by teams of experts, or in a crowd-based environment, leading in time to a truly accessible record of the cultural heritage embedded in the historical manuscripts held at libraries and archives around the world.

More information about TEITOK is available from the project website: http://www.teitok.org, as well as instruction on how to obtain the software.

References

1. Janssen, M.: TEITOK: text-faithful annotated corpora. In: Proceedings of the 10th International Conference on Language Resources and Evaluation, LREC 2016, pp. 4037–4043 (2016)
2. Bilansky, A.: TypeWright: an experiment in participatory curation. Digit. Hum. Q. **9**(4) (2015)
3. Evert, S., Hardie, A.: Twenty-first century corpus workbench: updating a query architecture for the new millennium. In: Corpus Linguistics 2011 (2011)

User Interaction

Predicting Retrieval Success Based on Information Use for Writing Tasks

Pertti Vakkari[1] [iD], Michael Völske[2]([✉]) [iD], Martin Potthast[3] [iD],
Matthias Hagen[4], and Benno Stein[2]

[1] University of Tampere, 33014 Tampere, Finland
pertti.vakkari@uta.fi
[2] Bauhaus-Universität Weimar, 99423 Weimar, Germany
{michael.voelske,benno.stein}@uni-weimar.de
[3] Leipzig University, 04109 Leipzig, Germany
martin.potthast@uni-leipzig.de
[4] Martin-Luther-Universität Halle-Wittenberg, 06108 Halle, Germany
matthias.hagen@informatik.uni-halle.de

Abstract. This paper asks to what extent querying, clicking, and text editing behavior can predict the usefulness of the search results retrieved during essay writing. To render the usefulness of a search result directly observable for the first time in this context, we cast the writing task as "essay writing with text reuse," where text reuse serves as usefulness indicator. Based on 150 essays written by 12 writers using a search engine to find sources for reuse, while their querying, clicking, reuse, and text editing activities were recorded, we build linear regression models for the two indicators (1) number of words reused from clicked search results, and (2) number of times text is pasted, covering 69% (90%) of the variation. The three best predictors from both models cover 91–95% of the explained variation. By demonstrating that straightforward models can predict retrieval success, our study constitutes first step towards incorporating usefulness signals in retrieval personalization for general writing tasks.

1 Introduction

In assessing information retrieval effectiveness, the value of search results to users has gained popularity as a metric of retrieval success. Supplementing established effectiveness indicators like topical relevance [1–3], the worth [4], utility [2], or usefulness [1] of information depends on the degree to which it contributes to accomplishing a larger task that triggered the use of the search system [2,4,5]. Despite the growing interest in information usefulness as a retrieval success indicator, only a handful of studies have emerged so far, and they typically focus on perceived usefulness rather than on the actual usage of information from search results. Even fewer studies explore the associations between user behavior and information usage during task-based search. The lack of contributions towards

E. Méndez et al. (Eds.): TPDL 2018, LNCS 11057, pp. 161–173, 2018.
https://doi.org/10.1007/978-3-030-00066-0_14

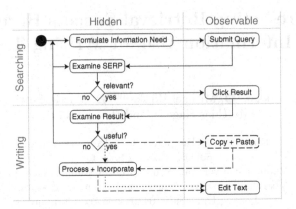

Fig. 1. User actions in the search and writing process: our study design involving text reuse (dashed lines) allows for more direct observation of users' usefulness assessments of search results than would be possible without reuse (dotted lines)

this important problem arises from the difficulty of measuring the usefulness of a search result with regard to a given task in a laboratory setting.

Focusing on essay writing, with this paper, we lift essay writing for the first time into the realm of fine-grained usefulness quantification. Using the task "essay writing with text reuse" as a surrogate, the usefulness of a search result becomes directly observable as a function of copy and paste (see Fig. 1). Furthermore, by analyzing a large corpus of essays with text reuse, where the search and writing behavior of their writers has been recorded, we identify two specific usefulness indicators based on text reuse behavior and build linear regression models that predict result usefulness based them. Keeping the limitations of our approach in mind, we believe that these results offer promising new directions for the development of search systems that support writing tasks at large.

In what follows, Sect. 2 reviews related work, Sect. 3 outlines our methodology, and Sect. 4 reports on our models. In Sect. 5 we discuss the consequences and potential limitations of our work.

2   Related Work

Search results are useful if information they contain contributes to the task that triggered information searching [5]. Users are expected to click, scan, and read documents to identify useful pieces of information for immediate or later use [6]. Only a few studies on the usefulness of search results focus on predicting the usefulness for some task [7–10], while most others are more interested in comparing relevance and usefulness assessments (e.g. [11]).

In most cases, usefulness is operationalized as perceived by users, not as the actual usage of information. Kelly and Belkin [7] explored the association between documents' display time and their usefulness for the retrieval task. Here, usefulness was operationalized as the degree of users' belief of how helpful the document was in understanding and completing a particular task. They found no association between usefulness and dwell time, regardless of the task type. Liu and Belkin [8] also studied whether the time spent on a clicked document was associated with its perceived usefulness for writing a journalistic article. They found a positive association between the dwell time on a document and its usefulness assessment. Users typically moved back and forth between the text they produced and the document informing their writing. In the context of the essay writing task of the scenario we explore, the copy-pasting of (parts from) search results may be an even more direct relation. Liu et al. [9] later modeled users' search behavior for predicting the usefulness of documents: they had users assess the usefulness of each saved page for an information gathering task, and employed binary recursive partitioning to identify the most important predictors of usefulness. In an ascending order, dwell time on documents, time to the first click, and the number of visits on a page were the most important predictors— the longer the dwell time, the more visits on a page and the shorter the time to first click, the more useful the page. Mao et al. [10] recently modeled the usefulness of documents for answering short questions by content, context, and behavioral factors, where usefulness was measured on a four-point scale. They found that behavioral factors were the most important in determining usefulness judgments, followed by content and context factors: the perceived usefulness of documents was positively correlated with dwell time and similarity to the answer, and negatively with the number of previous documents visited.

By comparison, Ahn [12] and He [13] evaluated the actual usefulness of information retrieved by measuring to what extent search systems support finding, collecting and organizing text extracts to help answer questions in intelligence tasks, with experts assessing the utility of each extract. Sakai and Dou [14] proposed a retrieval evaluation measure based on the amount of text read by the user while examining search results, presuming this text is used for some purpose during the search session.

3   Experimental Design

We base our investigation on a dataset of 150 essays and associated search engine interactions collected by Potthast et al. [15] as part of a large-scale crowdsourcing study with 12 different writers (made available to the research community as the Webis Text Reuse Corpus 2012,[1] Webis-TRC-12). We briefly review key characteristics of the Webis-TRC-12 and its collection process below, before detailing our conceptualization of document usefulness, and the variables we derive for our study.

3.1   Data

Each of the dataset's essays is written in response to one of 150 writing prompts derived from the topics of the TREC Web tracks 2009–2011. The writers were

---

[1] https://webis.de/data/webis-trc-12.html.

instructed to use the ChatNoir search engine [16] indexing the ClueWeb09 web crawl to gather material on their essay topic; all submitted queries and visited search results were recorded. For the purpose of writing the essay, the corpus authors provided an online rich text editor, which logged the interactions of writers with their essay texts in fine-grained detail, by storing a new revision of the text whenever a writer stopped typing for more than 300 ms.

The 12 writers were hired via the crowdsourcing platform Upwork, and were instructed to choose a topic to write about, and produce an essay of 5000 words using the supplied search engine and rich text editor. Writers were encouraged to reuse text from the sources they retrieved, paraphrasing it as needed to fit their essays. As reported by the corpus authors [15], the writers were aged 24 years or older, with a median age of 37. Two thirds of the writers were female, and two thirds were native English speakers. A quarter each of the writers were born in the UK and the Philippines, a sixth each in the US and India, and the remaining ones in Australia and South Africa. Participants had at least two and a median of eight years of professional writing experience.

The fine-grained data collection procedure, along with the intermeshing with established datasets like the TREC topics and ClueWeb09, has enabled the search and writing logs in the Webis-TRC-12 to contribute to several research tasks, including the study of writing behavior when reusing and paraphrasing text [15], of plagiarism and source retrieval [17], and of search behavior in exploratory tasks [18]. In the present study, we examine writers' search and text collection behavior to predict the usefulness of retrieved documents for the underlying essay writing task.

3.2   Operationalizing the Usefulness of Search Results

We limit our conception of usefulness to cover only information usage that directly contributes to the task outcome in form of the essay text, and exclude more difficult to measure "indirect" information usage from our consideration, such as learning better query terms from seen search results.

Usefulness implies that information is obtained from a document to help achieve a task outcome. In the following, we quantify usefulness by focusing only on cases where information is directly extracted from a document, not where it is first assimilated and transformed through the human mind to form an outcome. According to our definition, information is useful if it is extracted from a source and placed into an evolving information object to be modified. In the context of essay writing with text reuse, this means that information is copied from a search result and pasted in the essay to be written.

We measure the usefulness of documents for writing an essay in two dimensions, both based on the idea that a document is useful if information is extracted from it. First, we measure the number of words extracted from a document and pasted into the essay—this measure indicates the amount of text that has the potential to be transformed as a part of the essay. Second, we quantify usefulness as the number of times any text is pasted per clicked document.

Table 1. Means and standard deviations of study variables (n = 130).

Query Variables	μ	σ	Click Variables	μ	σ
Queries	46.5	41.9	Clicks per query	4.0	4.6
Unique queries	24.9	18.1	Click trails per query	1.8	1.4
Anchor queries	5.2	6.1	% Useful clicks per query	30.8	12.9
Querying time per query	53.1	46.1	Result reading time per paste	262.0	357.7
% Unique queries of all queries	62.6	20.8	... per click per query	48.9	30.4
% Anchor queries of all queries	10.5	7.1	... per major revision	172.4	141.7
Query terms per query	5.4	1.5	*Text Editing Variables*		
Unique query terms (UQT) per query	0.8	0.4	Writing time per major rev.	867.3	666.0
UQT from documents per query	0.6	0.3	Revisions per paste	175.7	225.7
% UQT from snippets	78.6	8.7	Writing time per paste	1270.4	1100.5
% UQT from docs	67.1	20.8	Words in the essay	4988.1	388.8
% UQT per query	15.9	7.6	*Other Independent Variables*		
UQT per unique query	1.3	0.4	Search sessions	7.4	4.1
Dependent Variables					
Words per useful click per query				325.0	420.8
Pastes per useful click per query				1.2	1.8

The limitations of these measures include that they do not reflect the possible synthesis of pasted information or the importance of the obtained passage of text. It is evident that the amount and importance of information are not linearly related, although users were allowed to use the pasted text directly for the essay without originality requirements. Our idealization excludes the qualitative aspects of information use; the presupposition that an increase in the amount of pasted text reflects usefulness directly resembles typical presuppositions in information retrieval research: for instance, Sakai and Dou [14] suppose that the value of a relevant information unit decays linearly with the amount of text the user has read. In general, a similar supposition holds for the DCG measure. These presuppositions are idealizations that we also apply in our analyses.

3.3   Independent Variables

For predicting the usefulness of search results, we focus on query, click, and text editing variables to build linear regression models. Temporally, querying and clicking precedes the selection of useful information, while the usage and manipulation of information succeeds it. Since we use aggregated data over all user sessions, we treat the search and writing process as a cross-sectional event, although querying, clicking, and text editing occur over several sessions. Since the editing of the essay text is connected with querying and clicking in a session, it is important to take into account also text editing variables in analyzing the usefulness of search results over all sessions. Therefore, we also select aggregated text editing variables for our models, although this solution is not ideal in every respect for representing the temporal order of the process.

Based on previous studies [9,10,18], we select 13 query variables, 6 click variables, 4 text editing variables, and 1 other variable, yielding the 24 variables

Independent Variable	Group	β	R^2 Change
Clicks per Q	C	0.38***	0.400
Revisions per paste	T	−0.15*	0.144
Unique queries	Q	−0.34***	0.081
Search sessions	O	−0.18**	0.037
Words in the essay	T	0.12*	0.019
Result reading time per paste	C	−0.24***	0.010

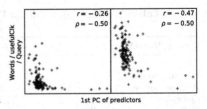

Fig. 2. Left: Regression model for the number of words pasted per useful click per query (n = 130), with predictors from the (Q)uery, (C)lick, (T)ext editing and (O)ther variables. Right: First principal component of this model's predictors and dependent variable; effect of the latter's log-transform on Pearson (r) and Spearman (ρ) correlation.

in total depicted in Table 1. Here, anchor queries refer to those queries repeatedly revisited throughout a session, in order to keep track of the main theme of the task; time spent querying, reading, and writing is measured in seconds; a click trail begins on the search result page, potentially following further links in the result document. We build regression models for both dependent variables and apply a stepwise entering method of predictors [19].

Regression analysis requires linearity between independent and dependent variables but in our case, the associations of both measures of usefulness with the major independent variables turned out to be non-linear—as evidenced by a large discrepancy between the Pearson and Spearman correlation coefficients, shown in the right-hand plots in Figs. 2 and 3. Therefore, we logarithmically transformed both words per useful click per query, and pastes per useful click per query (base of 10), enhancing linearity notably (Fig. 2, right). The predictor result reading time per click per query still showed a non-linear association, and was log-transformed as well (Fig. 3, right).

While the writers were instructed to produce essays of about 5000 words, some essays were notably shorter or longer [18]. We excluded essays shorter than 4000 and longer than 6000 words from the analysis, as well as four essays with missing variables, yielding 130 observations in total.

4   Results

Based on the variables we derive from the dataset, we investigate two linear regression models of document usefulness—each predicting one of the dependent variables at the bottom of Table 1. The first model uses the number of words pasted per useful click per query as dependent variable, using the amount of text extracted as an indicator of a document's usefulness, whereas the second model quantifies usefulness as the number of times text was extracted, using the number of pastes per useful click per query as dependent variable.

Independent Variable	Group	β	R^2 Change
Clicks per Q	C	0.33***	0.484
Writing time per paste (sec)	T	−0.42***	0.210
Queries	Q	−0.18**	0.162
Unique queries	Q	−0.24***	0.013
% Useful Clicks per Q	C	0.18***	0.008
Result reading time per C per Q	C	−0.15***	0.015
% Anchor Q of all Q	Q	0.12***	0.006
% Unique Q of all Q	Q	0.14**	0.007
% UQT from snippets of all UQT	Q	0.07*	0.004

Fig. 3. Left: Regression model for the number of pastes per useful click per query ($n = 129$; groups as in Fig. 2, left). Right: Scatter plots of "Result reading time per click per query" against the dependent variable, each before and after log-transform.

4.1 The Number of Words Pasted per Document

The model is significant ($R^2 = .703$; Adj $R^2 = .688$; $F = 48.4$; $p < .000$) consisting of six predictors. It explains 68.8% of the variation in the number of words pasted per useful click per query. The tolerance of all variables is greater than .60. The four strongest predictors—the number of clicks per query, the number of revisions per paste, the number of unique queries and the number of search sessions—cover 66.2% points of the variation in the number of words pasted (Fig. 2, left). The remaining two variables cover 2.9% points of it. Limiting the model to the four major factors, it is possible to reach an accuracy of two thirds in predicting document usefulness.

As per the model coefficients shown in Fig. 2, left, the more clicks users make per query, and the less time they spend reading result documents per paste, the more words are pasted per click per query. The number of revisions per paste reduces the number of words pasted. Increases in the number of search sessions and in the number of unique queries reduce the amount of text pasted, while an increase in the number of words in the essay increases the amount of text pasted. The number of unique queries and the number of search sessions are partially correlated, but contribute to the model in this case. Further, fewer clicks per query, more time reading documents, and a greater number of revisions per paste are associated with a smaller amount of text pasted. We hypothesize that difficulties in formulating pertinent queries lead to voluminous querying, and to a greater number of search sessions, which lead to fewer clicks, to longer dwell times per paste, and to a greater number of revisions per paste, all contributing to a smaller number of words pasted.

4.2 The Number of Pastes per Document

The regression analysis produces a model with nine variables significantly predicting the number of pastes per useful click per query. The model is significant ($R^2 = .908$; Adj $R^2 = .902$; $F = 131.2$; $p < .000$) and covers 90.2% of the variance in the number of pastes per document (Fig. 3, left). The three most important predictors—the number of clicks, the writing time per paste, and the number of queries—together explain 85.6% (R^2 Change) of the variation in the number of pastes; the remaining six predictors cover 4.6% points of variation.

The direction of effect in click, query and text editing variables differs: Increasing values of click variables—except reading time—increase the chance that documents provide material for the essay. The query variables both increase and decrease the chance of finding useful documents, while an increase in writing time per paste decreases that chance. Compared to the previous model, click variables have a proportionally smaller contribution compared to query variables, while the relative contribution of text editing variables remains on about the same level. The direction of effect in predictors remains similar; the content of the model essentially resembles the previous one, although some predictors change: writing time per paste resembles revisions per paste, while result reading time per click per query resembles result reading time per paste. The proportions of anchor queries and unique queries are new predictors compared to the previous model.

The model indicates that the more clicks per query, the larger the proportion of useful clicks of all clicks and the shorter the dwell time in clicked documents per query, the more useful the retrieved documents are. Increases in the number of queries and unique queries decrease the usefulness of clicked documents, while increases in the proportion of anchor queries and unique queries of all queries increase the chance that documents are useful.

Multicollinearity tolerance is the amount of variability of an independent variable (0–1) not explained by the other independent variables [19]. Five out of the nine predictors in the model were query variables. Tolerances of the number of queries (.240), the number of unique queries (.291) and the proportion of unique queries (.394) indicate that they depend quite heavily on other variables in the model. Therefore, leaving only the number of queries to represent querying would be reasonable and make the model more parsimonious.

We may conjecture that a smaller number of unique queries with good keywords from snippets produce a good result list. This contributes to a proportionally larger number of useful documents that require less dwell time for obtaining needed information for the essay. The information pasted is pertinent, not requiring much time to edit to match the evolving text. Naturally, the validity of this hypothetical process remains for later studies to test.

4.3   Comparing the Models

The explanatory power of the model predicting the number of words pasted is weaker, covering 68.8% of the variation in document usefulness, while the model for pastes covered about 90% of the variation.

The contributions of query, click and text editing variables vary between the models (Table 2). The relative effect ($\sum R^2$ Change) of click variables is notably greater in both models compared to query and text editing variables. Text editing variables have a somewhat greater role compared to query variables

Table 2. Summary of models: number of predictors, and relative importance ($\sum R^2$ change), per variable group for both models of search result usefulness.

Characteristics	Number of words	Number of pastes
Adj R^2	0.688	0.902
# Variables ($\sum R^2$ change)	6	9
Query	1 (0.08)	5 (0.19)
Click	2 (0.41)	3 (0.51)
Text editing	2 (0.16)	1 (0.21)
Other (sessions)	1 (0.04)	-

in predicting usefulness as indicated by the number of words pasted. Also the number of search sessions and the number of words in an essay have a minor impact on potential document usefulness. The models have only two predictors in common: the number of unique queries and the number of clicks per query.

In each model, three variables cover over 90% of the explained variation in document usefulness, one of them being a query, one a click and one a text editing variable. The most powerful variable is clicks per query in both models. Thus, one could predict each type of document usefulness by a very simple model.

In both models, the number of queries and the number of unique queries have a negative effect on document usefulness, while all proportional query variables have a positive effect. Clicks per query have a positive contribution to usefulness, while dwell time has a negative contribution. Number of revisions and writing time per paste both have a negative effect on document usefulness. In the model for the number of pasted words, the number of sessions has a negative effect on usefulness, while the number of words in the essay has a positive effect.

Altogether, it seems that clicked result documents are more useful, if: the user issues fewer queries over fewer sessions, makes more clicks per query, but with shorter dwell time on individual documents, makes fewer revisions to the essay per pasted text snippet, and writes a longer essay. Although regression analysis does not indicate associations between independent variables, we conjecture that users who issue fewer queries have better result lists, click more per query, spend less time reading documents, all this producing more useful documents per click per query. This hypothetical process remains to be tested in future work.

5   Discussion and Conclusions

Our study is one of the first attempts to analyze the usefulness of clicked search results based on information usage, instead of measuring perceived usefulness by asking the user (cf. [12,13]). The results extend our knowledge about factors predicting the actual usefulness of documents—and thus the user's success at finding useful sources—in the context of longer-lasting tasks like essay writing. Usefulness itself, we model by the number of pastes per useful click (indicating whether a documents contain information used in composing the essay), and by the number of words pasted per click (indicating the potential amount of useful information in documents).

Our regression models cover about 90% of the variation in pastes from clicked search results and about 69% of the variation in the number of words pasted per clicked search result. We argue that the number of pastes and the number of pasted words reflect the actual usefulness of search results fairly validly: for the writers we study, pasting precedes usage in the final essay [18].

We also observe that increased search result usefulness is associated with decreasing effort to edit the pastes for the essay. This is likely a result of the fact that writers were explicitly permitted to reuse text from the sources they found, without having to think about originality requirements. Hence, provided they found appropriate sources, writers could place passages from search results directly as a part of their essays. If the usefulness indicators reflect authors finding sources that require little editing, they should be correlated with less editing of pastes. To test this hypothesis, we measure the proportion of reused words out of all words in the essay (authors annotated the text they reused themselves, as part of the original study); it can be reasonably assumed that the higher this proportion, the less the pasted text is edited. We find that Spearman correlations of the proportion of reused words with the number of pastes ($\rho = .27^{**}$) and the number of words pasted ($\rho = .18^{*}$) are significant. Thus, decreasing effort in editing pasted text reflects the usefulness of pastes in composing the essay.

Further, our models indicate that the fewer queries a user makes, the more clicks per query, and the less text editing takes place, the more useful the search results are. This matches well with previous findings: an increase in clicks has been shown to correlate with search satisfaction [20] and the perceived usefulness of documents [9]. However, our results also show that an increase in dwell time decreases search result usefulness. This contradicts many earlier findings that dwell time is positively associated with usefulness [8–10]. We believe that this difference is due to the study design underlying the dataset we used: First, previous studies have restricted task time considerably, while in the essay writing of the Webis-TRC-12 there was no time limit. Second, the required length of the essays is notably longer than in similar studies. Third, the writers of the essays in the Webis-TRC-12 were encouraged to reuse text from search results without originality requirements. These factors likely encouraged authors to copy-and-paste from search results, potentially editing the text later.

In a previous study, Liu and Belkin [8] observed that users kept their search result documents open while moving back and forth between reading documents and writing text. In their scenario, increased usefulness thus comes with increased dwell time. In the case of Webis-TRC-12 instead, many writers first selected the useful pieces from some search result, pasted them into their essay, and modified them later [15]. Thus, the actual dwell time on useful search results is lower in the Webis-TRC-12. Furthermore, the selection of useful text fragments likely resembles relevance assessments. It has been shown that it takes less time to identify a relevant document compared to a borderline case [21,22]; essay writers likely needed less time to identify useful text passages in search results containing plenty of useful information, compared to documents with less such information. This can also further explain the negative association between dwell time and usefulness in the scenario of our study.

Our study places users into a simulated web-search setting using the ClueWeb corpus, but we believe our results regarding information use for writing tasks apply also to the wider Digital Library context; while our experimental setting excludes access modalities such as a catalog or a classification system, and limits writers to retrieving sources using a full-text keyword search interface, the latter is clearly a major mode of access in modern, large digital libraries [23]. That said, it is worth exploring if the correlations we observe hold also for other modes of digital library search.

We further believe that our results can be generalized to arbitrary writing tasks of long texts: In essay writing, it is likely that querying and result examination behavior is similar regardless of originality requirements, while text editing will vary by originality. An interesting future research question is how search and text editing contribute to document usefulness in the form of information use, in the presence of stricter originality requirements. In the paragraphs above, we conjecture processes that could explain the associations between the predictors and the usefulness measures. While our regression models do not allow us to test these conjectures, such an analysis could form a promising future direction.

In both our regression models, three predictors cover 91–95% of the explained variation. In both cases, one of these is a query variable, one is a click variable, and one is a text editing variable. Thus, all three variable types are required for an accurate prediction of usefulness based on information usage. Click variables have the strongest effect on usefulness compared to query or text editing variables. However, it is essential to include also the latter ones in the models, as they cover a notable proportion of variation in usefulness. Consequently, personalization in real-world retrieval systems based on information use should include the major factors in these three variable groups, due to their strong effects.

We consider retrieval personalization based on actual information use a promising proposition: the query and click variables we measure are already logged in standard search logs. Beyond that, modern web search engines tend to be operated by companies that also offer writing support tools, and may very well be able to measure text editing variables, as well. By showing that writers' aggregate retrieval success can be predicted by a simple model consisting of three variables, our study takes a first tentative step in this direction. Directly predicting the utility of individual candidate documents for a particular writing task will be important future work, in order to apply this idea in practice.

References

1. Belkin, N., Cole, M., Liu, J.: A model for evaluating interactive information retrieval. In: SIGIR Workshop on the Future of IR Evaluation, 23 July 2009, Boston (2009)
2. Hersh, W.: Relevance and retrieval evaluation: perspectives from medicine. J. Am. Soc. Inf. Sci. **45**(3), 201–206 (1994)
3. Järvelin, K., et al.: Task-based information interaction evaluation: the viewpoint of program theory. ACM Trans. Inf. Syst. **33**(1), 3:1–3:30 (2015)
4. Cooper, W.S.: On selecting a measure of retrieval effectiveness. JASIST **24**(2), 87–100 (1973)
5. Vakkari, P.: Task based information searching. ARIST **1**, 413–464 (2003)
6. Yilmaz, E., Verma, M., Craswell, N., Radlinski, F., Bailey, P.: Relevance and effort: an analysis of document utility. In: Proceedings of the CIKM 2014, pp. 91–100. ACM (2014)
7. Kelly, D., Belkin, N.J.: Display time as implicit feedback: understanding task effects. In: Proceedings of the SIGIR 2004, pp. 377–384. ACM (2004)
8. Liu, J., Belkin, N.J.: Personalizing information retrieval for multi-session tasks: the roles of task stage and task type. In: Proceedings of the SIGIR 2010, pp. 26–33. ACM (2010)
9. Liu, C., Belkin, N., Cole, M.: Personalization of search results using interaction behaviors in search sessions. In: Proceedings of the SIGIR 2012, pp. 205–214. ACM (2012)
10. Mao, J., et al.: Understanding and predicting usefulness judgment in web search. In: Proceedings of the SIGIR 2017, pp. 1169–1172. ACM, New York (2017)
11. Serola, S., Vakkari, P.: The anticipated and assessed contribution of information types in references retrieved for preparing a research proposal. JASIST **56**(4), 373–381 (2005)
12. Ahn, J.W., Brusilovsky, P., He, D., Grady, J., Li, Q.: Personalized web exploration with task models. In: Proceedings of the WWW 2008, pp. 1–10. ACM, New York (2008)
13. He, D., et al.: An evaluation of adaptive filtering in the context of realistic task-based information exploration. IP & M **44**(2), 511–533 (2008)
14. Sakai, T., Dou, Z.: Summaries, ranked retrieval and sessions: a unified framework for information access evaluation. In: Proceedings of the SIGIR 2013, pp. 473–482. ACM (2013)
15. Potthast, M., Hagen, M., Völske, M., Stein, B.: Crowdsourcing interaction logs to understand text reuse from the web. In: Proceedings of the ACL 2013, pp. 1212–1221. Association for Computational Linguistics, August 2013
16. Potthast, M., et al.: ChatNoir: a search engine for the ClueWeb09 corpus. In: Proceedings of the SIGIR 2012, p. 1004. ACM, August 2012
17. Hagen, M., Potthast, M., Stein, B.: Source retrieval for plagiarism detection from large web corpora: recent approaches. In: CLEF 2015 Evaluation Labs, CLEF and CEUR-WS.org, September 2015
18. Hagen, M., Potthast, M., Völske, M., Gomoll, J., Stein, B.: How writers search: analyzing the search and writing logs of non-fictional essays. In: Kelly, D., Capra, R., Belkin, N., Teevan, J., Vakkari, P. (eds.) Proceedings of the CHIIR 2016, pp. 193–202. ACM, March 2016
19. Hair, J.F., Black, W.C., Babin, B.J., Anderson, R.: Multivariate Data Analysis. Prentice-Hall, New Jersey (2010)

20. Hassan, A., Jones, R., Klinkner, K.: Beyond DCG: user behavior as a predictor of a successful search. In: Proceedings of the WSDM 2010, pp. 221–230. ACM (2010)
21. Gwizdka, J.: Characterizing relevance with eye-tracking measures. In: Proceedings of the 5th Information Interaction in Context Symposium, pp. 58–67. ACM (2014)
22. Smucker, M., Jethani, C.: Time to judge relevance as an indicator of assessor error. In: Proceedings of the SIGIR 2012, pp. 1153–1154. ACM (2012)
23. Weigl, D.M., Page, K.R., Organisciak, P., Downie, J.S.: Information-seeking in large-scale digital libraries: strategies for scholarly workset creation. In: Proceedings of the JCDL 2017, pp. 1–4, June 2017

Personalised Session Difficulty Prediction in an Online Academic Search Engine

Vu Tran[✉] and Norbert Fuhr

University of Duisburg-Essen, Duisburg, Germany
{vtran,fuhr}@is.inf.uni-due.de

Abstract. Search sessions consist of multiple user-system interactions. As a user-oriented measure for the difficulty of a session, we regard the time needed for finding the next relevant document (TTR). In this study, we analyse the search log of an academic search engine, focusing on the user interaction data without regarding the actual content. After observing a user for a short time, we predict the TTR for the remainder of the session. In addition to standard machine learning methods for numeric prediction, we investigate a new approach based on an ensemble of Markov models. Both types of methods yield similar performance. However, when we personalise the Markov models by adapting their parameters to the current user, this leads to significant improvements.

Keywords: Session difficulty prediction · Search behaviour
Evaluation · User modelling

1 Introduction

Search session difficulty is an important problem for many users. While most search tasks can be carried out in short time (e.g. typical Web searches), more complex tasks require a substantial effort and take much longer.

In the work presented here, we look at real searches carried out in an academic search engine. In contrast to lab studies usually based on a set of predefined tasks, we regard the whole range of search sessions occurring in the system, and then focus at sessions lasting for at least five minutes.

While previous research on the prediction of search difficulty has been restricted to single queries (see next section), we regard the complete user-system interaction during a session. Thus, we also consider the other types of actions carried out by users (like e.g. query (re)formulation, scanning snippets or reading documents); moreover, instead of system-oriented estimates, we perform user-specific ones, which implicitly allow for considering a users' behaviour and competence (like e.g. reading speed, ability to formulate good queries, subjective relevance).

In this paper, we regard the time for finding the next relevant document (time to relevant, TTR) as a measure of search session difficulty, which varies by two orders of magnitude in our study. The potential benefits of estimating TTR are

© Springer Nature Switzerland AG 2018
E. Méndez et al. (Eds.): TPDL 2018, LNCS 11057, pp. 174–185, 2018.
https://doi.org/10.1007/978-3-030-00066-0_15

not fully explored yet. An obvious application would be informing the user about the expected time and guide the user to minimize her search time – similar to the estimated arrival time and the shortest path suggestion in a car navigation system. A system might also behave in an adaptive way by offering/suggesting specific functions (or even human assistance) for more complex searches.

In the following, we start with a survey of related work (Sect. 2), then characterise the data used for this study in Sect. 3, before we first describe the general problem of TTR estimation and the application of numeric prediction methods for this purpose (Sect. 4). In Sect. 5, we present our new approach based on ensembles of Markov models. Experimental results are discussed in Sect. 6, and the final section summarises the findings and gives an outlook on future work.

2   Related Work

Session difficulty prediction is a difficult task, and thus this topic has not been addressed directly. When working with search sessions, many works deal with categorising search sessions from log data. Russell-Rose et al. [18] used human judgements to distinguish between different types of search sessions. In this small study (60 sessions), most of the participants used searcher's interactions to group sessions together. Other works [15,23] showed that interaction data is indeed a good feature set for grouping search sessions using clustering techniques. Guo and Agichtein [8] extracted a number of features, not just from the query log but also from more fine-grained user interactions such as mouse movements and query context in order to detect web searcher goals. Their work comprises a study with 10 subjects and an analysis of search logs from 440 users of a university library.

Other works focused on *query difficulty* and *task difficulty* prediction. There are many approaches on predicting query difficulty, where the prediction takes place either pre-retrieval or post-retrieval. Pre-retrieval prediction methods [3, 6,10] focus on query terms, while post-retrieval models [5,19] analyse features like retrieval scores, the robustness and the clarity of the result list in order to estimate the difficulty of the current query. Guo et al. [9] also combined query, search, and interaction features, for predicting query performance.

While most of these approaches on predicting query difficulty are system-oriented, works on predicting task difficulty are mainly user-based. Liu et al. [13] and Arguello [1] built task difficulty predictive models based on behavioural measures, which showed fairly good prediction performance. They later extended the models [12] to predict search task difficulty at different search stages.

When working on user-oriented approaches, building user models is crucial in order to understand and model user behaviour. Azzopardi [2] presented Search Economic Theory (SET). With SET, user effort is measured via a cost function. Using simulated interactions with cognitive load as the cost, Azzopardi compared a variety of search strategies, examining the cost of interaction for a given level of expected output, or gain. Kashyap et al. [11] define a cost model for browsing facets to minimise the cost of interaction, and thereby increasing the usefulness of the interface. Zhang and Zhai [25] presented the card model as a theoretical framework for optimising the contents of the screen presented in a specific situation. As optimising criterion, they used information gain, which in terms of the IPRP can be regarded as a heuristic approach for estimating the difference between cost and benefit, based upon the Interactive Probability Ranking Principle (IPRP) [7].

In [20], we combined eyetracking data with system logs to model the search process as a Markov chain, where a searcher would transition between a variety of different states, including (re)formulating a query, examining the attractiveness of snippets, the examination of documents, and selecting relevant documents. With this Markov chain, we were able to estimate values for the IPRP with effort as the time spent on each state, and benefit saved wrt. the TTR. We then extended the Markov model to a more detailed one [21], where each result rank has its own state. By estimating the expected benefit for each state, we were able to determine the rank where the user should reformulate the query (instead of going further down the result list). Recently, we estimated the TTR for individual searchers by observing them for 100 s only [22]; however, this study was performed as a lab experiment, with predefined tasks.

3 Data

3.1 Log Data

The data set used for our experiments is derived from the transaction log of the web-based academic search engine sowiport (www.sowiport.de), which is a digital library for the social sciences (for a more detailed description of the data set see [14]). The data was collected over a period of one year (between March 2014 and June 2015). The web server log files and specific Javascript-based logging techniques were used to capture the usage behaviour within the system. In the original logs, there are 58 different types of actions a user can perform, including all types of activities and pages that can be carried out/visited within the system (e.g. typing a query, visiting a document, selecting a facet, etc.). For each action, a session id, the time stamp and additional information (e.g. queries, document ids, and result lists) are stored. The session id is assigned via a browser cookie and allows tracking user behaviour over multiple searches. Based on the session id and time stamp, the step in which an action is conducted and the length of the action is included in the data set as well. The data set contains 646,235 individual search sessions and a total of more than 8 million log entries. The average number of actions per search session is 8.

3.2 Activities Mapping

For the experiments described in this paper, we map the 58 different action types onto only 4 basic actions: **query, snippet, document** and **relevant**. This was necessary due to the limited amount of training data available (see below) - later studies with larger training samples might distinguish more actions. However, as we are mainly interested in the basic feasibility of the different estimation methods considered, this limitation to four action types is not a severe restriction.

Activities like *search*, *search_advanced*, *query_form*, etc. are mapped to **query**. Actions *view_record*, *view_description*, *view_references* and *view_citation* were mapped to **document**. **snippet** corresponds to *resultlistids*, where all the ids of documents in the result list are logged. Actions like *export_bib*, *to_favorites*, *save_to_multiple_favorites*, etc. are relevance signals, thus these actions were mapped to **relevant**. Other navigational activities which are not part of the actual search process like *goto_create_account*, *goto_login*, *goto_team*, etc. were discarded from the log file. Table 1 shows the complete list of activities and their mapping.

Table 1. Activities mapping

Action	Activities
query	*goto_advanced_search*, *goto_advanced_search_recon*, *goto_last_search*, search, *search_advanced*, *search_change_facets*, *search_change_nohts*, *search_change_nohts_2*, *search_change_only_fulltext*, *search_institution* *search_change_only_fulltext_2*, *search_change_sorting*, *query_form*, *search_keyword*, *search_person*, *search_from_history*, *search_thesaurus* *search_with_CTS_possiblity*, *select_from_CTS*, *goto_last_search*
snippet	*resultlistids*
document	*view_record*, *view_description*, *view_references*, *view_citation*
relevant	*export_bib*, *export_cite*, *export_mail*, *export_search_mail*, *goto_fulltext*, *goto_google_books*, *goto_google_scholar*, *goto_Local_Availability* *save_to_multiple_favorites*, *to_favorites*
[discarded]	*delete_comment*, *goto_about*, *goto_contribute*, *goto_create_account*, *goto_delete_account*, *goto_edit_password*, *goto_favorites*, *goto_history*, *goto_home*, *goto_impressum*, *goto_login*, *goto_partner*, *goto_sofis*, *goto_team*, *goto_thesaurus*, *goto_topic − feeds*, *goto_topic − research*, *goto_topic − research − unique*, *purge_history*, *save_search*, *save_search_history*

3.3 Data Filtering

The next step in our pipeline is to remove data not suitable for our experiments. First, we excluded sessions that contain no query actions at all. They were initiated from external search engines where information about the preceding interactions is not available. In these sessions, searchers typically stayed for a very short time, and then left. After this step, the amount of search sessions was reduced from 646,235 to 18,970.

As we aimed to predict the TTR, we only kept sessions which contain at least one action of the type relevant. Furthermore, short sessions that lasted less than 300 s were removed, as they did not contain enough information that would allow the models to produce feasible predictions (and also evaluate them for the remainder of the session). After this data filtering process, we had 1967 sessions to work with.

4   TTR Prediction

In this section, we introduce the test setting for TTR prediction, and then describe the application of well-known numeric prediction methods for this problem.

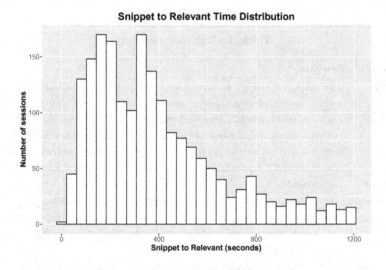

Fig. 1. Time distribution on snippet-to-relevant from the log data

Ideally, a system should be able to predict TTR at any point in time during a search session, and especially in any search situation. In fact, Markov models allow for this kind of prediction without further modification (see next Section). In contrast, the other prediction methods would have to be modified to consider the current search situation.

For the purpose of evaluation (and comparison of the different methods), here we focus only on a standard type of prediction, namely the search time needed per relevant document found. The most obvious definition for this time would be the span from query formulation until finding the first relevant document. However, after finding a relevant document, users often go to the next snippet, without formulating a new query. For this reason, we define TTR as the time from the first snippet after a query or the next snippet after locating a relevant document, until the next relevant document is found.

Figure 1 shows the distribution of the TTR values in our data set. As these values vary by two orders of magnitude, predicting TTR for a specific session obviously is a very challenging task.

An important research question is the relationship between the amount of session-specific training data and resulting quality of the predictions. To investigate this issue, we vary the training time per user, starting from 0 s (no user-specific information at all) in steps of 100 s up to a maximum of 1000 s. We then evaluate the quality of the predictions on the remainder of each session.

For TTR prediction based on numeric prediction, we considered three popular methods: M5P model tree, K-Nearest-Neighbours, and Support Vector Regression (SVR). They were performed using implementations of Weka 3.8.0. We used the standard M5P model tree with the improvements by Wang [24], based on the original M5 algorithm developed by Quinlan [17]. For kNN, we set k = 7, which is considered the best k evaluated by Weka, when using Euclidean distance as the distance function. SVR is an adaptation of Support Vector Machines for regression, which is considered state of the art in numeric prediction in other research fields [16]. We used SVR with a Polynomial Kernel.

From our sessions, we extracted the following statistical information as features: number of queries, number of snippets, number of documents and number of relevant documents. We also considered depth of the search, which is the lowest rank that the user clicked at. Interesting features like the average query duration, average snippet duration and average document duration were also used. Furthermore, we think the time proportions for each type of interaction are a valuable signal, thus we use the proportion of time that searchers spend on query, snippet and document as features. For example, on average people spent 6.5% of the time on queries, 44.8% on snippets and 48.7% on documents.

5   Markov Models for TTR Prediction

For modelling the user-system interaction as captured by the log data, a natural way is the application of Markov models. More specifically, we develop a model based upon a discrete time, discrete state Markov chain with four states (see Fig. 2). We consider the query state as the point from which a searcher focuses on the query box to submit a query. Examining a document is interpreted as the duration between displaying the document to the user until she either judges the document as relevant, or leaves the document. Snippet time is considered as the duration the subject spent examining a snippet. We assume that users look sequentially through the snippets; when the user clicks on a document at a specific rank, the time spent for inspecting the list of snippets is distributed evenly over the snippets up to this rank. Based upon this assumption, we created the corresponding number of snippet events. In case no snippet was clicked, we created artificial snippet events with the average duration per snippet derived from the observed clicks.

The transition probability between any two states s_i and s_j is estimated using maximum likelihood estimation:

$$P_{r(s_i,s_j)} = \frac{N_{s_i,s_j}}{N_{s_i}}$$

where N_{s_i,s_j} is the number of times we saw a transition from state s_i to state s_j, and N_{s_i} is the total number of times we saw state s_i in the training data. In a similar way, the expected time spent for each state (Query, Snippet, Document) is computed as the average of the observed times in these states, respectively.

Figure 2 shows two Markov models for simple vs. complex sessions. In the second case, searchers not only look at more snippets, they also spend a lot more time on each state. The query formulation took four times longer and the other states took twice as much time. The transition probabilities between the states snippet, document and relevant also show that it is harder to reach the relevant state.

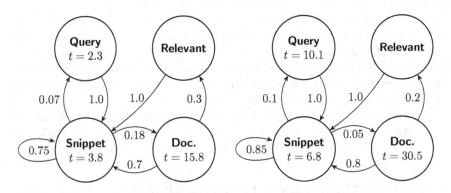

Fig. 2. The two Markov models for the shortest (left) vs. longest (right) TTR time intervals. The diagram also gives the average times spent per state and the transition probabilities.

For a given Markov model, TTR can be estimated via the so-called *mean first passage time*, which is the expected time from one state to another. Let us denote the four Markov states by q, s, d and r, the time in these states by t_q, t_s, t_d and t_r, and the transition probability from state x to state y as p_{xy}.

The expected times T_q, T_s and T_d for reaching the Relevant state from the query, snippet or document stage respectively can then be computed via the following linear equation system.

$$T_q = t_q + p_{qs}T_s$$
$$T_s = t_s + p_{sq}T_q + p_{ss}T_s + p_{sd}T_d$$
$$T_d = t_d + p_{ds}T_s$$

As TTR is defined as the time from the first/next snippet to the next relevant document, this corresponds to the T_s values estimated by our model - on which we focus our evaluation. As we can also compute T_q and T_d estimates, this allows us to perform predictions for any search state (after mapping it to one of the states of our Markov model).

With the method described above, a single Markov model would predict a constant TTR value for any search session. In order to be able to make more specific predictions, we apply two strategies:

1. Defining an ensemble of Markov models, and computing a session-specific weighted average of the predictions of the models.
2. Deriving personalised (session-specific) Markov models.

For defining an ensemble of Markov models, we group the training sessions by their average TTR values. After preliminary tests with varying numbers of groups, we found that five groups (i.e. five different Markov models) gave the best results. Since we only have a limited number of sessions, a larger number of groups would leave us with less training data per model, and thus lead to overfitting.

We calculated the average TTR of each session using the following formula: $TTR = T_{session}/N_r$, where $T_{session}$ is the session length and N_r is the total number of times we saw state *relevant* in the training data. Then, the training dataset was divided into 5 subsets using equal frequencies binning, which resulted in the following TTR boundaries: 7 – 158 – 269 – 386 – 581 – 1195.

When testing, the first task is to consider which Markov model should be used to provide the prediction for the current session. Given an interaction sequence, we calculate for each of the five Markov models the likelihood that the sequence was generated by this model. For example, the probability that the sequence $qsdm$ belongs to the current model is: $P_{qs} \cdot P_{sd} \cdot P_{dm}$. So in the end, each of the five models gives us a probability that the sequence belongs to this model. For making the predictions, we apply each of the five models and then weigh its output by the corresponding probability. Let us denote the time predictions of model 1 to 5 by t_1 to t_5, the probabilities of these models by pr_1 to pr_5 and the sum of them by pr, then we have the output prediction:

$$P = \frac{t_1 \cdot pr_1 + t_2 \cdot pr_2 + t_3 \cdot pr_3 + t_4 \cdot pr_4 + t_5 \cdot pr_5}{\sum_{i=1}^{5} pr_i}$$

For the second prediction method, we build personalised Markov models from observation data of the individual session (until cutoff time), in combination with the general models derived from the training data. For building these models after a short observation time, we face the problem of parameter estimation: some transitions or states even may not yet have been observed for a specific subject. For the states, we use the following Bayesian formula to estimate the time: $T_x = (\overline{T}_x v + Cm)/(v + m)$, where $\overline{T}$ is the time of the global models multiplied by their probabilities at the given point of time, and v is the total number of observations until that point. C is the mean time of that state across the entire session, and m is the weight given to the prior estimate that is based on the distribution of average times derived from the entire session.

As for the probabilities for our personalised models, even a few observed events will not lead to good estimates using the standard maximum likelihood technique. Instead, we use Bayes' estimates with beta priors where the parameters of the beta distribution are derived from the overall distribution of probabilities in the training sample via the *method of moments* [4].

6   Results

In our experiments, we split the set of sessions into ten folds, and then performed 10-fold cross-validation. For that, each of the methods was first trained on the 9 training folds for deriving a model. Then, for each session in the testing set, features were collected until cutoff time, which were used as input to the model for making TTR predictions. For the global Markov ensemble, the sequence observed until cutoff time was used for computing a session-specific weighted average of the predictions; in the case of personalised Markov models, also the parameters of these models were derived from the observed data.

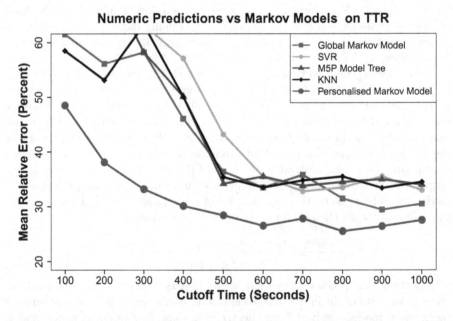

Fig. 3. The mean relative absolute error of the predictions for each Markov chain model over the cutoff times (refer to Sect. 4). Note that the relative error is compared over the various interaction data cutoff times.

The TTR predictions T_s (snippet to relevant) of the various approaches were compared to the actual values. These times $\hat{T}_s$ are calculated as $\hat{T}_s = (\hat{T}_{lR} - \hat{T}_{fS})/|R|$. Here $\hat{T}_{lR}$ is the timestamp of the last relevant document in the session.

Since we are making predictions for the remainder of a session at specific cutoff times, $\hat{T}_{fS}$ is the timestamp of the first snippet seen for which we have not reached a relevant document yet. $|R|$ denotes number of documents judged as relevant in the remainder of the session.

As quality metric, we regard the mean (absolute) relative error, which is defined as $|\hat{T}_s - T_s|/\hat{T}_s$. Note that due to the large variation of the $\hat{T}_s$ values (7s...1195s), the relative error can also exceed 100%.

The mean relative errors of the various models investigated are shown in Fig. 3, with various cutoff times. All approaches consider the snippet-to-relevant times for relevant documents occurring after the cutoff time. For the user models, user-specific parameters are derived from the observations occurring before the cutoff time (i.e. these models are trained for some time, allowing them to make predictions for sessions remainder). Significance tests are performed using 2-tailed paired t-tests, with $p < 0.05$. As mentioned above, we applied 10 fold cross-validation for all tests.

Despite the high variance in the data, all three numeric prediction methods performed reasonably well (given the high variance in the actual TTR values, as shown in Fig. 1). For the first 500 s, the errors are quite high, although they seem to get better over time. After 500 s, the performance stabilizes at about 35%.

Comparing these numeric predictions to our Markov models, we can see that the global Markov model performs similar for most of the time. After 800 s, our global Markov model gives better results, although this improvement is not significant. The personalised Markov model, on the other hand, performs significantly better than all the other methods.

When looking at the Markov models, we can see that the user-specific models are significantly better than the global ones. They also need a certain amount of data to reach their full potential, in this case about 500 s of interaction data.

Overall, we can see that (i) our global Markov models and the three numeric predictors perform similar (ii) user-specific models outperform the global ones and (iii) after 500 s, more training time does not improve the predictions (although more detailed representations might lead to further improvements in this case).

7   Conclusion and Future Work

In this paper, we have studied the new task of predicting session difficulty. We took a user-oriented approach and regarded real sessions in a social sciences digital library. As a user-oriented measure of session difficulty, we defined the time for finding the next relevant document. Besides applying standard methods for numeric prediction, we developed a new approach based on an ensemble of Markov models. While the variant based on global models showed a performance similar to that of the standard methods, significant improvements were achieved by adapting the Markov models to the specific user.

In our work, we have focused on interaction data alone, where we have to observe a user for at least 300 s for getting reasonable predictions. This time can certainly be reduced in case we have data about the same user from previous sessions. Also, we have not regarded the content of the searches, which has been the subject of many studies on query/topic difficulty. The two approaches obviously complement each other. Our method is able to adapt to the capabilities of the specific user, but needs training data from this user. For very short sessions, the content-oriented approaches might be more useful. Thus, a combination of the two approaches seems a promising direction for further research.

Another interesting research issue is the exploitation of the predictions, e.g. as information for the user, or for adapting the system behaviour.

References

1. Arguello, J.: Predicting search task difficulty. In: de Rijke, M., et al. (eds.) ECIR 2014. LNCS, vol. 8416, pp. 88–99. Springer, Cham (2014). https://doi.org/10.1007/978-3-319-06028-6_8
2. Azzopardi, L.: The economics in interactive information retrieval. In: Proceedings of the 34th International ACM SIGIR Conference on Research and Development in Information Retrieval. SIGIR 2011, pp. 15–24. ACM, New York (2011)
3. Bashir, S.: Combining pre-retrieval query quality predictors using genetic programming. Appl. Intell. **40**(3), 525–535 (2014)
4. Bowman, K.O., Shenton, L.R.: The beta distribution, moment method, Karl Pearson and RA Fisher. Far East J. Theor. Stat. **23**(2), 133 (2007)
5. Collins-Thompson, K., Bennett, P.N.: Predicting query performance via classification. In: Gurrin, C., et al. (eds.) ECIR 2010. LNCS, vol. 5993, pp. 140–152. Springer, Heidelberg (2010). https://doi.org/10.1007/978-3-642-12275-0_15
6. Yom-Tov, E., Carmel, D.: Estimating the Query Difficulty for Information Retrieval. Morgan & Claypool Publishers, San Rafael (2010)
7. Fuhr, N.: A probability ranking principle for interactive information retrieval. Inf. Retr. **11**(3), 251–265 (2008)
8. Guo, Q., Agichtein, E.: Ready to buy or just browsing?: detecting web searcher goals from interaction data. In: Proceedings of the 33rd International ACM SIGIR Conference on Research and Development in Information Retrieval. SIGIR 2010, pp. 130–137. ACM, New York (2010)
9. Guo, Q., White, R.W., Dumais, S.T., Wang, J., Anderson, B.: Predicting query performance using query, result, and user interaction features. In: Adaptivity, Personalization and Fusion of Heterogeneous Information. RIAO 2010, pp. 198–201. Le Centre de Hautes Etudes Internationales d'Informatique Documentaire, Paris (2010)
10. Hauff, C., Murdock, V., Baeza-Yates, R.: Improved query difficulty prediction for the web. In: Proceedings of the 17th ACM Conference on Information and Knowledge Management. CIKM 2008, pp. 439–448. ACM, New York (2008)
11. Kashyap, A., Hristidis, V., Petropoulos, M.: Facetor: cost-driven exploration of faceted query results. In: Proceedings of the 19th ACM International Conference on Information and Knowledge Management. CIKM 2010, pp. 719–728. ACM, New York (2010)
12. Liu, C., Liu, J., Belkin, N.J.: Predicting search task difficulty at different search stages. In: Proceedings of the 23rd ACM International Conference on Conference on Information and Knowledge Management. CIKM 2014, pp. 569–578. ACM, New York (2014)

13. Liu, J., Liu, C., Cole, M., Belkin, N.J., Zhang, X.: Exploring and predicting search task difficulty. In: Proceedings of the 21st ACM International Conference on Information and Knowledge Management. CIKM 2012, pp. 1313–1322. ACM, New York (2012)

14. Mayr, P.: Sowiport user search sessions data set (SUSS) (2016). https://doi.org/10.7802/1380

15. Chen, H.M., Cooper, M.D.: Using clustering techniques to detect usage patterns in a web-based information system. J. Am. Soc. Inf. Sci. Technol. **52**, 888–904 (2001)

16. Ouyang, Y., Li, W., Li, S., Qin, L.: Applying regression models to query-focused multi-document summarization. Inf. Process. Manag. **47**(2), 227–237 (2011)

17. Quinlan, R.J.: Learning with continuous classes. In: 5th Australian Joint Conference on Artificial Intelligence (1992)

18. Russell-Rose, T., Clough, P., Toms, E.G.: Categorising search sessions: some insights from human judgments. In: Proceedings of the 5th Information Interaction in Context Symposium. IIiX 2014, pp. 251–254. ACM, New York (2014)

19. Shtok, A., Kurland, O., Carmel, D., Raiber, F., Markovits, G.: Predicting query performance by query-drift estimation. ACM Trans. Inf. Syst. **30**(2), 11:1–11:35 (2012)

20. Tran, V., Fuhr, N.: Using eye-tracking with dynamic areas of interest for analyzing interactive information retrieval. In: Proceedings of the 35th International ACM SIGIR Conference on Research and Development in Information Retrieval. SIGIR 2012, pp. 1165–1166. ACM, New York (2012)

21. Tran, V., Fuhr, N.: Markov modeling for user interaction in retrieval. In: SIGIR 2013 Workshop on Modeling User Behavior for Information Retrieval Evaluation (MUBE 2013), August 2013

22. Tran, V., Maxwell, D., Fuhr, N., Azzopardi, L.: Personalised search time prediction using Markov chains. In: Proceedings of the ACM SIGIR International Conference on Theory of Information Retrieval. ICTIR 2017, pp. 237–240. ACM, New York (2017)

23. Wang, P., Wolfram, D., Zhang, J.: Modeling web session behaviour using cluster analysis: a comparison of three search settings. In: American Society for Information Science and Technology (2008)

24. Witten, I.H., Wang, Y.: Induction of model trees for predicting continuous classes. In: Poster Papers of the 9th European Conference on Machine Learning (1997)

25. Zhang, Y., Zhai, C.: A sequential decision formulation of the interface card model for interactive IR. In: Proceedings of the 39th International ACM SIGIR Conference on Research and Development in Information Retrieval. SIGIR 2016, pp. 85–94. ACM, New York (2016)

User Engagement with Generous Interfaces for Digital Cultural Heritage

Robert Speakman[1] , Mark Michael Hall[2(✉)] , and David Walsh[1]

[1] Edge Hill University, Ormskirk, Lancashire, UK
{robert.speakman,david.walsh}@edgehill.ac.uk
[2] Martin-Luther-Universität Halle-Wittenberg,
Von-Seckendorff-Platz 1, 06120 Halle, Germany
mark.hall@informatik.uni-halle.de

Abstract. Digitisation has created vast digital cultural heritage collections and has spawned interest in novel interfaces that go beyond the search box and aim to engage users better. In this study we investigate this proposed link between generous interfaces and user engagement. The results indicate that while generous interfaces tend to focus on novel interface components and increasing visit duration, neither of these significantly influence user engagement.

Keywords: Digital cultural heritage · Generous interfaces
User engagement

1 Introduction

Cultural Heritage institutions have embraced efforts to digitise their extensive collections, making them available to the general public and opening up our cultural heritage. These efforts have created large digital collections [12], but access remains through the search box. However, there has been interest in more "generous" interfaces that go beyond simply providing "1–10 of 10000" results.

These "generous" interfaces are primarily aimed at non-expert users who often find the search box and the need to formulate and interpret queries a major obstacle [16,17]. As a result digital cultural heritage (DCH) websites often have very high bounce rates with up to 60% of users leaving in the first ten seconds. The argument for the new interfaces is that by generously offering up the available content, the novice user will engage more with the content and the interface. Unlike traditional search interfaces, which are quite heavily standardised, generous interfaces demonstrate a large amount of variation and it is unclear how this variation affects their ability to engage users.

In the study presented here we investigate user engagement with three different generous interfaces and attempt to determine how different user interface elements affect user engagement. The remainder of this paper is structured as follows: Sect. 2 discusses the current state of generous interfaces and user engagement, Sect. 3 presents the experiment, Sect. 4 discusses the results, and Sect. 5 concludes with recommendations for future research.

ⓒ Springer Nature Switzerland AG 2018
E. Méndez et al. (Eds.): TPDL 2018, LNCS 11057, pp. 186–191, 2018.
https://doi.org/10.1007/978-3-030-00066-0_16

2   Generous Interfaces and User Engagement

To support the novice user in accessing large collections, alternative interfaces have been developed that focus primarily on browsing and visualisation to create richer user experiences [2] that are preferred by non-expert users [7,15]. In the field of DCH the labels "generous interfaces" [16] and "rich prospect browsing" [13] have been attached to this type of interface. While both terms are relatively new, they trace their core ideas to the concept of "overview first, zoom and filter, then details on demand" developed in the 90s [14].

In general these kinds of interface initially provide the user with a sample of the content available in the collection or an overview visualisation that highlights the available types of content. This enables the novice user to learn about the collection as a whole [6] and then through browsing and visualisation explore the collection in order to gradually build up a more detailed understanding of the content [5,8].

While user testing has shown that generous interfaces support users in their interaction with DCH collections [3], it is unclear whether they actually manage to engage users in the way they claim. User engagement takes into account both the usability of the interface, but also the users' sense of captivation with the task and system [1,10] and depth of interaction [11]. A number of metrics have been defined for measuring user engagement, but in the study presented here the User Engagement Scale is used in its short-form (UES-SF) [9].

3   Experiment

3.1   Methodology

To investigate user engagement with generous interfaces we developed an on-line experiment and tested three different generous user interfaces. The experiment used a standard interactive information retrieval setup, initially acquiring demographics data, then letting each participant use one of the three interfaces, and finally assessing their experience and engagement.

The first step acquired information on participants' age, gender, education, employment status, and cultural heritage experience. In the second step participants were randomly assigned one of the three tested interfaces. To test the generous interfaces' open-ended exploration support we used the open-ended task instructions from [4], which instructed participants to freely explore until they had enough. The experiment automatically tracked the time participants spent in the system. While showing participants more than one interface was considered, this would have increased the experiment duration and from our experience in on-line settings this would lead to high drop-out rates.

After completing step two, participants were asked to rate how much use they made of the available user interface components. Then participants were asked to assess whether they had used similar interfaces before, whether the initial set of items they saw was interesting, and whether they looked at items that they would not normally be interested in. Finally the UES-SF was administered.

The experiment was piloted with ten participants and after correcting issues it was made available on-line over a two-week period in November 2017.

3.2   Tested Generous Interfaces

As generous interfaces come in a wide range of styles, the decision was made
to test three existing, live, production-quality systems, enabling the study to
assess engagement in a realistic context. The three interfaces (see Fig. 1) were:
Discover the Queenslander[1] (DtQ) [16], the Rijksmuseum's Rijksstudio[2] (RS),
and the Tyne and Wear museum's Collections Dive[3] (CD) [3].

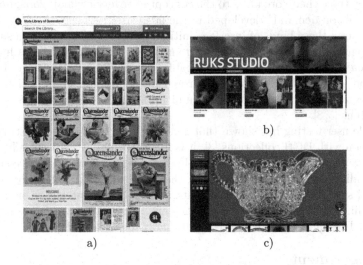

Fig. 1. Screenshots showing the initial view presented by each of the three tested
interfaces: (a) Discover the Queenslander, (b) Rijksstudio, (c) Collections Dive.

The DtQ system contains a collection of illustrations from the Queenslander
newspaper between the years 1866 and 1939 [16]. It uses an image grid to give
the user a generous overview over the data and help them explore it. Additionally it supports exploration via colour and a time-line. It is representative of
generous interfaces that aim to surface a very specific collection by providing a
visualisation and browsing-based interface.

The RS system lets users curate their own galleries out of the Rijksmuseum's
digital collections, which can then be explored via browsing. Unlike the other
systems, it includes a curated aspect with galleries created by the Rijksmuseum's
curators. Like the DtQ it also supports exploration via colour. The RS interface
is heavily influenced by social sharing sites (Tumblr, Pinterest, ...). It is included
as an example of a mixed expert-driven and crowd-sourced generous interface.

The CD interface is the most novel interface in that it allows exploration of
the collection simply by scrolling down the page. It initially shows a randomly
selected set of related items and when the user scrolls down, the system either

---

[1] http://www.slq.qld.gov.au/showcase/discover-the-queenslander.
[2] https://www.rijksmuseum.nl/en/rijksstudio.
[3] http://www.collectionsdivetwmuseums.org.uk/.

shows more similar items (slow scrolling) or switches to showing very different items (fast scrolling). It is representative of generous interfaces that attempt to provide a very different interaction pattern and user experience.

3.3 Participants

Participants were recruited via social media (Facebook, Twitter, LinkedIn) and via physical and electronic noticeboards at Edge Hill University. A total of 620 participants were recruited of which 56 completed the experiment (9%). Most drop-outs occurred on the initial landing page, a common scenario when inviting participants via social media.

Of the 56 participants 32 were male, 23 female, and 1 undisclosed. While the largest group is from the student age-bracket 18–25 (24), there is a good distribution across the other age brackets as well 26–36 (9), 36–45 (12), 46–55 (7), 56–65 (3), over 65 (1). 24 participants identified as students, 27 as in employment, and the remaining 5 were undeclared. All participants undertook the experiment on-line, using their own devices in their own environment.

4   Results and Discussion

Participants were automatically balanced across the three interfaces. To ensure that this did not introduce any accidental biases, we tested for potential biases due to CH experience and previous exposure to generous interfaces and found no significant influence by either of these aspects.

Table 1. Participants' visit duration, whether they found the initial items interesting, discovered novel content, and their engagement. Values are formatted [mean (standard deviation)] and, excepting the visit duration, are on a scale of 1 (low) to 5 (high).

	DtQ	RS	CD
Visit duration	4 m13 s (6 m29 s)	2 m30 s (4 m7 s)	2 m18 s (2 m1 s)
Item interest	2.58 (1.26)	3.38 (1.2)	3 (1.37)
Novel items	3.47 (1.35)	3.52 (1.47)	3.7 (0.96)
User engagement	3.43	4.24	4.12

Table 1 shows the main results. All three interfaces are successful in introducing participants to the collections and showing them items they had previously not seen (*novel items*). While the *visit duration* is higher for the DtQ, the difference is not statistically significant. Likewise, there is no statistically significant difference on the initial *item interest* either. However, the DtQ is significantly less engaging than either the RS or the CD (Wilcoxon rank-sum $p < 0.05$).

An interesting result is how long some participants spent on the DtQ, even though its engagement score is significantly lower. Clearly visit duration is not a predictor for engagement, even though longer visit times are frequently noted as an aim for generous interfaces.

Potentially the engagement difference is due to the type of data in the three collections. Although initial item interest is not significantly different across the three interfaces, for the RS and the CD there is a significant correlation between initial item interest and engagement (RS: $\rho = 0.6, p < 0.01$, CD: $\rho = 0.75, p = 0.001$), which does not exist for the DtQ. The exact nature of this effect needs further study.

The CD's use of scrolling to navigate is novel and clearly manages to engage users. At the same time it has the lowest average visit duration and the lowest standard deviation. We believe that this is because, while the interface engages, the lack of control leads to a relatively consistent point in time where the user has had enough, and since they cannot focus their exploration, they leave.

The generous interface literature focuses on the impact of interface components on engagement, but, with two exceptions, we find no significant correlation between component use and user engagement. For the CD use of scrolling weakly correlates with engagement ($\rho = 0.53, p = 0.04$), as does use of the image viewing component in the DtQ ($\rho = 0.47, p = 0.04$).

5   Conclusion

Generous interfaces have been put forward as a solution to the high bounce rates experienced by DCH sites. In this paper we presented a study of user engagement with three such interfaces. The central result is that while work on generous interfaces tends to focus on increasing visit durations and developing novel interfaces and visualisations, the results of our study show little influence of these aspects on user engagement. However, what items a user initially sees does significantly correlate with engagement for two of the interfaces.

This strongly indicates that the research focus needs to change from time and novel interfaces to actually understanding novice users' needs and information journeys and then developing interfaces that can support them through these, with a particular focus on the initial interaction moments and the data the users see at that point.

The study has some limitations due to its nature. Three live systems were tested, which did not allow us to track participants' interactions with the interfaces, instead we relied on participants' self-assessment regarding which aspects of the interfaces they used. Additionally the relatively small sample sizes of between 16 and 21 participants per interface limits the strength of our conclusions.

References

1. Attfield, S., Kazai, G., Lalmas, M.: Towards a science of user engagement (position paper). In: WSDM Workshop on User Modelling for Web Applications (2011). http://www.dcs.gla.ac.uk/~mounia/Papers/engagement.pdf
2. Bates, M.J.: What is browsing-really? a model drawing from behavioural science research, no. 4 (2007)

3. Coburn, J.: I dont know what lm looking for: better understanding public usage and behaviours with Tyne & Wear Archives & Museums online collections. In: MW2016: Museums and the Web 2016 (2016). http://mw2016.museumsandtheweb.com/paper/i-dont-know-what-im-looking-for-better-understanding-public-usage-and-behaviours-with-tyne-wear-archives-museums-online-collections/
4. Gäde, M., et al.: Overview of the SBS 2015 interactive track (2015)
5. Giacometti, A.: The texttiles browser: an experiment in rich-prospect browsing for text collections. Ph.D. thesis, Faculty of Arts (2009)
6. Hibberd, G.: Metaphors for discovery : how interfaces shape our relationship with library collections. In: The Search Is Over! Exploring Cultural Collections with Visualization International Workshop in conjunction with DL 2014, September 2014
7. Lopatovska, I., Bierlein, I., Lember, H., Meyer, E.: Exploring requirements for online art collections. Proc. Assoc. Inf. Sci. Technol. **50**(1), 1–4 (2013)
8. Mauri, M., Pini, A., Ciminieri, D., Ciuccarelli, P.: Weaving data, slicing views: a design approach to creating visual access for digital archival collections. In: ACM International Conference Proceeding Series CHItaly, pp. 1–8 (2013). https://doi.org/10.1145/2499149.2499159
9. OBrien, H.L., Cairns, P., Hall, M.: A practical approach to measuring user engagement with the refined user engagement scale (UES) and new UES short form. Int. J. Hum.-Comput. Stud. **112**, 28–39 (2018). https://doi.org/10.1016/j.ijhcs.2018.01.004,, http://www.sciencedirect.com/science/article/pii/S1071581918300041
10. Peters, C., Castellano, G., de Freitas, S.: An exploration of user engagement in HCI. In: Proceedings of the International Workshop on Affective-Aware Virtual Agents and Social Robots, pp. 9:1–9:3 (2009). https://doi.org/10.1145/1655260.1655269
11. Peterson, E.T., Carrabis, J.: Measuring the immeasurable: visitor engagement, p. 54 (2008)
12. Petras, V., Hill, T., Stiller, J., Gäde, M.: Europeana - a search engine for digitised cultural heritage material. Datenbank-Spektrum **17**(1), 41–46 (2017). https://doi.org/10.1007/s13222-016-0238-1
13. Ruecker, S., Radzikowska, M., Sinclair, S.: Visual Interface Design for Digital Cultural Heritage: A Guide to Rich-Prospect Browsing (Digital Research in the Arts and Humanities). Routledge, Abingdon (2016)
14. Shneiderman, B.: The eyes have it: a task by data type taxonomy for information visualizations. In: Proceedings 1996 IEEE Symposium on Visual Languages, pp. 336–343 (1996). https://doi.org/10.1109/VL.1996.545307
15. Walsh, D., Hall, M., Clough, P., Foster, J.: The ghost in the museum website: investigating the general public's interactions with museum websites. In: Kamps, J., Tsakonas, G., Manolopoulos, Y., Iliadis, L., Karydis, I. (eds.) TPDL 2017. LNCS, vol. 10450, pp. 434–445. Springer, Cham (2017). https://doi.org/10.1007/978-3-319-67008-9_34
16. Whitelaw, M.: Generous interfaces for digital cultural collections. DHQ: Digit. Humanit. Q. **9**(1) (2015)
17. Wilson, M.L., Elsweiler, D.: Casual-leisure searching: the exploratory search scenarios that break our current models. In: Proceedings of HCIR, pp. 28–31 (2010)

Resources

Peer Review and Citation Data in Predicting University Rankings, a Large-Scale Analysis

David Pride[(✉)] and Petr Knoth

The Knowledge Media Institute, The Open University, Milton Keynes, UK
{david.pride,petr.knoth}@open.ac.uk

Abstract. Most Performance-based Research Funding Systems (PRFS) draw on peer review and bibliometric indicators, two different methodologies which are sometimes combined. A common argument against the use of indicators in such research evaluation exercises is their low correlation at the article level with peer review judgments. In this study, we analyse 191,000 papers from 154 higher education institutes which were peer reviewed in a national research evaluation exercise. We combine these data with 6.95 million citations to the original papers. We show that when citation-based indicators are applied at the institutional or departmental level, rather than at the level of individual papers, surprisingly large correlations with peer review judgments can be observed, up to $r <= 0.802, n = 37, p < 0.001$ for some disciplines. In our evaluation of ranking prediction performance based on citation data, we show we can reduce the mean rank prediction error by 25% compared to previous work. This suggests that citation-based indicators are sufficiently aligned with peer review results at the institutional level to be used to lessen the overall burden of peer review on national evaluation exercises leading to considerable cost savings.

1 Introduction

Since the late 20th century there has been a seismic shift in many countries in how research is funded. In addition to traditional grant or patronage funding, there is growing use of Performance-based Research Funding Systems (PRFS) in many countries. These systems fall largely into two categories; those that focus on peer review judgments for evaluation and those that use a bibliometric approach. The UK and New Zealand both have systems heavily weighted towards peer review. Northern European countries other than the UK tend to favour bibliometric methodologies whereas Italy and Spain consider both peer review judgments and bibliometrics. Research Evaluation Systems overall have dual and potentially dichotomous ends, firstly identifying the best quality research but also, in many cases, the distribution of research funds. There is, however, a large variance in the level of institutional funding granted based on the results of these exercises. The UK's Research Councils distribute £1.6 billion annually

© Springer Nature Switzerland AG 2018
E. Méndez et al. (Eds.): TPDL 2018, LNCS 11057, pp. 195–207, 2018.
https://doi.org/10.1007/978-3-030-00066-0_17

entirely on the basis of the results of the Research Excellence Framework (REF) which is the largest single component of university funding. At the other end of the scale, the distribution of funds based on the results of the Finnish PRFS is just 3% of the total research budget. Furthermore, the PRFS in Norway and Australia are both used for research evaluation but are not used for funding distribution [1]. Peer-review based PRFS are hugely time-consuming and costly to conduct. In this investigation we ask how well do the results of peer-review based PRFS correlate with bibliometric indicators at the institutional or disciplinary level. A strong correlation would indicate that metrics, where available, can lessen the burden of peer review on national PRFS leading to considerable cost savings, while a weak correlation would suggest each methodology provides different insights.

To our knowledge, this is the first large-scale study exploring the relationship between peer-review judgments and citation data at the institutional level. Our study is based on a new dataset compiled from 190,628 academic papers in 36 disciplines submitted to UK REF 2014, article level bibliometric indicators (6.95m citations) and institutional/discipline level peer-review judgments. This study demonstrates that there is a surprisingly strong correlation between an institutions' *Grade Point Average* (GPA) ranking for its outputs submitted to the UK Research Excellence Framework for many Units of Assessment (UoAs) and citation data. We also shows that this makes it possible to predict institutional rankings with a degree of accuracy in highly cited disciplines.

2   Related Work

There has long been wide ranging and often contentious discussion regarding the efficacy of both peer review and bibliometrics and whether one or other, or both should be used for Research Evaluation. Several other studies have specifically investigated the correlation between the results of different nations' peer review focused Performance-based Research Funding Systems and bibliometric indicators. Anderson [2] finds only weak to moderate correlation with results from the New Zealand PRFS and a range of traditional journal rankings. The highest correlation is $r = 0.48$ with the Thomson Reuters Journal Citation Report. However Anderson states that this may be due to the much broader scope of research considered by PRFS processes and the additional quality-related information available to panels. Contrary to Anderson, Smith [3] used citations from Google Scholar (GS) and correlated these against the results from the New Zealand PRFS in 2008. He found strong correlation, $r = 0.85$ for overall PRFS results against Google Scholar citation count.

A comprehensive global PRFS analysis was conducted by Hicks in 2012. Hicks states there is convincing evidence that when PRFS are used to define league tables this creates powerful incentives for institutions to attempt to 'game' the process, whether in regards to submission selection or staff retention and recruitment policies [1]. A UK government funded report, The Metric Tide, was published in 2015 and gave a range of recommendations for the use of metrics in research evaluation exercises. The Metric Tide study had access to the anonymised scores for the individual submissions to the REF and was therefore directly able to compare on a paper by paper basis the accuracy of a range of bibliometric indicators. This study tested correlations with a range of different bibliometric measures and found correlation with rankings for REF 4* and 3* outputs for some UoAs. Metrics found to have moderately strong correlations with REF scores for a wide range of UoAs included: number of tweets; number of Google Scholar citations; source normalised impact per paper; SCImago journal rank and citation count [4].

However, The Metric Tide study used different citation metrics and citation data sources from our approach. It is at the institutional UoA level that our study reveals some of the strongest correlations, higher than previously shown. In a related study, Mryglod et al. [5] used departmental h-index aggregation to predict REF rankings. Their work was completed before December 2014 when the REF results were published and contained ranking predictions based on their model with some degree of success. They also experimented by normalising the h-index for each year between 2008 and 2014 but surprisingly found little evidence that timescale played a part in the strength of the correlations they found. An ad hoc study by Bishop [6] also found a moderate to strong correlation between departmental research funding based on the results of the UK's Research Assessment and Evaluation (RAE) exercise conducted in 2008, and departmental h-index. Mingers [7] recently completed an investigation that collected total citation counts from Google Scholar (GS) for the top 50 academics[1] from each UK institute and he found strong correlations with overall REF rankings. To our knowledge, ours is the first large-scale in-depth study that investigates the correlation between citation data and peer review rankings by discipline at the institutional level, taking into account all papers submitted to REF.

3 Results

For this study we used data from the UK's Research Excellence Framework (REF). The last REF exercise undertaken in the UK in 2014 was the largest overall assessment of universities' research output ever undertaken globally. These experiments focus on the academic outputs (research papers) component of the REF, for which the metadata are available for download from the REF website. The REF 2014 exercise peer reviewed and graded approximately 191,000 outputs from 154 institutions and in 36 Units of Assessment (UoAs) from zero to four stars. The grading for each submission was determined according to *originality, significance and rigour*. The peer review grades for the individual submissions were aggregated for each UoA to produce a *Grade Point Average* for each institute. The rankings are of critical importance to the institutes as approximately £1.6 billion in QR funding from central government is distributed annually entirely on the basis of the REF results [8].

---

[1] If there were not 50 academics then the total number of academics on GS for that institute was used.

Each of the REF peer review panels individually chose whether or not to use citation data to inform their decisions. Eleven out of 36 selected to do so and were provided with citation data from Elsevier Scopus to assist their decision making. For each area and age of publication they were given the number of citations required to put the paper in the top 1%, 5%, 10% or 25% of papers within its area. This gave REF reviewers a subject-level benchmark against which to consider the citation data [9].

Whereas the aggregate GPA ranking for all UoAs and all institutes is publicly available, it is now not possible to obtain a direct comparison between citation data and the individual rankings for each submission as HEFCE state that these data were destroyed. The rationale behind this was to preempt any requests for this data under the Freedom of Information Act. [10].

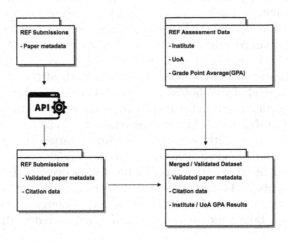

Fig. 1. Citation enrichment workflow used in dataset creation.

3.1   Dataset

The dataset creation procedure is depicted in Fig. 1. We first downloaded the REF 2014 submission list [9]. For each output, the list contains; publication title, publication year, publication venue, name of institute and UoA. These fields were fully populated for 190,628 out of 190,963 submissions to the 'outputs' category of the REF process.

We decided to utilise the Microsoft Academic Graph (MAG) to enrich the REF submission list with citation information. At the time of the experiment MAG contained approximately 168m individual papers and 1.15 billion citation pairs. This decision was motivated by the fact that while Scopus, operated by Elsevier, was used to provide citation data to the REF process, the free version of the Scopus API service is limited to 20,000 requests per week. It would have therefore taken almost two months to gather the required data which was not practical as this was more than 10 times slower than using MAG. Additionally, studies by [11,12] have recently confirmed how comprehensive the MAG citation data are. We could not utilise Google Scholar as it does not offer an API and prohibits 'scraping' of data.

We systematically queried the MAG Evaluate API for each submission using a normalised version of the publication's title (lower case, diacritics removed). This returned a set of MAG IDs which were potential matches of the article. We subsequently queried the MAG Graph Search API to validate each of the potential matches. We accepted as a match the most similar publication title that had at least 0.9 cosine similarity. This threshold was set by manually observing about one hundred matches. Using this process we successfully matched 145,415 REF submissions with 6.95 million citations, corresponding to a recall of 76% of the total initial REF submission list.

Table 1. UoAs with the highest mean citations per paper (MCPP).

UoA/Subject	Outputs	% in MAG	Citations	MCPP
Public health	4,881	94.61%	505,950	109.56
Clinical medicine	13,394	90.78%	1,278,810	105.17
Physics	6,446	84.51%	491,151	90.15
Biological sciences	8,608	92.20%	620,009	78.12
Earth systems/environment	5,249	91.64%	315,429	65.58
Chemistry	4,698	87.71%	246,361	59.78
Allied health professions	10,358	89.35%	402,033	43.43
Ag. Vet. and food science	3,919	90.76%	150,959	42.44
Comp. science and informatics	7,645	89.22%	284,815	41.76
Economics and econometrics	2,600	88.81%	95,591	41.4

Table 2. Dataset statistics

Number of Units of Assessment (UoAs)	36
Number of institutes	154
Number of UoAs/institution pairs	1,911
Number of submissions (papers)	190,628
Number of submissions (papers) in MAG	145,415
Number of citations	6,959,629

Table 1 is ordered by the mean citations per paper (MCPP) and shows total number of submissions, percentage of these submissions available in MAG and the total citations of these submissions.

Additionally, as described in Fig. 1, we downloaded the Assessment Data from the REF 2014 website. These data contain the GPA, calculated by aggregating the peer review assessment results of individual papers for each given institution per UoA. We then joined these data with the enriched REF submission list by institution name and UoA. By doing so, we obtained 1,911 UoA/institution pairs together with their peer assessment information (GPA) and corresponding lists of submissions and their citation data (Table 2).

The full dataset used in our experiments and all results can be downloaded from Figshare.[2]

3.2 How Well Do Peer Review Judgments Correlate with Citation Data at the Institutional Level?

Once we assembled the full dataset, we extracted the following overall citation statistics: mean citations in December 2017 (mn_{2017}), median citations in December 2017 (med_{2017}), mean citations at the time of the REF exercise (mn_{2014}), and median citations at the same point (med_{2014}). These data were then used to test the correlation between citation data and REF GPA rankings for outputs for every institute in every UOA. The ten highest and ten lowest measured correlations by UoA are shown in Table 3. The citation data (cd) column denotes whether the REF judging panels considered citation data in their deliberations. While we attempted to run correlations with other similar aggregate functions, these are not shown in this table as they have far lower correlations with GPA.

Strong positive correlations can be observed at the discipline level for a large proportion of the UoAs, particularly for median citation count in 2017. Whilst the correlation was most often stronger for those UoAs that had used citation data in the REF peer review process, this was not always the case. Aeronautical and Mechanical engineering and Social work & Policy are two disciplines, which did not use citation data yet, show very strong correlations with GPA results. At the lower end of the scale, there was little correlation between GPA ranking and citation data, notably for those subjects covered by REF panels C and D. [4]. Lack of coverage in many of these areas is, however, understandable as these are disciplines which do not always produce journal articles, conference proceedings and other digitally published and highly citable artifacts as their main type of output. There is, however, clear delineation between the more highly correlated UoAs and those less correlated. The UoAs with the lowest are distinct from the rest, they are having a very weak or no correlation ($r <= 0.159, n = 37, p < 0.001$). Those above this level have a medium to strong correlation ($r > 0.353, n = 37, p < 0.001$). The low correlation for mean citations for Biological Sciences is explained by a single paper which was the most highly cited paper in the UoA. This paper received 4,626 citations, 58% more than next cited paper and nine times as many citations as all other submissions for that institute combined. Furthermore, this paper came from second lowest ranked (by

---

[2] https://figshare.com/s/69199811238dcb4ca987.

Table 3. Correlation between REF GPA output rankings and citation data

	UoA	mn_{2017}	med_{2017}	mn_{2014}	med_{2014}	cd
1	Chemistry	0.663	**0.802**	0.637	0.738	Y
2	Biological sciences	0.188	**0.797**	0.288	0.785	Y
3	Aero. mech. chem. engineering	**0.771**	0.758	0.745	0.760	N
4	Social work and policy	0.697	**0.752**	0.629	0.635	N
5	Comp. sci. and informatics	0.715	**0.743**	0.720	0.678	Y
6	Economics	0.750	0.737	0.760	**0.770**	Y
7	Earth systems and enviro. sciences	0.472	**0.707**	0.512	0.686	Y
8	Clinical medicine	0.654	**0.677**	0.666	0.662	Y
9	Public health and primary care	0.535	**0.674**	0.607	0.653	Y
10	Physics	0.600	**0.666**	0.627	0.605	Y
...						
27	Comm. cultural and media studies	**0.369**	0.355	0.334	0.267	N
28	Philosophy	0.352	**0.353**	0.268	0.270	N
29	Law	0.318	0.159	**0.365**	0.136	N
30	Theology and religious studies	0.404	0.154	**0.439**	0.153	N
31	English language and literature	−0.168	**0.102**	−0.192	0.094	N
32	Art and design	0.157	0.075	**0.187**	0.118	N
33	Anthropology and dev. studies	0.062	−0.009	**0.222**	0.145	N
34	Modern languages and linguistics	0.141	−0.069	**0.182**	0.188	N
35	Classics	0.155	−0.07	0.079	**0.285**	N
36	Music drama dance & perf. arts	0.046	−0.094	0.051	**0.039**	N

GPA) of 44 institutes. Had this paper been discounted from the correlations, the prediction results would have been far more clearly aligned with the other UoAs (mn2017 = 0.782, mn2014 = 0.766).

The variance of citation data coverage across UoAs led us to explore whether there could be a relationship between the strength of the correlations GPA and citation data correlation with the coverage of citation in a given UoA. Figure 2 plots this for both the UoAs that used citation data and those that did not. While the graph confirms that the highly cited UoAs in MAG are those UoAs that used citation data, it indicates that a few UoAs that did not also exhibit strong correlations. Unsurprisingly, the plot suggests that there might be a small bias exhibited by extra correlation strength in UoAs that utilised citation data. However, given the small number of UoAs, this is not statistically significant.

Table 4. Rankings by GPA and predictions produced using med_{2017} and med_{2014} respectively for the three most highly correlated UoAs.

REF UoA/Rank	GPA	med_{2017}	mc2017	rdiff	med_{2014}	mc2014	rdiff
Chemistry							
Liverpool	3.44	Liverpool	64	0	Liverpool	26	0
Cambridge	3.42	Cambridge	54	0	Lancaster	25	+8
Oxford	3.32	Warwick	53	+3	Oxford	22	0
UEA	3.29	Bath	51	+12	Cambridge	22	−2
Bristol	3.26	Oxford	50	−2	Queen Mary	20	+2
Bio sciences							
ICR	3.44	ICR	77	0	ICR	31	0
Newcastle	3.33	Queen Mary	66	+15	Sheffield	26	+5
Dundee	3.3	Imperial	59	+1	Imperial	25	+1
Imperial	3.26	Sheffield	56	+3	Leeds	24	+27
Oxford	3.26	Edinburgh	55	+4	Edinburgh	23	+4
Aero. mech.							
Cambridge	3.34	Cambridge	25	0	Cambridge	9	0
Imperial	3.12	Imperial	23	0	Imperial	8	0
UCL	3.06	Sheffield	19	+2	Brighton	7	+13
Cranfield	3.01	Brighton	18	+12	Manchester	6	+4
Sheffield	3.01	Manchester	17	+3	Sheffield	6	0

Table 5. Rank prediction quality for top 10 UoAs with the highest mean citations per paper.

UoA	HEIs	rdiff	nrdiff	MAP rt = 3	MAP rt = 5	MAP rt = 10	MAP rt = 10%	MAP rt = 20%	MAP rt = 30%
Comp sci.	89	12.39	0.139	0.19	0.32	0.50	0.46	0.75	0.87
Ag. vet.	29	4.02	0.139	0.45	0.65	0.86	0.45	0.68	0.86
Clinical med.	31	4.38	0.141	0.51	0.70	0.93	0.51	0.77	0.93
Allied H.	83	12.03	0.145	0.20	0.30	0.55	0.43	0.72	0.86
Economics	28	4.07	0.145	0.57	0.71	0.92	0.57	0.78	0.92
Chemistry	37	5.51	0.149	0.54	0.56	0.83	0.54	0.78	0.86
Earth systems	45	7.24	0.161	0.40	0.51	0.77	0.51	0.68	0.84
Public health	32	5.18	0.162	0.50	0.62	0.84	0.50	0.68	0.84
Bio. science	44	7.59	0.173	0.34	0.52	0.72	0.52	0.66	0.79
Physics	41	7.36	0.180	0.36	0.53	0.78	0.43	0.73	0.80
All (mean)	45	6.98	0.153	0.41	0.54	0.77	0.49	0.72	0.86

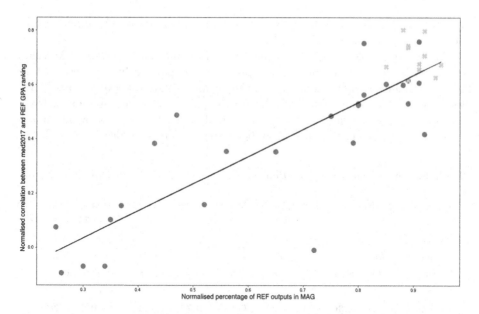

Fig. 2. Correlation between med_{2017} citations per UoA and GPA against the coverage of REF submissions in MAG for all UoAs. An 'o' represents a non-citation based UoA whilst and 'x' denotes a UoA that used citations.

3.3 How Well Can Citation Data Predict Peer Review Based Institutional Rankings?

Tables 4 shows the top five institutions for Chemistry, Biological Sciences and Aeronautical and Mechanical Engineering as ranked in the REF by GPA and predictions of ranking using med_{2017} and med_{2014} respectively. $mc2017$ and $mc2014$ show the median citation count for that institute. $Rdiff$ shows the rank difference when ranked by a particular citation metric. The prediction performance indicated in these tables is not unique, in four of the five top UoAs by correlation strength the highest ranked institute is predicted correctly by both med_{2014} and med_{2017}.

Table 5 demonstrates the effectiveness of predicting based on med_{2014} for the 10 most highly cited UoAs. To compare the prediction error, expressed by $rdiff$, across UoAs, we calculated the mean rank difference normalised by number of institutions ($nrdiff$). To express overall prediction accuracy, we used Mean Average Precision (MAP). The HEI column denotes the number of institutes submitting to that UoA. The parameter rt denotes the prediction rank tolerance. For example, $rt = 3$ indicates that a prediction within 3 positions of the original assessment result will be considered as correct. Given the simplicity of the prediction method, this is a strong indication of the power of citation data in this task. One could reasonably expect that further improvements can be made by employing more sophisticated indicators. However, as the predictions are not as good for UoAs that have lower than average mean citations per paper, we would restrain from recommending the use of citation data unaccompanied by peer review assessments in those UoAs.

We wanted to compare our prediction performance to the study of Mryglod et al. [5]. In order to conduct a fair and exact comparison, it was necessary to parse a number of institutions from our input data. Mryglod et al. reported they were unable to obtain citation indicators for all institutions in a given UoA. Their study covered three of the top ten highly cited UoAs, we show in Table 6 that our predictions are significantly better across all categories.

Table 6. Comparison of the prediction performance of our study with Mryglod et al. [5]

UoA	HEIs	rdiff	nrdiff	MAP rt = 3	MAP rt = 5	MAP rt = 10	MAP rt = 10%	MAP rt = 20%	MAP rt = 30%
Mryglod [5]									
Chemistry	29	4.89	0.169	0.37	0.82	0.82	0.37	0.82	0.82
Physics	32	8.63	0.270	0.28	0.40	0.65	0.28	0.46	0.65
Bio Science	31	8.38	0.270	0.22	0.38	0.70	0.22	0.51	0.64
All (mean)	**31**	**7.30**	**0.24**	**0.29**	**0.53**	**0.72**	**0.29**	**0.60**	**0.70**
Pride & Knoth (this study)									
Chemistry	29	4.00	0.138	0.68	0.72	0.89	0.68	0.72	0.86
Physics	32	5.68	0.178	0.34	0.59	0.90	0.34	0.75	0.90
Bio science	31	7.16	0.231	0.35	0.45	0.74	0.35	0.51	0.71
All (mean)	**31**	**5.61**	**0.18**	**0.46**	**0.59**	**0.84**	**0.46**	**0.66**	**0.82**
Improvement		**23%**	**25%**	**59%**	**11%**	**17%**	**59%**	**10%**	**17%**

4   Discussion

It has been shown in [4,13] and that many bibliometric indicators show little correlation with peer review judgments at the article level. This study, and those by [2,3,7], demonstrate that some bibliometric measures can offer a surprisingly high degree of accuracy when used at the institutional or departmental level. Our work has been conducted on a significantly larger dataset and our prediction accuracy is higher than shown in previous studies, despite deliberately using fairly simplistic indicators. Several studies including The Metric Tide [4], The Stern Report [14] and the HEFCE pilot study [15] all state that metrics should be used as an additional component in research evaluation, with peer review remaining as the central pillar. Yet, peer review has been shown by [16–18] amongst others to exhibit many forms of bias including institutional bias, gender/age related bias and bias against interdisciplinary research. In an examination of one of the most critical forms of bias, that of publication bias, Emerson [19] noted that reviewers were much more likely to recommend papers demonstrating positive results over those that demonstrated null or negative results.

All of the above biases exist even when peer review is carried out to the highest international standards. There were close to 1,000 peer review experts recruited by the REF, however the sheer volume of outputs requiring review calls into question the exactitude of the whole process. As an example the REF panel for UoA 9, Physics, consisted of 20 members. The total number of outputs submitted for this UoA was 6,446. Each paper is required to be read by two referees. This increases the overall total requirement to read 12,892 paper instances. Therefore each panel member was required to review, to international standards, an average of 644 papers in a little over ten months. If every panel member, worked every day for ten months, each member would need to read and review 2.14 papers *per day* to complete the work on time. This is, of course, in addition to the panelist's usual full-time work load. Moreover, Physics is not an unusual example and many other UoAs tell a similar story in terms of the average number of papers each panel member was expected to review; Business and Management Studies (1,017 papers), General Engineering (868 papers), Clinical Medicine (765 papers). The burden placed on the expert reviewers during the REF process was onerous in the extreme. Coles [20] calculated a very similar figure of 2 papers per day, based on an estimate before the data we now have was available. 'It is blindingly obvious,' he concluded, 'that whatever the panels do will not be a thorough peer review of each paper, equivalent to refereeing it for publication in a journal'. Sayer [21] is equally disparaging in regards to the volume of papers each reviewer was required to read and also expresses significant doubts about the level of expertise within the review panels themselves.

In addition to the potential pitfalls in the current methodologies, there is also the enormous cost to be considered. This was estimated to be £66m for the UK's original PRFS, the Research Assessment Exercise (RAE) in 2008. This rose markedly to £246m for the 2014 Research Excellence Framework. This is comprised of £232M in costs to the higher education institutes and around £14M in costs for the four UK higher education funding bodies. The cost to the institutions was approximately £212M for preparing the REF submissions for the three areas; outputs, impact and environment, with the cost for preparing the outputs being the majority share of this amount. Additionally, there were costs of around £19M for panelists' time [22]. If bibliometric indicators can in any way lessen the financial burden of these exercises on the institutions this is a strong argument in favour of their usage.

5   Conclusion

This work constitutes the largest quantitative analysis of the relationship between peer reviews (190,628 paper submissions) and citation data (6.9m citation pairs) at an institutional level. Firstly, our results show that citation data exhibit strong correlations with peer review judgments when considered at the institutional level and within a given discipline. These correlations tend to be higher in disciplines with high mean citations per paper. Secondly, we demonstrate that we can utilise citation data to predict top ranked institutions with a surprisingly high precision. In the ten UoAs with the highest number of mean citations per paper we achieve 0.77 MAP with prediction rank tolerance 10 with respect to the REF 2014 results. In four out of five top UoAs by correlation strength, the highest ranked institute in the REF results was predicted correctly. It is important to note that these predictions are based on citation data that were available at the time of the REF exercise.

While our analysis does not answer whether using citation-based indicators we can predict institutional rankings better than by relying on a peer review system, our results evidence that the REF peer review process led to highly similar results as those that could have been predicted automatically using citation data. The 11 REF UoAs with the highest mean citations per paper in MAG are the identical UoAs in which the peer review panels used citation data to inform their decisions. We argue that if peer-review is conducted in the way it was conducted in the REF, then it would have been more cost effective to save a significant proportion of the £246m spent on organising the peer review process [22] and carry out the institutional evaluation purely using citation data, particularly in UoAs with high mean citations per paper.

This has wide implication for PRFS globally. The countries whose PRFS still have a peer review component should carefully consider the way in which the peer review process is conducted. Thus ensuring that the peer review results add a new dimension to the information over that which can be obtained by predictions based on citation data alone. However, this advice only applies when the goal of the PRFS is to rank institutions, as it is the case in the UK REF, rather than individual papers or researchers.

Acknowledgements. This work has been funded by Jisc and has also received support from the scholarly communications use case of the EU OpenMinTeD project under the H2020-EINFRA-2014-2 call, Project ID: 654021.

References

1. Hicks, D.: Performance-based university research funding systems. Res. Policy **41**(2), 251–261 (2012)
2. Anderson, D.L., Smart, W., Tressler, J.: Evaluating research-peer review team assessment and journal based bibliographic measures: New Zealand PBRF research output scores in 2006. NZ Econ. Pap. **47**(2), 140–157 (2013)
3. Smith, A.G.: Benchmarking Google Scholar with the New Zealand PBRF research assessment exercise. Scientometrics **74**(2), 309–316 (2008)
4. HEFCE: The Metric Tide: Report of the Independent Review of the Role of Metrics in Research Assessment and Management (2015). http://www.hefce.ac.uk/pubs/rereports/year/2015/metrictide/
5. Mryglod, O., Kenna, R., Holovatch, Y., Berche, B.: Predicting results of the research excellence framework using departmental h-index. Scientometrics **102**(3), 2165–2180 (2015). https://doi.org/10.1007/s11192-014-1512-3
6. Bishop, D.: An alternative to REF2014? (2013). http://deevybee.blogspot.co.uk/2013/01/an-alternative-to-ref2014.html

7. Mingers, J., O'Hanley, J.R., Okunola, M.: Using Google Scholar institutional level data to evaluate the quality of university research. Scientometrics **113**(3), 1627–1643 (2017)
8. HEFCE: Research Excellence Framework 2014: Overview report by Main Panel A and Sub-panels 1 to 6; (2015). http://www.ref.ac.uk/2014/media/ref/content/expanel/member/Main
9. HEFCE: Research Excellence Framework - Results and Submissions (2014). http://results.ref.ac.uk/Results
10. HEFCE: Annex A - Summary of additional information about outputs (2014). http://www.ref.ac.uk/2014/media/ref/content/pub/panelcriteriaandworkingmethods/01_12a.pdf
11. Herrmannova, D., Knoth, P.: An analysis of the Microsoft academic graph. D-Lib Mag. **22**(9/10) (2016)
12. Hug, S.E., Brändle, M.P.: The coverage of Microsoft academic: analyzing the publication output of a university. Scientometrics **113**(3), 1551–1571 (2017)
13. Baccini, A., De Nicolao, G.: Do they agree? Bibliometric evaluation versus informed peer review in the Italian research assessment exercise. Scientometrics **108**(3), 1651–1671 (2016)
14. Stern, N., et al.: Building on success and learning from experience: an independent review of the research excellence framework. UK Government, Ministry of Universities and Science, London (2016)
15. HEFCE: Report on the pilot exercise to develop bibliometric indicators for the research excellence framework (2016)
16. Hojat, M., Gonnella, J.S., Caelleigh, A.S.: Impartial judgment by the "gatekeepers" of science: fallibility and accountability in the peer review process. Adv. Health Sci. Educ. **8**(1), 75–96 (2003)
17. Lee, C.J., Sugimoto, C.R., Zhang, G., Cronin, B.: Bias in peer review. J. Assoc. Inf. Sci. Technol. **64**(1), 2–17 (2013)
18. Smith, R.: Peer review: a flawed process at the heart of science and journals. J. R. Soc. Med. **99**(4), 178–182 (2006)
19. Emerson, G.B., Warme, W.J., Wolf, F.M., Heckman, J.D., Brand, R.A., Leopold, S.S.: Testing for the presence of positive-outcome bias in peer review: a randomized controlled trial. Arch. Internal Med. **170**(21), 1934–1939 (2010)
20. Coles, P.: The apparatus of research assessment is driven by the academic publishing industry (2013). https://bit.ly/2EfNMeV
21. Sayer, D.: Rank Hypocrisies: The Insult of the REF. Sage, Thousand Oaks (2014)
22. Technopolis. REF Accountability Review: Costs, Benefits and Burden (2015)

The MUIR Framework: Cross-Linking MOOC Resources to Enhance Discussion Forums

Ya-Hui An[1,3]([✉]), Muthu Kumar Chandresekaran[1], Min-Yen Kan[1,2], and Yan Fu[3]

[1] Web IR/NLP Group (WING), National University of Singapore, Singapore, Singapore
[2] Smart Systems Institute, National University of Singapore, Singapore, Singapore
[3] Web Sciences Center, School of Computer Science and Engineering, University of Electronic Science and Technology of China, Chengdu, China
anyahui.120@gmail.com

Abstract. New learning resources are created and minted in Massive Open Online Courses every week – new videos, quizzes, assessments and discussion threads are deployed and interacted with – in the era of on-demand online learning. However, these resources are often artificially siloed between platforms and artificial web application models. Facilitating the linking between such resources facilitates learning and multimodal understanding, bettering learners' experience. We create a framework for MOOC Uniform Identifier for Resources (MUIR). MUIR enables applications to refer and link to such resources in a cross-platform way, allowing the easy minting of identifiers to MOOC resources, akin to #hashtags. We demonstrate the feasibility of this approach to the automatic identification, linking and resolution – a task known as Wikification – of learning resources mentioned on MOOC discussion forums, from a harvested collection of 100K+ resources. Our Wikification system achieves a high initial rate of 54.6% successful resolutions on key resource mentions found in discussion forums, demonstrating the utility of the MUIR framework. Our analysis on this new problem shows that context is a key factor in determining the correct resolution of such mentions.

Keywords: Digital library · MOOC · Learning resource
Unique resource identifier · DOI · MUIR

1 Introduction

Digital libraries for open knowledge goes beyond the scholarly library and extends into the pedagogical one [9]. While participation in Massive Open Online Courses (MOOCs) and online learning has expanded [5,8,13,14], the methods by which learners participate in these classes has still been confined to the limitations of the Learning Management Systems (LMS) [4,6]. Such LMSes often

© Springer Nature Switzerland AG 2018
E. Méndez et al. (Eds.): TPDL 2018, LNCS 11057, pp. 208–219, 2018.
https://doi.org/10.1007/978-3-030-00066-0_18

have separated and distinct views of each form of learning resource – discussion forums, lecture videos, problem sets, homeworks – where cross-linking resources is difficult or impossible to achieve. Learners "cannot see the forest for the trees" when concepts are siloed and easy cross-referencing is impeded.

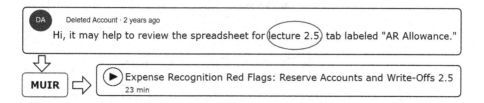

Fig. 1. Crosslinking a lecture resource mention in a discussion forum.

A concrete instance of this is in the discussion forum, where both instructors and students co-construct arguments to support critical thinking and knowledge [2,7]. Students often reference a certain quiz, this week's lecture or a particular slide, as in Fig. 1. Automatically hyperlinking such mentions to the target resource brushes and links the two endpoints, facilitating the contextualization of course materials across disparate views. To address this, we introduce and reduce to practice a pipeline that adds appropriate hyperlinks to natural language mentions of MOOC resources in discussion forums – a task known as *Wikification*, named after the same task which was first applied to Wikipedia.

In addressing this challenge, we needed to also propose an important standalone contribution: a framework for MOOC Uniform Identifier for Resources, which we name MUIR[1]. The MUIR framework is a two-component framework that pairs a transparent, guessable URL syntax for learning resources with a best-effort resolver that connects MUIR identifiers to their target resource. Best thought of as a hybrid between bibliographic records that identify a scholarly work, and the Digital Object Identifier that gives a resolution, our MUIR framework facilitates the cross-linking functionality that allows for the Wikification of natural language mentions in learner and instructor discourse.

MUIR also facilitates resource discovery. As a central harvester, the MUIR resolver components crawls MOOC platforms for resources and can expose related course material across different providers, formulating a MOOC domain Linked Open Data (LOD) [3], which creates typed links between data from different sources. This helps to address learning resource reuse, a problem that has been exacerbated with exponential success of MOOCs [18]. Without an aggregation service like MUIR, each MOOC LMS platform is siloed: having its own resource identifier schema that is non-portable, opaque and non-interpretable.

We demonstrate the use of the MUIR framework for the application of Wikification. In this case study, our Wikification application recognizes mentions

---

[1] MUIR refers to **M**OOC **U**niform **I**dentifier for **R**esources as well to the eponymous framework that creates such identifiers.

to publicly exposed resources, and generates short form references to those resources which the framework resolves and forwards links.

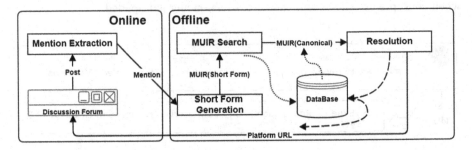

Fig. 2. MUIR system architecture: (l) online system, (r) offline harvester and resolution components.

2   Related Work

The MUIR framework contributes to both the topics curation and indexing, as well as identification schemes. We review these areas in turn.

Curation and Indexing. Both MOOCs and Learning Resource collection and indexing have prior work. MOOC List[2] curates a commercial, faceted indexing website to find current MOOC offerings. More general and academically inclined, MERLOT[3] achieves broader goals for thousands of learning resources for K–12 and tertiary education, for learners, educators, and faculty development for specific discipline. It acts as both an aggregator of submitted content for peer curation as well as a focal point for gathering the community concerning these resources [11]. MERLOT allocates a unique identifier to each material submitted as a pairing of a unique 'materialId' and an 'entryType'. More recently, the OpenAIRE project [1] aggregates metadata about scholarly research – projects, publications, people, organizations, etc. – into a central information space .

Identifier Schemes. Wikification uses MUIR to cross-link resources, creating a MOOC domain-specific form of Linked Open Data (LOD) [3,10]. It is a method of publishing to create and publish typed links between data entities from different sources, so that the data can be interconnected and put to better use. The MUIR scheme aims to aggregate resources across platforms and should be persistent, transparent and resolvable for various providers. We are informed of the design by related resource identifiers such as PURL, DOI, Dublin Core and general bibliographic metadata.

A Persistent Uniform Resource Locator [15] (PURL) provides a single layer of indirection built over the standard URL protocol for web addressing. PURLs

---

[2] https://www.mooc-list.com/.

[3] or "Multimedia Educational Resource for Learning and Online Teaching", https://www.merlot.org/.

> **I. MUIR (Short Form, Transparent (sample)):** `www.example.org/`
> `accounting-analytics/Week 2/lecture/2-5`
>
> **II. MUIR (Canonical Transparent):** `www.example.org/Coursera/accounting-`
> `analytics/1480320000000/Brian J Bushee&Christopher D. Ittner/Videos/`
> `expense-recognition-red-flags-reserve-accounts-and-write-offs-2-5`
>
> **III. MUIR (Opaque):** `www.example.org/id/1239jdn3oni3123s`
>
> **IV. Coursera URL:** `www.coursera.org/learn/accounting-analytics/lecture/`
> `1UzkX/expense-recognition-red-flags-reserve-accounts-and-write-offs-2-5`

Fig. 3. A Coursera learning resource URI in MUIR's threefold identifier scheme.

solve the problem of transitory URIs through their indirection, but omit any guidelines or enforcement of the identifier minting schema; the choice of identifier is up to the minting agent, somewhat akin to custom URL shorteners such as `bit.ly` and `tiny.cc`. The Digital Object Identifier [12] (DOI) schema goes further, not bound by any dependent protocols (e.g., HTTP for PURLs) and admits different authorities (e.g., different journal publishers) and distributed and hierarchical resolution via its use of the handle system. Our MUIR proposal is technically a PURL service, where our effort has been to create strong guidelines for the identifier portion of the schema.

Both Dublin Core [17] (DC) and bibliographic metadata are flexible containers that specify preferred (or mandatory) metadata attribute–value fields for different types of materials, such as *title* or *contributor*. Unlike PURL and DOI which are opaque, MUIR opts for transparent identifiers, taking the cue from DC and bibliographic metadata. The components of a MUIR encode the metadata values directly as part of the URL syntax for the identifier, and uniformly across various LMS providers.

3   The MUIR Framework

"When we try to pick out anything by itself, we find it hitched to everything else in the universe"—John Muir

Our uniform identifier scheme for MOOC learning resources, embodies the American naturalist John Muir's insight that everything is interconnected. In creating MUIR, our aim is to objectify MOOC resources so that they can be inventoried, referenced and subsequently better "hitched" to other resources, in the spirit of LOD, creating a densely tangled web of knowledge crucial for the contextualization of learning. We discuss the desiderata for our MUIR schema, while relating it to the practices of related work.

We motivate this section by working through the elements of a hypothetical MUIR associated with a learning resource from Coursera representing a specific lecture on accounting analytics:

1. Indirection. MUIRs provide two layers of indirection over actual resolvable resources such as a Coursera discussion forum, or a quiz hosted on a course on EdX. The first layer serves as a semantically transparent, short form where fields can be omitted and the search functionality of MUIR invoked to form the best-effort resolution to the canonical form. Similar to the simplicity of #hashtags, the MUIR short form encourages direct use by humans, later to be resolved to a canonical form or directly to the platform URL via best-guess relevance search.

The second layer of indirection (from the canonical form to the platform URL) provides both a uniform access mechanism to the resources that is platform-/provider-independent. As with PURLs, it also lends itself to preservation, having a single authority for resolution. Both the canonical form and the opaque form map one to one to the platform instance.

2. Transparent. Unlike traditional schema that use succinct opaque identifiers to serialize and identify objects, MUIR takes the cue from bibliographic systems that admit multiple, value–attribute fields to name resources. Much like how Dublin Core mandates certain fields be specified, MUIR also splits fields into required (*Resource Title, Resource Type, Course Name, Session Date, Instructor(s), Institution, Source Platform*) and optional categories (*Other Elements*). The short form MUIR invokes search by the resolution system to find the most appropriate learning resource, akin to search in a web search engine or an online public access catalog.

3. Comprehensive. MUIR's resource type categorizes the most common learning resources exposed in MOOCs. We survey learning resources provided on 29 worldwide MOOC platforms to inventory the common learning resources exposed, and map these forms to MUIR's *Resource Type* (Table 1). *Videos* present the lecture content. *Slides* provide the lecture content for download and separate review, often aligned to those in the video. *Transcripts* of the videos are sometimes available for various languages, often for other languages than the one used in the video. *Assessments* capture any form of assessments, exercises, homeworks and assignments that aim to self-diagnose the learners' knowledge commitment of the course content. *Exams* evaluate the knowledge and/or skills of students, including quizzes, tests, mid-exams and final examinations. *Readings* optionally provide a list of other learning resources provided by courses. *Additional Resources* help to catch other materials made available for specialized discipline-specific courses. For example, computer programming courses can provide program files for reference.

4. Stable. In addition to standard descriptor-like identifier structure, MUIR also has an alternate serial identifier syntax that is opaque and succinct, permitting short references that are permanent, as in the final MUIR opaque identifier in Fig. 3. Thus there can be many MUIR short form, transparent descriptors that map to a single unique opaque identifier.

Table 1. Prevalence of resource types exposed on global MOOC platforms. 'Scale' indicates # of courses/# of learners. The subsequent columns on top represents videos, slides, exams, quizzes, transcript, homeworks, assignments, assessments, exercises, readings, articles, programming scripts and additional materials, respectively. Each resource type is mapped to one of MUIR's canonical resource types (bottom row).

No. platform	Country	Scale (C /L)	V.	S.	E.	Q.	Tr.	HW.	Asg.	Ass.	Ex.	Re.	Art.	Pro.	Add.
1. Coursera	US	2000+ /25M+	✓	✓	✓	✓	✓	✓	✓	✓			✓		✓
2. edX	US	950+ /14M+	✓	✓	✓	✓	✓	✓	✓	✓		✓	✓		✓
3. Udacity	US	200+ /4M+	✓			✓	✓								
4. FutureLearn	UK	400+ /6.5M+	✓		✓				✓		✓		✓		✓
5. iversity	GER	50+ /0.75M+	✓						✓						
6. Open2Study	AU	45+ /1.1M+	✓	✓		✓	✓		✓						
7. Acumen+	US	34+ /0.3M	✓				✓		✓				✓		✓
8. P2PU	US	200+ /—	✓	✓	✓	✓	✓	✓	✓	✓	✓			✓	✓
9. Academic Earth	US	600+ /5.8M+	✓	✓	✓	✓	✓	✓	✓	✓	✓	✓	✓	✓	✓
10. Alison	IE	1000+ /11M+	✓												
11. Athlete Learning Gateway	CH	27+ /14K+	✓				✓					✓			
12. Canvas Network	US	200+ /0.2M+	✓			✓	✓		✓						
13. Course Sites	US	493+ /—	✓			✓			✓				✓		✓
14. KhanAcademy	US	— /57M+	✓			✓	✓				✓	✓			
15. Open Learning	JP	30+ /—	✓	✓	✓	✓	✓	✓	✓	✓	✓	✓	✓	✓	✓
16. OpenupEd	EU	190+ /—	✓	✓	✓	✓	✓	✓	✓	✓	✓	✓	✓	✓	✓
17. Saylor	US	100+ /—	✓		✓	✓					✓		✓	✓	
18. Udemy	US	— /20M+	✓		✓										✓
19. CNMOOC	CN	600+ /—	✓		✓										
20. Complexity Explorer	US	11+ /—	✓			✓	✓	✓							✓
21. Ewant	TW	600+ /20K+	✓	✓		✓									
22. Janux	US	20+ /31K+	✓			✓	✓		✓						✓
23. Microsoft Virtual Academy	US	800+ /—	✓	✓		✓	✓								
24. NTHU MOOCs	TW	46 /—	✓			✓			✓						
25. Stanford Online	US	100+ /—	✓	✓	✓	✓		✓	✓	✓					✓
26. XuetangX	CN	1300+ /9M+	✓	✓	✓	✓		✓	✓	✓					✓
27. icourse163	CN	1000+ /—	✓		✓	✓		✓	✓						
28. FUN	FR	330+ /1M+	✓		✓						✓				
29. FX Academy	ZA	10+ /—	✓		✓							✓			✓
# of platforms w/ Resource Type			29	11	10	24	15	11	16	11	9	8	5	6	15
Mapping to MUIR's Resource Type			V.	S.	E.	T.		Ass.			Re.			Add.	

3.1 Collected Dataset

We operationalize our MUIR framework by creating a series of crawlers to proactively collect learning resources from MOOC platforms. In the remainder of the paper, we study using MUIR against a subset of crawled resources from Coursera as a proof of concept. Our Coursera corpus, collected at January 31, 2017, includes all posts and resources of 142 courses that had already completed, totalling 102,661 posts and 11,484 learning resources spanning all 7 resource types.

4   Discussion Forum Wikification

We operationalise the MUIR framework through the task of *discussion forum Wikification*. Our system for forum Wikification extracts and hyperlinks mentions of learning resources in student posts as shown in Fig. 1.

The skeptic might ask: Is Wikification meeting a real demand for crosslinking learning resources? To answer this, we wish to calculate the number of mentions that are actually present in discussion forum. Let us assume that mentions to the seven resource types do contain a descriptive keyword. While the presence of these keywords may not necessarily denote an actual mention (i.e., *"I have a question"*), the percentage of posts that contain the relevant keywords serves as an upper bound for the number of mentions. Restricting our examination to content subforums (excluding forums for socializing; e.g. *'Meet & Greet'* and *'General Discussion'*), we find that approximately $15,529/69,025 = 22.5\%$ posts contain one or more keywords. Restating, about 1 of 4 posts in discussion forums potentially have mentions that need Wikification. So there is a real need that we address with Wikification.

The process presumes that the MUIR system has proactively crawled and indexed MOOC resources, as previously discussed. We reduce the problem into 4 concrete phases as shown in Fig. 2: (1) Mention Extraction: mention identification, (2) Short Form Generation: MUIR short form construction, (3) MUIR Search: MUIR short form to canonical form resolution, (4) Resolution: forwarding the request to the platform URL. Note that the first two phases take place outside of the MUIR framework, in our Wikification application that processes discussion forums. We step through these four phases in turn to illustrate how the MUIR framework interacts with the Wikification process.

Phase 1: Mention Extraction. Wikification begins by identifying important mentions from a post of a course. As natural language mentions can occur in an infinite variety, in this initial study, we constrain the problem scope to identifying only **S**ingle, **C**oncrete, w**I**thin-course entities (or SCI). As counterexamples, references to collective entities (i.e., "the quizzes"), specific topics taught within a course (similar to keywords, i.e., "corporate risk") fall outside the scope of our SCI definition.

Analyzing actual SCI mentions in discussion forums, such as *"lecture 2.5"* in Fig. 1 and those in Fig. 4 show us that SCI entities do lend themselves to be captured by a simple regular expression matching with a keyword followed by a numeric offset. We thus programmatically find and delimit such mentions as spans for hyperlinking. This solution, although overly simplistic, serves well as a starting point for Wikification. We revisit this decision later in our evaluation.

Phase 2: Short Form Generation. For each mention, Wikification generates a MUIR short form programmatically. The short form is used to split the mention into component words, using which our algorithm maps them to fields in the MUIR short form. Inferrable missing components are added by the context of the hosted discussion forum. Continuing with our running example, this stage takes the mention *"lecture 2.5"* that appears in an *Accounting Analytics* course on Coursera, and constructs the short form I in Fig. 3, where the mention's text of { *"lecture"*, *"2"*, *"."* and *"5"*} constructs the s_4 and s_5 short form components: the relative block number (2–5 denotes module 2 lecture 5), and remaining components (s_2 and s_3) are inferred from context:

$$\underbrace{\texttt{www.example.org}}_{s_1} / \underbrace{\texttt{accounting-analytics}}_{s_2} / \underbrace{\texttt{Week2}}_{s_3} / \underbrace{\texttt{lecture}}_{s_4} / \underbrace{\texttt{2-5}}_{s_5}$$

Here, s_1 is the MUIR resolver host, s_2 is the course name, s_3 is the forum name (usually the week number) of the post, s_4 is the resource type and s_5 represents the relative block number.

Phase 3: MUIR Search. A click on a short form requests the resource from MUIR resolver. This search process is the first layer of indirection, combining the post information in the MUIR database from which MUIR obtains additional peripheral information (platform, session date and instructor(s) name) about the post that embeds the mention. The search process first utilizes the origin post data {source platform, s_2, session date and instructor(s) name} to locate the hosting course's context. The remainder of the short form (s_4 and s_5) are used to match the resource type and name in a full text search, where exact matches are favored. The resolver searches its index of canonical MUIRs using this custom search logic to match with the short form and deems the best match its resolution. As in the running example, this process matches the MUIR short form I to the MUIR canonical form II:

$$\underbrace{\texttt{www.example.org}}_{f_1} / \underbrace{\texttt{Coursera}}_{f_2} / \underbrace{\texttt{accounting - analytics}}_{f_3} / \underbrace{\texttt{1480320000000}}_{f_4} / \underbrace{\texttt{BrianJBushee\&Christopher}}_{f_5}$$

$$\underbrace{\texttt{D.Ittner}}_{f_5} / \underbrace{\texttt{Videos}}_{f_6} / \underbrace{\texttt{expense-recognition-redflags-reserve-accounts-and-write-offs-2-5}}_{f_7}$$

Here, f_1, f_3 and f_6 are migrated from the short form, and the remaining fields have been imputed from context: f_2, f_4, f_5 and f_7 give the source platform, the session date, instructors' names, and the slug name of the resource, respectively.

Phase 4: Resolution. This final phase is simple, as the canonical MUIR maps one-to-one with a platform URL, through a hash table lookup. This process maps the running example's canonical form II to the platform-specific URL IV through the second layer of indirection.

5   Wikification Evaluation

We believe the MUIR identifier framework is useful on its own right, but it is hard to evaluate its intrinsic utility. We instead evaluate extrinsically, assessing the utility of MUIR as a component within discussion forum Wikification. Specifically we ask ourselves the following research questions (RQ):

RQ1. What is the coverage rate for posts that actually contains mentions?

Table 2. Mention extraction coverage.

Annotator ID	# of posts	# of posts identified as having mentions	# extracted by our wikifier	# correct	Coverage
Annotator 1	1,087	156	5	5	14.4%
Annotator 2	1,087	175	5	5	16.1%
Overall	1,087	196 (Union)	5	5	18.0%

> **YES:** $\langle m_1 \rangle$ Is it just me or were some questions on Quiz 2 a surprise? There were a few questions that were not discussed in the lesson plan.
> **YES:** $\langle m_2 \rangle$ Hello, I just would like to note that on 12:30 in the answer to question 3 in the lecture 2.4 it says that the network is deadlock-free, whereas ...
> **NO:** $\langle m_3 \rangle$ The last item, that is "Probability Models for Customer-Base Analysis.pdf", in the Resources > Additional Readings by Week section for Week 3 is not accessible.
> **NO:** $\langle m_4 \rangle$ I'm working on the programming assignment for ML, week 2. I successfully submitted answers to the obligatory questions.
> **NO:** $\langle m_5 \rangle$ At around 5:00 in the lecture, we see that the regularization term in the cost function is summed from 1 to L-1. Shouldn't this be 2 to L?
> **NO:** $\langle m_6 \rangle$ Hello. I wanted to use "e" as a number for ex.2/week3. It didn't work, and I didn't find useful help with "help exponent".

Fig. 4. Actual resource mentions in our 1,087 sample sized dataset, illustrating the variety of expressions. Our Wikification currently handles the first two mentions.

RQ2. How accurate is the resolution for different resource types?

RQ1: Mention Coverage. With a full annotation of the dataset we could conclusively measure the coverage of our regular expressions in capturing actual natural language mentions to SCI. However, the effort for full annotation is infeasible, and instead we randomly sample ~1,000 posts to check the actual coverage of our Wikifier syntax. We note that it can be unintuitive for annotators to identify whether a word, phrase or sentence is a mention, so we employed two independent annotators to reduce bias. Results for this sample annotation are shown in Table 2.

In our 1K sample of posts, 18% of posts or more contain mentions to learning materials. This is significant, as it shows that there is much potential to better interlink resources, even just for the silo of discussion forums. In these sampled posts, our Wikifier matched 5 mentions, which were all actual mentions (correct). This result shows that our $<$"*keyword*" + number$>$ pattern has high precision but suffers from low recall, covering only about 2.6% of possible mentions.

How can we improve mention extraction coverage? We examine the causes for the coverage disparity, where the parenthetical percentage is determined over the same sampled data.

Table 3. Resolution accuracy evaluation. Only mentions to 4 MUIR types are present in our Coursera subset. P_I represents precision of Annotation I and P_II is for Annotation II.

Resource	# of instances	P_I	P_II
Videos	89	71.9%	57.3%
Slides	27	74.1%	33.3%
Exams	718	83.0%	53.3%
Assessments	12	50.0%	25.0%
Total	846	81.1%	54.6%

1. **Implicit Contextual Knowledge (∼45% of errors).** In sequential posts, posters often refer to the content from the previous posters, and refer using demonstrative pronouns such as '*this*', '*that*' or '*the*'. Without context knowledge, our prototype simply does not capture such mentions, such as in '*that video you mentioned*'.
2. **Named Reference (∼30% of errors).** Direct use of the resource name – especially for videos, slides and quizzes – makes such mentions impossible to capture, without predicating prior MUIR lookup (cf m_3 in Fig. 4 or '*the problem "Hashing with chains"*').
3. **Informal Expressions (∼15% of errors).** Colloquial expressions abound (Fig. 4's m_4 and m_6) and fall outside the current scheme. Adding regular expressions to capture these would improve coverage at the cost of precision.

RQ2: MUIR Resolution Accuracy. The other component that needs evaluation is Phase 3, MUIR Search. Given the short forms that are generated by Wikification, MUIR Search connects the short form to a (hopefully correct) platform URL.

We offer two evaluations that give complementary data on the resolution accuracy, shown in Table 3. Comparing P_I against P_II, the accuracy of Annotation I is generally better than Annotation II. That is because Annotation I is generated only by depending on the information of mentions and the limited relevant information of posts, foregoing the implicit contextual knowledge of the previous and subsequent posts. This gives an upper-bound for how well mentions are actually resolved by our simple search logic. But in Annotation II when we annotate the ground truth test data, we consider all of the context of the mentions including the content around the mentions and other posts in the same thread. This is a realistic evaluation on the full complexity of the problem.

The results are best analyzed jointly. We see that the mentions we capture are easy to extract (higher performance on Annotation I), but hard to resolve without context (lower performance on Annotation II). The accuracies for four *Resource Types* have different degrees of reduction. But the results are encouraging: our prototype, even with its simple logic, can already handle almost 55% of learning resources.

As we did for RQ1, we further categorized a rough cause to the errors in the resolution process:

1. **Mentions needing context to resolve against multiple matches** (~20% **of errors**): Learners may write mentions such as *"lecture 4.5"*, where *"4"* and *"5"* are used by MUIR Search but could refer to different lectures that both have textual components "4" and "5" in their slug name.
2. **Multiple potential targets** (~70% **of errors**): Even considering context, certain mentions are still ambiguous. If a mention states *"question 3"* but there are multiple quizzes within the context, all which have a Question 3, the target is ambiguous. MUIR can only guess in this case.
3. **Errors in mention extraction** (~10% **of errors**): These are cascaded from the Phase 1 process of mention extraction. Examples include *partial mention extraction* (*"lecture's 2 transcript"* may be written by a learner, but only "lecture 2" was detected) and *informal reference* (*cf* m_6 in Fig. 4).

Both RQ1 and RQ2 discussions clearly point forward in the direction of improving coverage, especially in Phase 1, as such errors cascade. A clear direction is to incorporate contextual knowledge: our current work thus aims to incorporate such knowledge by the machine reading of the posts, by leveraging recurrent neural network based learning models [16] currently making much impact in natural language processing research. This will help the Wikification process by both capturing more natural mention expressions and minting better Phase II MUIR short forms that better facilitate correct resolution downstream.

We note that mention extraction can also be facilitated by introducing linking conventions, similar to #hashtags. MUIR's short form can be further facilitated by the future learner's explicit triggering when writing their posts: i.e., *"I have a question about #video5"*, where mention identification are solved by the learner.

6   Conclusion

For a learner to see the forest for the trees requires seamless interlinking of learning resources. Discussion forum Wikification takes us closer towards this goal. Our prototype shows the feasibility of the approach for simple mention types, and further motivates research on better mention identification and search resolution of such mentions.

Underlying this development is our core contribution of the MUIR framework for identifying and referencing the burgeoning set of MOOC resources being generated by the community. Our solution hybridizes best practices among ease-of-use descriptions, search practices and the persistence and identification standards. Our work aims to catalyse work towards making linked open data a closer reality for the world's learners.

Acknowledgement. This research is funded in part by NUS Learning Innovation Fund – Technology grant #C-252-000-123-001, and also in part by the scholarship from China Scholarship Council (CSC), National Natural Science Foundation of China under

Grant Nos.: 61673085 and 61433014, and UESTC Fundamental Research Funds for the Central Universities under Grant No.: ZYGX2016J196. We also thank the anonymous reviewers for their useful comments.

References

1. Ameri, S., Vahdati, S., Lange, C.: Exploiting interlinked research metadata. In: Kamps, J., Tsakonas, G., Manolopoulos, Y., Iliadis, L., Karydis, I. (eds.) TPDL 2017. LNCS, vol. 10450, pp. 3–14. Springer, Cham (2017). https://doi.org/10.1007/978-3-319-67008-9_1

2. Andresen, M.A.: Asynchronous discussion forums: success factors, outcomes, assessments, and limitations. J. Educ. Technol. Soc. 12(1), 249 (2009)

3. Bizer, C., Heath, T., Berners-Lee, T.: Linked data-the story so far. Int. J. Semant. Web Inf. Syst. 5(3), 1–22 (2009)

4. Dalsgaard, C.: Social software: e-learning beyond learning management systems. Eur. J. Open, Distance E-Learn. 9(2) (2006)

5. Hew, K.F., Cheung, W.S.: Students and instructors use of massive open online courses (MOOCs): motivations and challenges. Educ. Res. Rev. 12, 45–58 (2014)

6. Mahnegar, F.: Learning management system. Int. J. Bus. Soc. Sci. 3(12) (2012)

7. Marra, R.M., Moore, J.L., Klimczak, A.K.: Content analysis of online discussion forums: a comparative analysis of protocols. Educ. Technol. Res. Dev. 52(2), 23 (2004)

8. Martin, F.G.: Will massive open online courses change how we teach? Commun. ACM 55(8), 26–28 (2012)

9. McAuley, A., Stewart, B., Siemens, G., Cormier, D.: The MOOC model for digital practice (2010)

10. Mendes, P.N., Jakob, M., García-Silva, A., Bizer, C.: DBpedia spotlight: shedding light on the web of documents. In: Proceedings of the 7th International Conference on Semantic Systems, pp. 1–8. ACM (2011)

11. Moncada, S.M.: Rediscovering MERLOT: a resource sharing cooperative for accounting education. J. High. Educ. Theory Pract. 15(6), 85 (2015)

12. Paskin, N.: Digital object identifier (DOI®) system. Encycl. Libr. Inf. Sci. 3, 1586–1592 (2010)

13. Peña-López, I., et al.: Giving knowledge for free: the emergence of open educational resources (2007)

14. Seely Brown, J., Adler, R.: Open education, the long tail, and learning 2.0. Educ. Rev. 43(1), 16–20 (2008)

15. Shafer, K., Weibel, S., Jul, E., Fausey, J.: Introduction to persistent uniform resource locators. In: INET 1996 (1996)

16. Sutskever, I., Vinyals, O., Le, Q.V.: Sequence to sequence learning with neural networks. In: Advances in Neural Information Processing Systems, pp. 3104–3112 (2014)

17. Weibel, S., Kunze, J., Lagoze, C., Wolf, M.: Dublin core metadata for resource discovery. Technical report (1998)

18. Zemsky, R.: With a MOOC MOOC here and a MOOC MOOC there, here a MOOC, there a MOOC, everywhere a MOOC MOOC. J. Gen. Educ. 63(4), 237–243 (2014)

Figures in Scientific Open Access Publications

Lucia Sohmen[2]([✉])[iD], Jean Charbonnier[1][iD], Ina Blümel[1,2][iD],
Christian Wartena[1][iD], and Lambert Heller[2][iD]

[1] Hochschule Hannover, Expo Plaza 12, 30539 Hannover, Germany
[2] Technische Informationsbibliothek, Welfengarten 1B, 30167 Hannover, Germany
Lucia.Sohmen@tib.eu

Abstract. This paper summarizes the results of a comprehensive statistical analysis on a corpus of open access articles and contained figures. It gives an insight into quantitative relationships between illustrations or types of illustrations, caption lengths, subjects, publishers, author affiliations, article citations and others.

Keywords: Open access · Scientific figures · Statistical analysis

1 Motivation and Target

Researchers often reuse figures from other publications for their own work, for example presentations or articles. In order to find those images, it is useful to have a search engine that finds figures from scientific articles.

The goal of the NOA (*Nachnutzung von Open Access Bildern*, Reuse of Open Access Images) project is to build a freely accessible corpus of figures from open access articles, providing links to the original article as well [3]. A first version of a search engine allowing for filtering and searching is available at http://noa.wp.hs-hannover.de/. In order to secure access to the images after project completion, they will be uploaded to Wikimedia Commons (commons.wikimedia.org). As a side effect of the mentioned extraction of figures from papers, we use the built-up corpus of images linked to corresponding articles for various analyses and relations to other quantitative data/article such as citations. This paper summarizes the results of a comprehensive statistical analysis on our corpus and gives an insight into quantitative relationships between illustrations or types of illustrations, subjects, publishers, journals, article citations and others.

2 Related Work

Over the years, there have already been attempts at creating search engines for scientific images. So far, all of these have used some subset of articles from the life sciences. FigSearch [7], developed in 2004, claims to be the first of these applications. The Yale Image Finder [9] was developed in 2008 Another search

© Springer Nature Switzerland AG 2018
E. Méndez et al. (Eds.): TPDL 2018, LNCS 11057, pp. 220–226, 2018.
https://doi.org/10.1007/978-3-030-00066-0_19

Table 1. Publishers (including aggregators), number of papers, figures, percentage of papers with figures and years included in the dataset.

Publisher	# Articles	# Figures	% With figures	Years included
Copernicus	9 592	85 720	71,7	2014–2017
Springer	78 418	310 214	98,0	2003–2018
Hindawi	147 848	1 172 657	80,3	2008–2017
Frontiers	57 621	217 897	83,3	2009–2017
PMC	747 839	2 796 271	81,3	1848–2017
All	1 041 318	4 582 759	80,7	

engine is Figuresearch [1] from 2009. Viziometrics [6] from 2016 is the newest application that allows users to directly search for images. Their dataset contains 650 000 articles and 4.8 million images from the PubMedCentral (PMC) corpus. Their search engine is the only one that is still available to search in at viziometrics.org.

Several statistical analyses of article corpora containing images have been done. [6] analyzes the Viziometrics corpus. [4] extracted 6.4 million figures from 1 million papers in computer science and biomedicine. They found that, over time, figure counts and their captions lengths have increased. There was a small positive correlation between the figure count and the number of citations to a paper. [5] looked at 1133 psychology papers to find out what factors influence the number of citations to a paper. The authors found that the number of graphs had a negative correlation while the number of tables and models had a positive correlation with the citations. [2] analyzed 5180 articles from six journals in different domains to analyze the figure use of multiple authors versus single authors and found that multiple authors use more figures per article.

3   Corpus and Analysis Method

Our corpus includes figures from open access articles from different sources. Criteria for inclusion were accessibility (difficulty of downloading a large set of articles), format (easy to parse, like XML) and license (suitable for reuse and upload to Wikimedia Commons). A big part of the corpus is a subset from PubMedCentral (PMC) which stores millions of articles from the life sciences. Other articles were downloaded from the publishers as a dump or via API.

All the articles that we downloaded have the XML format with most of them using the JATS-XML specification that is required by PMC. After download, the articles were parsed with a Java program that was developed within our project. It extracts all the relevant data from the documents (for example article metadata, figure URLs and captions) and writes it to the project database. Furthermore, this data has been enhanced with additional information, including journal discipline, corresponding Wikipedia categories and citation data from Crossref. This makes up the dataset on which we base our statistics.

We found 3 million figures in 1 million articles, including articles with zero figures. We counted everything that was embedded in a "figure" tag in the XML form of an article. These do not usually include tables and equations. See Table 1 for an overview of the different publishers and their image count in our dataset.

4   Results

4.1   Licences and Figures with Source Reference

The license type of the figures is of interest for re-usability. CC-BY clearly dominates the corpus: CC-BY-4.0 came to a number of 351694, −3.0 to 75729, −2.5 to 30036 and −2.0 to 216472. CC0 was only assigned 1986 times. Although we did not filter out CC-BY-SA type licenses, none of the articles in the corpus are under that license type. 7878 times no license was found.

To identify figures that were reused from an external source and are therefore not under the same license as the article, we spotted keywords in the captions to find out whether an external source is cited. This algorithm identified about 5% of all images. Manual inspection revealed that roughly 8/9 of those results were false positives, so the actual rate of reused images is about 0.55%. Recall was valued over precision to avoid violation of copyright.

4.2   Figure Types

Table 2 shows the average number of charts (including charts and graphics) and images (including photos, microscopy and other imaging methods) per paper for disciplines with 2000 or more papers. The often much higher proportion of charts is noticeable in almost all disciplines, especially in the subjects belonging to the field of Engineering and Technology[1]. In total, Engineering and Technology subjects contain the highest number of figures, followed by Natural Sciences and Medical and Health Sciences. All disciplines with less than 2000 papers can be derived from the underlying raw data [8].

4.3   Figure Caption Length

Since the captions are usually the most important source for information about an image, we determined the caption length for all images. In Table 3 we can see that there are large differences in the average caption length per discipline. While life sciences usually have long captions, mathematics and technical sciences tend to use shorter captions. In Fig. 1 we see the distribution of caption lengths.

---

[1] We refer to the Revised Field of Science and Technology (FOS) classification at http://www.oecd.org/sti/inno/38235147.pdf..

Table 2. Average number of charts and images for disciplines with 2000 or more papers.

Discipline	#Papers	Charts/paper	Images/paper
All	932542	3.6	0.7
Medicine	432424	2.4	0.8
Biology	136655	3.9	0.6
Chemistry and pharmacy	78525	3.7	0.3
Mathematics	34668	4.8	0.4
Physics	29900	5.7	0.8
Geosciences	25845	2.2	0.1
Process engineering, biotechnology	24019	4.6	1.4
Science in general	21779	6.4	1.1
Computer science	19563	5.9	0.4
Electrical engineering	14648	7.0	0.8
Energy, environmental protection	13321	4.7	0.8
General technology	11587	9.7	1.0
Measurement and control engineering	14648	7.0	0.8
Mechanical engineering	11052	8.6	3.2
Materials science	11052	8.6	3.2
Agriculture and forestry	12444	2.6	0.5
Nuclear engineering	13297	4.7	0.8
Earth sciences	7388	6.5	0.7
Psychology	5755	2.0	0.3
General engineering	3375	6.7	1.3
Sports	3144	1.5	0.1
Architecture, civil engineering and surveying	2774	12.7	1.5
Education	2736	1.4	0.1
Economics	2337	3.3	0.1

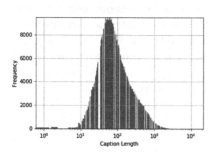

Fig. 1. Distribution of caption length on a logarithmic scale.

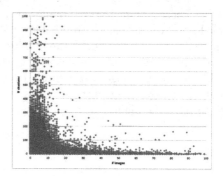

Fig. 2. Count of references.

Table 3. Caption length in characters for disciplines with over 10.000 figures. Disciplines are counted according to assignment of journals. Figures from journals assigned to more than one discipline are counted for each of these disciplines.

Discipline	n	Mode	Median	Mean
All	2963059	54	265	411.9
General technology	124131	52	81	119.8
Mathematics	179023	52	84	126.3
Architecture civil engineering and surveying	36931	70	89	117.6
Electrical eng., measurement and control eng.	115415	50	101	141.7
Energy, environmental protection, nuclear eng.	74368	68	116	175.5
Mechanical eng., materials science	137878	69	125	174.0
Computer science	126198	43	133	243.5
Geosciences	58875	111	140	159.5
General engineering	27018	83	198	269.4
Agriculture and forestry	29942	59	201	291.7
Earth sciences	54480	86	220	294.2
Chemistry and pharmacy	319335	111	228	416.5
Physics	199369	75	274	468.0
Psychology	13142	117	338	443.9
Process eng., biotechnology	144268	123	355	440.3
Medicine	1374680	69	357	471.8
Science in general	162051	47	513	697.1
Biology	615226	330	524	652.8

4.4 Citations

We investigated whether the number of figures correlates with the citations to an articles as suggested by [5,6]. This information was added using the Crossref API. Those numbers were compared with other services. Although they were a bit lower overall, they correlated strongly. We assumed that more figures lead to more readers. Interestingly, the number of figures in an article does not correlate with the number of citations it has received (correlation: $6.19 \cdot 10^{-3}$, Fig. 2). This does not change considerably even after excluding all outliers with over 20 figures and over 100 citations (Table 4). However, articles with a figure count of 6–10 have the highest median citation count of 4. See [8] for details.

Table 4. Number of images and related citation counts

Articles in set (f = figures, c = citations)	number of papers	Median citation count	Mean citation count	Correlation between citation count and figure count
All	1048575	3	8,3	0,006192715
0–20 f., 0–100.c	984284	3	7	0,037702209
0 f	211441	1	5,3	Not possible
1–5 f	519924	3	9	0,022292513
6–10 f	238525	4	9,8	−0,008327417
11–678 f	78688	2	6,1	−0,008684956

5 Discussion

The study gives an insight into a large data set based exclusively on open access articles. The dataset consists of articles with CC-BY-licenses that were available for mass download in an XML-format. The majority of figures within our corpus are charts. This figure type often visualizes research results and can range from the very standardized form of a graph with an x- and y-axis to drawings that can show abstract concepts in different formats. These figures could be used for research in the field of automatic information extraction. Images, on the other hand, are the more likely candidates for reuse since they usually do not show numbers that are only relevant for one paper. Researchers that work in analyzing images should consider the average caption length in each discipline. Our paper shows a clear trend towards shorter captions in technology and longer captions in the life sciences. This could mean that captions in the life sciences generally contain more information and are therefore a better source for analysis than captions in other disciplines. However, it could also mean that this field needs more words to explain a single concept. Our results on the citation numbers do not match what [6] found. These differences could be explained by our inclusion of different disciplines or the slightly different way of ordering the numbers. This invites more study into the question whether figure use is a predictor for scientific impact, possibly with a focus on different disciplines. The result of our study is that the number of figures in a paper is not a good predictor for scientific impact. However, it seems like papers with between 1 and 10 figures, which are the most common, receive the most citations. Further research should include a more faceted classification of figure types and how they relate to different disciplines and citations.

Acknowledgment. This research was funded by the DFG under grant no. 315976924.

References

1. Agarwal, S., Yu, H.: FigSum: automatically generating structured text summaries for figures in biomedical literature, 6–10 (2009). https://www.ncbi.nlm.nih.gov/pmc/articles/PMC2815407/
2. Cabanac, G., Hubert, G., Hartley, J.: Solo versus collaborative writing: discrepancies in the use of tables and graphs in academic articles. J. Assoc. Inf. Sci. Technol. **65**(4), 812–820 (2014)
3. Charbonnier, J., Sohmen, L., Rothman, J., Rohden, B., Wartena, C.: NOA: a search engine for reusable scientific images beyond the life sciences. In: Pasi, G., Piwowarski, B., Azzopardi, L., Hanbury, A. (eds.) ECIR 2018. LNCS, vol. 10772, pp. 797–800. Springer, Cham (2018). https://doi.org/10.1007/978-3-319-76941-7_78
4. Clark, C., Divvala, S.: PDFFigures 2.0: mining figures from research papers. In: Proceedings of the 16th ACM/IEEE-CS on Joint Conference on Digital Libraries. JCDL 2016, pp. 143–152. ACM, New York (2016). https://doi.org/10.1145/2910896.2910904
5. Hegarty, P., Walton, Z.: The consequences of predicting scientific impact in psychology using journal impact factors **7**(1), 72–78. https://doi.org/10.1177/1745691611429356
6. Lee, P., West, J., Howe, B.: Viziometrics: analyzing visual information in the scientific literature. IEEE Trans. Big Data **4**, 117–129 (2018). https://doi.org/10.1109/TBDATA.2017.2689038
7. Liu, F., Jenssen, T.K., Nygaard, V., Sack, J., Hovig, E.: FigSearch: a figure legend indexing and classification system **20**(16), 2880–2882. https://doi.org/10.1093/bioinformatics/bth316
8. Sohmen, L., Charbonnier, J., Blümel, I., Wartena, C., Heller, L.: Figures in scientific open access publications - underlying data (2018). https://doi.org/10.5281/zenodo.1295579
9. Xu, S., McCusker, J., Krauthammer, M.: Yale image finder (YIF): a new search engine for retrieving biomedical images **24**(17), 1968–1970. https://doi.org/10.1093/bioinformatics/btn340

Information Extraction

Finding Person Relations in Image Data of News Collections in the Internet Archive

Eric Müller-Budack[1,2]($\boxtimes$) , Kader Pustu-Iren[1] , Sebastian Diering[1],
and Ralph Ewerth[1,2]

[1] Leibniz Information Centre for Science and Technology (TIB), Hannover, Germany
{eric.mueller,kader.pustu,ralph.ewerth}@tib.eu,
diering@stud.uni-hannover.de
[2] L3S Research Center, Leibniz Universität Hannover, Hannover, Germany

Abstract. The amount of multimedia content in the World Wide Web is rapidly growing and contains valuable information for many applications in different domains. The Internet Archive initiative has gathered billions of time-versioned web pages since the mid-nineties. However, the huge amount of data is rarely labeled with appropriate metadata and automatic approaches are required to enable semantic search. Normally, the textual content of the Internet Archive is used to extract entities and their possible relations across domains such as politics and entertainment, whereas image and video content is usually disregarded. In this paper, we introduce a system for person recognition in image content of web news stored in the Internet Archive. Thus, the system complements entity recognition in text and allows researchers and analysts to track media coverage and relations of persons more precisely. Based on a deep learning face recognition approach, we suggest a system that detects persons of interest and gathers sample material, which is subsequently used to identify them in the image data of the Internet Archive. We evaluate the performance of the face recognition system on an appropriate standard benchmark dataset and demonstrate the feasibility of the approach with two use cases.

Keywords: Deep learning · Face recognition · Internet Archive
Big data application

1 Introduction

The World Wide Web with its billions of web pages and related multimedia content includes valuable information for many academic and non-academic applications. Therefore, the *Internet Archive* (www.archive.org) and national (digital) libraries have been capturing the (multimedia) web pages with time-stamped snapshots in huge archives since the mid-nineties. This serves as a playground

E. Méndez et al. (Eds.): TPDL 2018, LNCS 11057, pp. 229–240, 2018.
https://doi.org/10.1007/978-3-030-00066-0_20

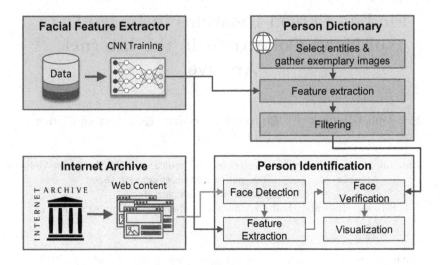

Fig. 1. Workflow of the proposed person identification framework

for researchers and analysts in different domains such as politics, economics, and entertainment. One of the main challenges is to make the available unstructured data, which is rarely enriched with appropriate metadata, accessible and explorable by the users. For this reason, it is necessary to develop (semi-)automatic content analysis approaches and systems to extract metadata that can be subsequently used for semantic search and information visualization in order to provide users with relevant information about a given topic. In recent years, many tools like *BabelNet* [17], *Dandelion* [3], and *FRED* [6] have been introduced that aim to track entities and their relations using textual information. However, we argue that text does not cover every entity in general and that image (and video) data can contain additional information. Visual and textual content can be complementary and their combination can serve as a basis for a more complete entity recognition system. However, approaches that exploit image or video data in the *Internet Archive* are rare. In this paper, we present a system (Fig. 1) that enables researchers and analysts to find and explore media coverage and relations of persons of interest in the image content of news articles in the Internet Archive for a given domain such as politics, sports and entertainment. A number of sample images is crawled for every entity using an image search engine like *Google Images*. Due to noise in the retrieved data of this web-based approach, we investigate two strategies to improve the quality of the sample dataset. A state of the art convolutional neural network (CNN) is used to learn a robust feature representation in order to describe the facial characteristics of the entities. Based on the sample dataset the trained deep learning model is applied to identify the selected entities in the image content of all German web pages in the *Internet Archive*. All required processing steps in the pipeline are designed to match the requirements of this big data application in terms of the computational efficiency. The performance of our CNN-based feature representation is evaluated on the Labeled Faces in the Wild (LFW) dataset [9]. Finally, we evaluate the performance of our system by presenting two use cases along with appropriate graphical representations. To the best of our knowledge, this is the first approach to identify entities in the *Internet Archive* using solely image data.

The remainder of the paper is organized as follows. A brief overview of related work for face recognition and entity extraction is given in Sect. 2. In Sect. 3, we introduce our deep learning system to identify persons in image content of the *Internet Archive*. Experimental results for the face recognition approach as well as some use cases are presented in Sect. 4. The paper concludes in Sect. 5 with a summary and areas of future work.

2   Related Work

In recent years a number of very powerful tools for named entity recognition like *BabelNet* [16,17], *Dandelion* [3], *FRED* [6] and *NERD* [21] have been developed. The recognition of entities depicting public personalities already achieves very good results. But especially online news articles are often provided with photos, which potentially show additional entities that are not mentioned in the text. Furthermore, possible disambiguations could be resolved using the visual content. For this purpose, face recognition approaches can be applied to predict persons in the images. Face recognition has been a well studied computer vision task for decades and the performance significantly improved since convolutional neural networks [10] as well as huge public data collections like *CASIA-WebFace* [24] and Microsoft-Celebrity-1M (*MS-Celeb-1M*) [7] have been introduced. DeepFace [20] has been one of the first CNN-based approaches that treat face recognition as a classification approach and subsequently uses the learned feature representation for face verification. While in general face recognition approaches based on deep neural networks benefit from learning a robust face representation, new loss functions like the contrastive loss [19], triplet loss [18] and angular softmax [11] have been introduced to enhance the discriminative power. To improve the robustness against pose variation some approaches [12–15,25,26] aim to frontalize the face using 3D-head models or synthesize new views of the face to augment the training dataset with all available poses, respectively. Another widely used technique to increase the robustness to poses and occlusions are approaches [5,23] that use several image patches around facial landmarks as input for the CNN network training. Finally, other approaches [2,22] are suggested to overcome variations due to the aging of faces.

3   Person Identification in Archived Web News

In this section, a system for the identification of interesting persons in images of archived web news is introduced. First, a CNN is trained to learn a robust representation for faces (Sect. 3.1). In Sect. 3.2, we describe a way to define a lexicon of persons and to automatically gather sample images from the Web for them, i.e., to build an entity dictionary for a given domain like politics or entertainment. In this context, we explain how to reduce noise in the sample dataset which is caused by the web-based approach. The proposed framework retrieves image data from the *Internet Archive* according to a predefined search space (Sect. 3.3). Section 3.4 presents the complete workflow for identifying persons in images of the *Internet Archive* based on the predefined dictionary. Furthermore, single and joint occurrences of persons in images are explored and visualized. The workflow of the approach is illustrated in Fig. 1.

3.1   Learning a Feature Representation for Faces

A CNN is trained to learn a reliable representation for person identification in the subsequent steps. Given a dataset of face images such as *MS-Celeb-1M* [7] or *CASIA-WebFace* [24] covering n individual persons, a model with n classes is trained for classification. During training the cross-entropy loss is minimized given the probability distribution C for the output neurons and the one-hot vector $\hat{C}$ for the ground-truth class:

$$E(C, \hat{C}) = -\sum_i \hat{C}_i log(C_i). \tag{1}$$

Removing the fully-connected layer that assigns probabilities to the predefined classes of faces transforms the model to a generalized feature extractor. Thus, for a query image the model outputs a compact vector of facial features. In this way a query image can be compared with the facial features of entities in the predefined dictionary, which is presented in the next section.

3.2   Creating a Dictionary of Persons for a Domain

First, the necessary steps to automatically define entities and gather sample images for them from the web are explained. Second, the process of defining a compact representation for every entity is described. In this context, two strategies for filtering inappropriate facial features are introduced and discussed.

Defining Entities and Gathering Sample Images: In order to automatically define relevant persons of a given domain of interest, a knowledge base such as the *Wikipedia* encyclopedia is queried for persons associated with the target group, e.g., politicians. To retrieve the most relevant entities for the selected group, the query is further constrained to persons whose pages were viewed most frequently in a given year and who were born after 1920. Sample images for the selected persons are retrieved in an unsupervised manner employing a web-based approach. Given the names of the selected entities, an image search engine such as *Google Images* is crawled to find a given number of k sample images for each person. However, the collected images do not necessarily always or only depict the target person but involve some level of noise which should be eliminated in the following steps.

Extraction and Filtering of Feature Vectors: To extract face regions in an image, we use *dlib* face detector based on the *histogram of oriented gradients* (HOG) [4]. Though not able to detect extreme facial poses, this face detector ensures efficiency in terms of computational speed when it comes to the large-scale image data of news pages gathered by the *Internet Archive* (Sect. 3.4). For each detected face i associated to person p the feature vector $f_i \in F_p, i = 1, \ldots, |F_p|$ is computed using the CNN model (Sect. 3.1). Since the detected faces can depict the target person as well as other individuals in the corresponding image or entirely different persons, a data cleansing step on the extracted facial features F_p is conducted. The cosine similarity of each feature vector f_i to a target feature vector f_t representing the individual p is computed. For the choice of the target feature vector, we propose (1) using the mean of all feature vectors in F_p, or (2) selecting one representative example vector among those in F_p. Facial feature vectors that yield a similarity value smaller than a given cleansing threshold λ_c are removed. Choosing the mean feature vector is advantageous in the sense that it does not require supervision, unlike the manual selection of exemplary vectors for the dictionary entries. On the other hand, the average vector may not be meaningful if the images collected from the web contain a lot of noise. In addition, the selection of exemplary face vectors in a supervised manner unambiguously represents the target entities and thus ensures a more robust filtering of false positives. The evaluation of the proposed vector choices for filtering as well as the choice of threshold λ_c are discussed in Sect. 4.

Definition of the Final Dictionary: After the filtering step is applied, the set of the remaining facial features F_p is represented by the mean vector:

$$\overline{f_p} = \frac{1}{|F_p|} \sum_{i=1}^{|F_p|} f_i. \tag{2}$$

This is computationally advantageous for the subsequent steps in terms of numbers of comparisons. For the resulting dictionary $D_P = \{\overline{f_1}, \ldots, \overline{f_{|P|}}\}$ only one computation per person is required to verify if a query depicts the given entity, instead of comparing against the entire set of feature vectors describing an individual entity.

3.3   Retrieving Images from the Internet Archive

The *Internet Archive* contains an enormous amount of multimedia data that can be used to reveal dependencies between entities in various fields. Looking only at the collection of web pages, a large part of the multimedia content is irrelevant for person search, e.g., shopping websites. For this reason, we aim at selecting useful and interesting domains in which the entities from the dictionary are depicted. In particular, we retrieve image data of web pages such as the German domain welt.de for political subjects. Furthermore, the amount of images is restricted to image formats as JPEG and PNG, excluding formats like GIF can reduce the amount of possible spam. Another useful criterion for restricting the image search space is the publication date, which enables the exploration of events of a certain year.

3.4 Person Identification Pipeline

Using the components introduced in the preceding sections, persons in the image data of the *Internet Archive* can be automatically identified. First, a HOG-based face detector [4] extracts all available faces in the set of retrieved images of the *Internet Archive*. Based on the CNN described in Sect. 3.1, feature vectors for the faces are computed. Subsequently, each query vector is compared against the dictionary of persons (Sect. 3.2). The cosine similarity of a query vector to each entity vector is computed in order to determine the most similar person in the dictionary. Given the similarity value, the identification threshold λ_{id} determines whether the queried entity is part of the person dictionary, or an unknown person is depicted. Based on the results of person identification, visualizations based on single and joint occurrences of persons of interest in the selected *Internet Archive* data can be created.

4 Experimental Setup and Results

In this section, we evaluate the components of our person identification framework. We present details of the technical realization as well as experimental results on the learned face representation (Sect. 4.1) and the dictionary of persons (Sect. 4.2). Without loss of generality, the feasibility of our system is demonstrated on image data of the *Internet Archive* concentrating on a selection of German web content (Sect. 4.3). Finally, visualizations for relations among the persons of interest in the selected data are shown.

4.1 CNN for Facial Feature Extraction

Training of the CNN Model: We use the face images of MS-Celeb-1M [7] as training data for our CNN. Comprising 8.5M images of around 100K different persons, it is the largest publicly available dataset. A classification model is trained using the *ResNet* architecture [8] with 101 convolutional layers. The weights are initialized by a pre-trained ImageNet model. Furthermore, we augment the data by randomly selecting a region covering at least 70% of the image. The input images are then randomly flipped and cropped to 224 × 224 pixels. Similar to *DeepFace* [20], Stochastic Gradient Descend (SGD) is used with a momentum of 0.9 and an initial learning rate of 0.01, which is exponentially lowered by a factor of 0.5 after every 100,000 iterations. The model is trained for 500,000 iterations with a batch size of 64. Using two Nvidia Titan X graphics cards with 12 GB VRAM each, the training takes around 4 days. The implementation is realized using the TensorFlow library [1] in Python. The trained model is available at: https://github.com/TIB-Visual-Analytics/PIIA.

Evaluation of the CNN Model: The trained model is evaluated on the well-known LFW benchmarking set [9] in the verification task. Therefore, the similarity between two feature vectors using the cosine distance is measured. As suggested by the authors of LFW we perform a 10-fold cross validation in our experiments, where each fold consists of respectively 300 matched and mismatched face pairs of the test set. For each subset, the best threshold maximizing the accuracy on the remaining 9 subsets is calculated. Thus, the validation accuracy as well as threshold values are averaged for the 10 folds. The trained model obtains an accuracy of 98.0% with a threshold set to 0.757. Compared to the much more complex systems achieving state of the art results on the LFW benchmark[1], our model yields a satisfactory accuracy using a basis architecture as well as loss function and provides a good basis for the face verification step in our pipeline. Moreover, the estimated threshold has a standard deviation of 0.002. This demonstrates that the threshold value, indicating whether a query image depicts the same entity as a reference image, is very stable for the variety of input faces. In the following, the cleansing threshold for comparing a target vector against dictionary entity vectors for filtering (Sect. 3.2) is assigned to $\lambda_c = 0.757$.

4.2   Creating a Dictionary of Persons

Selecting Entities and Gathering Image Samples: The goal in our experimental setup is to recognize persons in the German web content of the *Internet Archive* and visually infer relations among them. Hence, people of public interest have to be selected for the dictionary. We exemplarily choose the groups of *politicians* and *actors*, for each of whom we create a dictionary according to the description in Sect. 3.2. In order to find relevant people in German media, we explicitly query the German *Wikipedia* for persons according to the selected occupations and further criteria specified in Sect. 3.2. The entity names are fetched via SPARQL queries to the Wikidata repository. Additionally, the relevance of an entity is determined by the number of page views. Thus, the person dictionary is limited to the $|P| = 100$ most frequently viewed entities[2]. Since Wikidata provides page views from mid 2015, we fetch the numbers for the year 2016. Given the sets of persons for the selected occupational groups, we crawl the *Google Images* search engine for a maximum of $k = 100$ images per entity.

Evaluating the Methods for Feature Vector Cleansing: In Sect. 3.2 two methods for selecting a target vector for filtering entity vectors are introduced. Using the cleansing threshold λ_c which was estimated on the LFW benchmark, we separately filter the entity vectors according to the average entity vector and a manually chosen reference vector. The methods are evaluated on an annotated subset of 1100 facial images covering 20 politicians of the dictionary.

Table 1 reports filtering results for both strategies in comparison to the unfiltered test set. As shown, the use of the mean vector boosts precision from initially

---

[1] Results can be found on: http://vis-www.cs.umass.edu/lfw/results.html.
[2] The entity list can be found at: https://github.com/TIB-Visual-Analytics/PIIA.

Table 1. Results of methods for the cleansing step of the entity dictionary on a subset of 20 politicians.

Method	Precision	Recall	F_1
No filtering	0.669	1	0.802
Mean vector	0.993	0.449	0.618
Reference vector	0.977	0.922	0.949

Table 2. Number of images and faces in news articles of the selected domains published in 2013.

Domain	Images	Faces
welt.de	648,106	205,386
bild.de	566,131	243,343

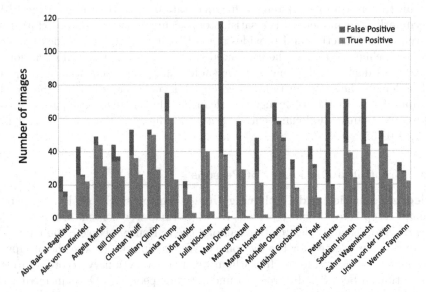

Fig. 2. Composition of true and false positives according to filtering methods on given entity test set. Grouped bars denote (a) unfiltered entity images, (b) filtering with a reference vector and (c) filtering according to the mean vector.

0.669 to 0.993, but recall is reduced to 0.449. As already hypothesized in Sect. 3.2, this is due to the noise in the exemplary images caused by the web-based approach and thus the strong distortion of the mean vector. Figure 2 illustrates that the average vector drastically discards false as well as true positives making this strategy impractical for our purposes. In comparison, the manually selected vector boosts recall to 0.922 and thus significantly reduces the false positive rate of entity images (see also Fig. 2). For most depicted entities such as *Malu Dreyer* or *Angela Merkel* almost every false positive is filtered out while the correct images are maintained. The method yields a slightly smaller precision of 0.977 and requires supervision, which we take into account for the subsequent steps due to the high F_1 score.

Evaluating a Global Threshold for Face Verification: After noisy vectors are filtered out, each entity is described by its mean vector. The use of a mean vector is plausible for our framework since we do not detect faces in extreme poses. Since the global face verification step for query images is carried out with the average of all vectors of a entity, a new identification threshold λ_{id} value different to λ_c has to be estimated. For this reason, a cross-fold validation in the same way as in Sect. 4.1 is performed based on the subset of politicians used for the evaluation of the dictionary cleansing. An accuracy of 96% is obtained. The threshold results in $\lambda_{id} = 0.833$ and shows a standard deviation of 0.002. In particular, the very small standard deviation implies that the use of the mean entity vector works very stable for the face verification task of our framework.

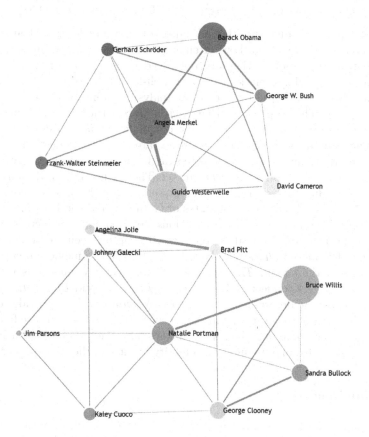

Fig. 3. Graph showing relations among an exemplary set of international politicians (top) and actors (bottom) using the domain welt.de and bild.de, respectively. The size of vertices encodes the occurrence frequency of the entity. The strength of edges denotes the frequency of joint occurrences.

4.3   Face Recognition in Image Collections of the *Internet Archive*

Selection of Image Data: We demonstrate our framework upon selected web content of the two German news websites welt.de and bild.de in the *Internet Archive*. The number of images and faces are shown in Table 2. While the former addresses political subjects, bild.de has a stronger focus on entertainment news as well as celebrity gossip. Therefore, we separately exploit welt.de for identifying politicians and bild.de for identifying the selected actors. We select image data of the year 2013, in which the German elections took place. Please note the minor offset compared to the selected entities using statistics of *Wikipedia* from 2016 (Sect. 4.2). However, the persons are still identifiable and relevant.

Visualization: To quantify the relevance of individuals and their relation to other entities, we count how often single entities appear in the selected image data and how often they are portrayed with persons of the dictionary. Figure 3 (top) visualizes relations between well-known heads of states and other politicians in 2013 inferred by our visual analysis system for the German news website welt.de. The graph shows that *Angela Merkel*, the German chancellor, as well as the former German minister of foreign affairs, *Guido Westerwelle*, appear most frequently in the image data and also share a strong connection. The most relevant international politician detected in the news images is *Barack Obama* with a strong connection to *Angela Merkel*. The connection of *Westerwelle* to *Steinmeier* is due to the transition of power in late 2013. Also, connections between former and new heads of states of Germany and the USA exist.

Figure 3 (bottom) visualizes connections between different actors in 2013. For example, the graph reveals strong connections between the actors *George Clooney* and *Sandra Bullock* who have both acted in the movie *Gravity*. More-over, actors of the sitcom *The Big Bang Theory* share connections with each other. Also a strong connection between *Angelina Jolie* and *Brad Pitt*, a famous actor couple can be determined. The actress *Natalie Portman* provides connections to all actors of the graph having the second strongest appearance frequency. This implies that there must be several images published in bild.de which depict her with colleagues, maybe due to a celebrity event like the Academy Awards.

5   Conclusions

In this paper, we have presented a system for the identification of persons of interest in image content of web news in the *Internet Archive*. For this task, a CNN-based feature representation for faces was trained and evaluated on the standard LFW benchmark set. Moreover, we introduced a semi-automatic web-based method for creating a dictionary of persons of interest for a given domain. In addition, two methods for filtering inappropriate images in the sample data were introduced and evaluated. In order to cope with the enormous amount of image content the *Internet Archive* provides, a constrained search domain was defined. The proposed system reliably detects dictionary entities and reveals relations between the entities by means of joint occurrences. In order to process the huge amount of image data, the system is realized using efficient and well scalable solutions.

In the future, we plan to further improve individual steps of the pipeline. In particular, we aim to improve our deep learning model using a more sophisticated loss function like the triplet loss [18] or preprocessing for more robustness against pose variation and aging. In addition, a face detector will be used that deals with arbitrary poses. The process of determining a ground truth vector can be automated by querying *Wikipedia* for a representative image of the entity. Finally, the framework will be extended to allow the exploration of relations of persons across different domains.

Acknowledgement. This work is financially supported by the German Research Foundation (DFG: Deutsche Forschungsgemeinschaft, project number: EW 134/4-1). The work was partially funded by the European Commission for the ERC Advanced Grant ALEXANDRIA (No. 339233, Wolfgang Nejdl).

References

1. Abadi, M., et al.: Tensorflow: large-scale machine learning on heterogeneous distributed systems. CoRR abs/1603.04467 (2016)
2. Best-Rowden, L., Jain, A.K.: Longitudinal study of automatic face recognition. Trans. Pattern Anal. Mach. Intell. **40**, 148–162 (2018)
3. Brambilla, M., Ceri, S., Della Valle, E., Volonterio, R., Acero Salazar, F.X.: Extracting emerging knowledge from social media. In: International Conference on World Wide Web, pp. 795–804. IW3C2 (2017)
4. Dalal, N., Triggs, B.: Histograms of oriented gradients for human detection. In: Conference on Computer Vision and Pattern Recognition, pp. 886–893. IEEE (2005)
5. Ding, C., Tao, D.: Trunk-branch ensemble convolutional neural networks for video-based face recognition. Trans. Pattern Anal. Mach. Intell. **40**, 1002–1014 (2017)
6. Gangemi, A., Presutti, V., Reforgiato Recupero, D., Nuzzolese, A.G., Draicchio, F., Mongiovì, M.: Semantic web machine reading with FRED. Semant. Web **8**(6), 873–893 (2017)
7. Guo, Y., Zhang, L., Hu, Y., He, X., Gao, J.: MS-Celeb-1M: a dataset and benchmark for large-scale face recognition. In: Leibe, B., Matas, J., Sebe, N., Welling, M. (eds.) ECCV 2016. LNCS, vol. 9907, pp. 87–102. Springer, Cham (2016). https://doi.org/10.1007/978-3-319-46487-9_6
8. He, K., Zhang, X., Ren, S., Sun, J.: Deep residual learning for image recognition. In: Conference on Computer Vision and Pattern Recognition, pp. 770–778. IEEE (2016)
9. Huang, G.B., Ramesh, M., Berg, T., Learned-Miller, E.: Labeled faces in the wild: a database for studying face recognition in unconstrained environments. Technical report 07–49, University of Massachusetts, Amherst (2007)
10. Krizhevsky, A., Sutskever, I., Hinton, G.E.: Imagenet classification with deep convolutional neural networks. In: Advances in Neural Information Processing Systems, pp. 1097–1105. NIPS (2012)
11. Liu, W., Wen, Y., Yu, Z., Li, M., Raj, B., Song, L.: Sphereface: deep hypersphere embedding for face recognition. In: Conference on Computer Vision and Pattern Recognition, vol. 1. IEEE (2017)

12. Masi, I., et al.: Learning pose-aware models for pose-invariant face recognition in the wild. Trans. Pattern Anal. Mach. Intell. (2018)

13. Masi, I., Hassner, T., Tran, A.T., Medioni, G.: Rapid synthesis of massive face sets for improved face recognition. In: International Conference on Automatic Face & Gesture Recognition, pp. 604–611. IEEE (2017)

14. Masi, I., Rawls, S., Medioni, G., Natarajan, P.: Pose-aware face recognition in the wild. In: Conference on Computer Vision and Pattern Recognition, pp. 4838–4846. IEEE (2016)

15. Masi, I., Tran, A.T., Hassner, T., Leksut, J.T., Medioni, G.: Do we really need to collect millions of faces for effective face recognition? In: Leibe, B., Matas, J., Sebe, N., Welling, M. (eds.) ECCV 2016. LNCS, vol. 9909, pp. 579–596. Springer, Cham (2016). https://doi.org/10.1007/978-3-319-46454-1_35

16. Moro, A., Raganato, A., Navigli, R.: Entity linking meets word sense disambiguation: a unified approach. Trans. Assoc. Comput. Linguist. **2**, 231–244 (2014)

17. Navigli, R., Ponzetto, S.P.: BabelNet: the automatic construction, evaluation and application of a wide-coverage multilingual semantic network. Artif. Intell. **193**, 217–250 (2012)

18. Schroff, F., Kalenichenko, D., Philbin, J.: FaceNet: a unified embedding for face recognition and clustering. In: Conference on Computer Vision and Pattern Recognition, pp. 815–823. IEEE (2015)

19. Sun, Y., Wang, X., Tang, X.: Deep learning face representation from predicting 10,000 classes. In: Conference on Computer Vision and Pattern Recognition, pp. 1891–1898. IEEE (2014)

20. Taigman, Y., Yang, M., Ranzato, M., Wolf, L.: DeepFace: closing the gap to human-level performance in face verification. In: Conference on Computer Vision and Pattern Recognition, pp. 1701–1708. IEEE (2014)

21. Van Erp, M., Rizzo, G., Troncy, R.: Learning with the web: Spotting named entities on the intersection of NERD and machine learning. In: Workshop on Making Sense of Microposts, pp. 27–30 (2013)

22. Wen, Y., Li, Z., Qiao, Y.: Latent factor guided convolutional neural networks for age-invariant face recognition. In: Conference on Computer Vision and Pattern Recognition, pp. 4893–4901. IEEE (2016)

23. Yang, S., Luo, P., Loy, C.C., Tang, X.: From facial parts responses to face detection: a deep learning approach. In: International Conference on Computer Vision, pp. 3676–3684. IEEE (2015)

24. Yi, D., Lei, Z., Liao, S., Li, S.Z.: Learning face representation from scratch. CoRR abs/1411.7923 (2014)

25. Yin, X., Yu, X., Sohn, K., Liu, X., Chandraker, M.: Towards large-pose face frontalization in the wild. CoRR abs/1704.06244 (2017)

26. Zhu, X., Lei, Z., Liu, X., Shi, H., Li, S.Z.: Face alignment across large poses: a 3D solution. In: Conference on Computer Vision and Pattern Recognition, pp. 146–155. IEEE (2016)

Ontology-Driven Information Extraction from Research Publications

Vayianos Pertsas[1(✉)] and Panos Constantopoulos[1,2]

[1] Department of Informatics, Athens University of Economics and Business, Athens, Greece
{vpertsas,panosc}@aueb.gr
[2] Digital Curation Unit, Athena Research Centre, Athens, Greece

Abstract. Extraction of information from a research article, association with other sources and inference of new knowledge is a challenging task that has not yet been entirely addressed. We present Research Spotlight, a system that leverages existing information from DBpedia, retrieves articles from repositories, extracts and interrelates various kinds of named and non-named entities by exploiting article metadata, the structure of text as well as syntactic, lexical and semantic constraints, and populates a knowledge base in the form of RDF triples. An ontology designed to represent scholarly practices is driving the whole process. The system is evaluated through two experiments that measure the overall accuracy in terms of token- and entity- based precision, recall and F1 scores, as well as entity boundary detection, with promising results.

Keywords: Information extraction from text · Ontology population
Linked data · Knowledge base creation

1 Introduction

Extracting and encoding the knowledge contained in a research article is a multi-dimensional challenge. For instance, detecting who has done what, their interests and goals, affiliations, etc., requires extracting, analyzing and mapping onto an appropriate schema information from the metadata of the article. Also, several kinds of named entities need to be recognized (e.g. method employed in an experiment) and linked to other relevant information. Established named entity recognizers offer pre-trained models that support "common" types of named entities such as: Location, Person, Organization, Money, Events, and 'miscellaneous' [1]. For "non-common" types of named entities (e.g. 'tools', 'methods'), a classifier needs to be trained using annotated corpora, specifically created by human annotators, an expensive, time consuming process. Furthermore, to capture the information contained in a publication about a scholarly activity, its context and outcomes, entities and relations of many different types have to be extracted, which differ considerably from named entities in that they extend over widely variable lengths of text, even in more than one sentences. Every possible aspect of context needs to be exploited: from surface/lexical form, to part of speech and deep syntactic role of each token in a sentence; and from discourse structure in sections and paragraphs, to the role and position of each sub-sentence in a sentence.

E. Méndez et al. (Eds.): TPDL 2018, LNCS 11057, pp. 241–253, 2018.
https://doi.org/10.1007/978-3-030-00066-0_21

Finally, the output of the above tasks needs to be aligned in a semantic framework for comparison or integration with other existing knowledge published as linked data.

In this paper we present Research Spotlight (RS), a system that extracts information from research articles, enriches it with relevant information from other Web sources, organizes it according to the Scholarly Ontology (SO) [2], and republishes it in the form of linked data. Existing information is leveraged by accessing SPARQL endpoints, scraping Web pages or through APIs. Harvested information is further used as background knowledge for training classifiers or for extracting information from semi-structured or unstructured texts. So, RS generates linked, contextualized, structured data describing research activities and their outcomes, thus addressing the growing need for integrated access to information scattered in different publications.

Knowledge bases created using RS can support researchers in finding details of relevant work without reading the articles; discovering uses of resources, processes and methods in particular contexts; promoting communities of interests; formulating future directions and project proposals. Besides, funders and research councils can get a "bird's eye view" of scholarly work useful for planning and evaluation.

2   Related Work

To the best of our knowledge the exact task of extracting information from scientific article and republishing it as Linked Data, as prescribed in this paper, has not been addressed yet. That said however, several past efforts aimed at extracting information from text based on an existing ontology: In [3] RDF triples are extracted from RSS feeds and published as Linked Open Data using mappings to DBpedia entities. The focus is on statistical methods and rules based on lexical form. Syntactic dependencies of tokens in the sentence that could allow better context understanding are not exploited. The DBpedia project itself [4] is a huge operation to automatically extract knowledge from Wikipedia pages and info-boxes involving various NLP and feature-matching extractors that create RDF triples as instances of the DBpedia ontology. Here predefined rules are based on the DBpedia schema, metadata mappings, statistics of page links or word counts, and a number of feature extractors that exploit xml/html tags. However, the lexical, syntactic or structural analysis of raw text is not supported. In [5] an ontology is used to guide the automatic creation of RDF triples from facts previously extracted from various Web pages and to publish linked data in a SESAME triple store. Here too, the methods employed exploit string features of noun phrases, distributions of text found around those phrases in other Web pages, and the HTML structure of the Web pages containing the noun phrases. In [6] a knowledge base is created with information extracted from French news wires by linking extracted entities to the instances of an ontology that unifies the models of GeoNames and Wikipedia and contains entities of type Person, Organization or Location retrieved from these sources. Common types of named entities are recognized and aligned with an existing database. In domain-specific endeavors, such as [7], an ontology is defined from fragments of CIDOC-CRM in order to describe the domain of Arts, on the basis of which knowledge is extracted from various Web pages in order to create personalized biographies of artists. Only common types of named entities are supported. In [8], a knowledge base is constructed by semi-automatic extraction of relations, based on the PRIMA ontology for risk management and a combination of machine learning techniques and predefined handcrafted rules. Syntactic dependencies that could yield patterns exploiting the deeper syntactic structure of sentences are not considered. In [9], an event ontology is used in order to guide NLP modules in extracting instances from unstructured texts in a semi-supervised manner based on shallow syntactic parsing. Finally, in [10] an ontology-based information extractor employs handcrafted rules in order to extract soccer-related entities from various Web sources and map them onto soccer-specific semantic structures. The recognition of named entities is based solely on named entity lists, thus not supporting recognition of entities that are absent from the lists.

In this context, the main contributions of RS are:

- An end-to-end solution for understanding "who has done what, how, why and with what results" from the text of research articles.
- A domain-independent procedure that automatically creates annotated corpora for training named entity recognizers, especially useful for entities of "non-common" type.
- A system that leverages semantic information, surface form as well as deep syntactic and structural text analysis in order to extract information using both machine learning modules and handcrafted rules.
- A workflow that combines information from metadata and linked data with knowledge extracted from text and republishes it as a knowledge base, adhering to linked data standards.

3   Conceptual Framework: The Scholarly Ontology

The conceptual model underlying RS is based on the Scholarly Ontology (SO) [2], a domain-independent framework for modeling scholarly activities and practices. The rationale behind SO is to support answering questions of the form "*who does what, when, and how*" in a given scholarly domain, so the ontology is built around the central notion of *activity* and combines three perspectives: the *agency* perspective, concerning actors and intentionality; the *procedure* perspective, concerning the intellectual framework and organization of work; and the *resource* perspective, concerning the material and immaterial objects consumed, used or produced in the course of activities. We here briefly review a subset of core SO concepts that constitute the RS schema guiding the extraction as well as the structuring of information (see Fig. 1).

Activity (e.g. an evaluation, a survey, an archeological excavation, a biological experiment, etc.) represents real events that have occurred in the form of intentional acts carried out by actors. Sequence of activities and composition from sub-activities are represented by the *follows* and *partOf* relations. The instances of the Activity class are real processes with specific results, as opposed to those of the *Method* class, which are specifications, or procedures for carrying out activities to address specific goals. *Actor* instances are entities capable of performing intentional acts they can be accounted or referenced for. Actors can participate in activities, actively or passively, in one or more roles. Subclasses of *Actor* are the classes *Person* and *Group* representing individual persons and collective entities respectively. Further specializations of *Group* are the classes *Organization* and *Research Team*. *Content Item* comprises information resources, regardless of their physical carrier, in human readable form (e.g. images, tables, texts, mathematical expressions, etc.). *Proposition* comprises assertions in affirmative or negative form, resulting from activities and *supportedBy* evidence provided by content items. Finally, the class *Topic* comprises thematic keywords expressing the subject of methods, the topic of content items, the research interests of actors, etc.

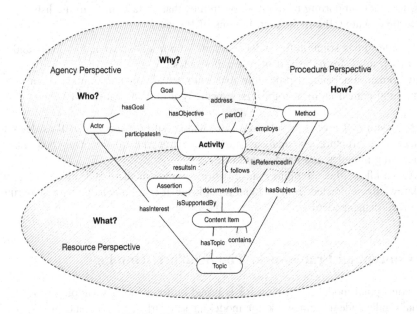

Fig. 1. The Scholarly Ontology core

4   Knowledge Base Creation

4.1   Process Overview

An overview of the knowledge creation process is given in Fig. 2. The input comprises published -open access- research articles retrieved from repositories or Web pages in the preferred html/xml format. The format is exploited in extracting the metadata of an article, such as authors' information, references and their mentions in text, legends of figures, tables etc. Entities, such as activities, methods, goals, propositions, etc., are extracted from the text of the article. These are associated in the relation extraction step, through various relations, e.g. *follows, hasPart, hasObjective, resultsIn, hasPartici-pant, hasTopic, has Affiliation*, etc. Encoded as RDF triples, these are published as linked data, using additional "meta properties", such as *owl:sameAs, owl:equiva-lentProperty, rdfs:Label, skos:altLabel*, where appropriate.

The entities targeted for extraction can be categorized into: (i) *named entities*, i.e. entities that have a proper name [1], such as instances of the SO classes: *ContentItem, Person, Organization, Method* and *Topic*; and (ii) *nameless*, or *non-named entities*, identified by their own description but not given a proper name, such as instances of SO classes *Activity, Goal* and *Proposition*.

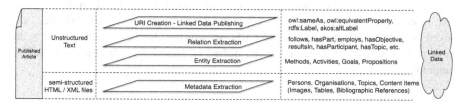

Fig. 2. Knowledge base creation. Left to right: input, processes, extracted entities and relations

Different modules handle entities of each category. Figure 3 shows the architecture of RS implementing the above process. The inputs of the system are:

(i)   SPARQL endpoints of various Web sources for creating Named Entities (NE) lists;
(ii) user search keywords indicating the type of named entity to be recognized; and
(iii) URLs (e.g. journal Web pages that can be scraped) or publishers' APIs.

The main output of the system is the knowledge base published as linked data. The knowledge base creation process consists of two phases: (1) Preprocessing, for creating named entities lists and training the NER classifier and (2) Main Processing for the actual information extraction and publishing.

In *Preprocessing*, information is retrieved from sources such as DBpedia in order to build lists of named entities through the *NE List Creation* module. Specific queries using these entities are then submitted to the sources via the *API Querying* module. Retrieved articles are processed by the *Text Cleaning* module and the raw text at the output is added to a training corpus through the *Automatic Annotation* module that uses the entries of *NE list* to spot named entities in the text. The annotated texts are used to train a classifier to recognize the desired type of named entities. For details regarding the pre-processing phase, see Sect. 4.2.

Main Processing begins with harvesting research articles from Web sources, either using their APIs or by scraping publication Web sites. The articles are scanned for metadata which are mapped to SO instances according to a set of rules. In addition, specific html/xml tags inside the articles indicating images, tables and references are extracted and associated with appropriate entities according to SO, while the rest of the unstructured, "raw" text is cleaned and segmented into sentences by the *Text Cleaning & Segmentation* module (Sect. 4.3). The unstructured, "raw" text of the article is then input into the *Named Entity Recognition* module, where named entities of specific types are recognized. The segmented text is also inserted into a dependency parser using the *Syntactic Analysis* module. The output consists of annotated text -in the form of dependency trees based on the internal syntax of each sentence- which is further processed by the *Non-Named Entities Extraction* module, so that text segments that contain other entities (such as Activities, Goals or Propositions) can be extracted (Sect. 4.4). The output of the above steps (named entities, non-named entities and metadata) is fed into the *Relation Extraction* module that uses four kinds of rules (Sect. 4.5): (i) syntactic patterns based on outputs of the dependency parser; (ii) surface form of words and POS tagging; (iii) semantic rules derived from SO; (iv) proximity constraints capturing structural idiosyncrasies of texts. Finally, based on the information extracted in the previous steps, URIs for the SO namespace are generated, and linked -when possible- to other strong URIs (such as the DBpedia entities stored in the named entities lists) in order to be published as linked data through a SPARQL endpoint.

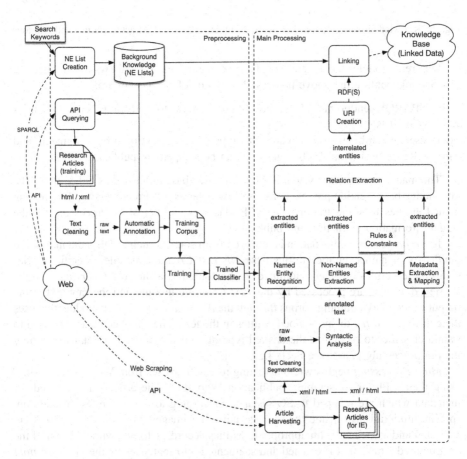

Fig. 3. Research Spotlight - system architecture

4.2 Preprocessing

In the Preprocessing phase (see Fig. 3), information is gathered from external sources in order to create a substantial amount of "background" knowledge. We currently use DBpedia, but other sources can be used as well. The use of background knowledge is twofold: (i) provide instances of the *Method* and *Topic* classes; and (ii) distant supervision for the creation of training data for named entity recognition (NER).

By querying DBpedia we create two NE lists, one for *Topics* and one for *Methods*. Research Methods is a named entity type not supported by existing NER models (they usually support common types of named entities such as: persons, organizations, locations, events and "miscellaneous"). We use the entries from the Methods List, for distant supervision so that training data from retrieved research articles can be created automatically. The benefits of this process are multiple: (1) being entirely automatic, it can help create a very large (noisy) training data set; (2) it can reduce the work of human annotators to correcting the already automatically annotated corpus. Here we employ the latter approach, in order to generate a dataset for the recognition of NEs of type *Method,* but in the same context, datasets of other types of NEs (such as *Tools, Persons, Locations* etc.) could be generated.

Along this line, the Methods List is used to generate training data for NER to recognize entities of type *Method.* Through the APIs of sources such as Springer and Elsevier we retrieve research articles that have Methods List entries as topic keywords, thus maximizing the likelihood of finding named entities of those specific types in the texts. Articles are segmented into sentences, which are scanned for entities from the Methods List and annotated by the automatic annotation module, using regular expressions for the name (and variants) of each entity in the NE list.

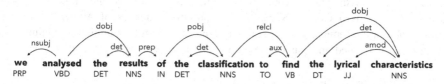

Fig. 4. Dependency tree

4.3 Metadata Extraction

By "metadata extraction" we mean the acquisition of all the structured information encoded in the article, either delivered in a separate format, such as Json, or embedded in the html/xml encoding of the document. Retrieved articles are parsed and entities of type *Person, Organization, Article* (subclass of Content Item) and *Topic* extracted from the xml tags. ORCID[1] is integrated through its API, so it can be used for duplicate

---

[1] https://orcid.org/.

detection and additional information. The html/xml encoding of the article is parsed using Beautiful Soup[2] to extract information about figures, tables and references. After extracting information from the html/xml encoding of the article, the *Text Cleaning* module is used to remove all the html/xml tags and the raw text is segmented into sentences and stored along with paragraph and section indicators.

4.4   Entity Extraction

Apart from "named entities" that can be identified using a NER (i.e. instances of *Method* class), we also need to extract "non-named" entities of highly variable length. Textual chunks indicating *Activities*, *Goals* and *Propositions* are detected using syntactic analysis in conjunction with rules that exploit lexico-syntactic patterns derived from the reasoning frame of SO [2]. A dependency tree containing POS tags and syntactic dependencies for each word in a sentence is obtained using Spacy[3]. Each sentence is further analyzed using the semantic definitions of SO classes, the surface form of words, their POS tag and their syntactic dependencies.

A sentence with verb in past or past/present perfect tense -in active or passive voice- containing no markers such as 'if' or 'that', quite likely describes an *Activity*, assuming the subject has the correct surface form ('we' or 'I' depending on the number of authors for active voice, no personal pronouns or determiners -to exclude vague subjects- for passive voice). Besides, 'that' following a verb can introduce a sub-sentence classified as *Proposition*, while a verb with dependent nodes with surface form 'to' or 'in order to' can introduce a sub-sentence classified as *Goal*. For example, consider the sentence

"We analyzed the results of the classification to find the lyrical characteristics". The syntactic analysis would yield the dependency tree of Fig. 4, from which "analysed the results of the classification" would be classified as an *Activity* and "find the lyrical characteristics" as a *Goal*. RS can detect multiple instances of *Activity*, *Goal*, or *Proposition* in the same sentence using the same rules with the addition of conjunction indicators. A detailed analysis of the employed algorithms can be found in [11].

4.5   Relation Extraction

The last step of information extraction involves detecting relations between previously extracted entities. SO semantics are employed for identifying the proper relation based on its domain and range. The organization of the text in sections and paragraphs induces proximity constraints enabling the inference of more complex, possibly inter-sentence, relations such as parthood and sequence of activities. The constraints used to identify relations are listed in Table 1. Relations marked with * are inherited from entity super-classes (Image, Table, Bib. Reference, Article from ContentItem; Person from Actor). The constraints for the *partOf* and *follows* relations (marked with **) can be relaxed in the presence of certain special indicators in the text (see Table 2).

---

[2] https://www.crummy.com/software/BeautifulSoup/.
[3] https://spacy.io/.

Parthood or sequence relations are assigned between the current and the last extracted activity either when a parthood or sequence indicator is detected, or by virtue of the relevant constraint. Figure 5 illustrates the extraction of sequence and parthood relations.

Table 1. Types of constraints per relation type

Relation Type	Semantic Constrains (derived from SO)	Proximity Constraints - $P\_C()$ (from text structure)
isSupportedBy*	Domain: Proposition Range: Image|Table|Bibl. Reference	XML/HTML pointers inside the Proposition chunk
partOf**	Domain: Activity Range: Activity	Co-occurrence with parent Activity in the same paragraph
follows**	Domain: Activity Range: Activity	Co-occurrence with last Activity in the same paragraph
contains*	Domain: Article Range: Image|Table| Bibl. Reference	Co-occurrence in the same Article
participatesIn*	Domain: Person Range: Activity	Co-occurrence in the same Article
employs	Domain: Activity Range: Method	Co-occurrence in the same sentence
resultsIn	Domain: Activity Range: Proposition	Co-occurrence in the same paragraph
hasObjective	Domain: Activity Range: Goal	Co-occurrence in the same sentence
addresses	Domain: Method Range: Goal	Co-occurrence in the same sentence
hasTopic*	Domain: Article Range: Topic	Co-occurrence in the same Article
hasSubject	Domain: Method Range: Topic	Co-occurrence in the same Article
hasInterest*	Domain: Person Range: Topic	Co-occurrence in the same Article
hasGoal*	Domain: Person Range: Goal	Co-occurrence in the same Article
isDocumentedIn*	Domain: Activity Range: Article	Co-occurrence in the same Article
isReferencedIn*	Domain: Method Range: Article	Co-occurrence in the same Article

Table 2. Sequence and Parthood indicators along with their surface forms

Sequence and parthood indicators	Surface forms
beginning_of_sequence	'first', 'initially', 'starting'
middle_of_sequence	'second', 'third', 'forth', 'fifth', 'sixth', 'then', 'afterwards', 'later', 'moreover', 'additionally', 'next'
end_of_sequence	'finally', 'concluding', 'lastly', 'last'
parthood_indicators	'specifically', 'first', 'concretely', 'individually', 'characteristically', 'explicitly', 'indicatively', 'analytically'

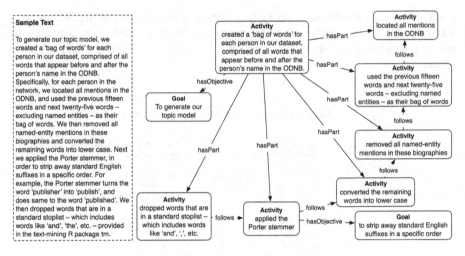

Fig. 5. Parthood and sequence relations

5   Evaluation

Metadata association exhibited very good performance since it relies solely on preconstructed mappings between fixed schemas. Few isolated incidents (less than 1%) of improper association were due to errors in xml/html tags. The Information Extraction Modules of RS were evaluated by comparing their output with a "gold standard" produced by human annotators. According to established practice [12–14], we generated the confusion matrices by comparing the output of the system with that of the human annotators and, using micro and macro-averaging, we calculated the precision, recall and F1 scores. We conducted two evaluation experiments: one "strict" and one "lenient", in which the confusion matrices were created based on "per-entity" and "per-token" calculations respectively.

5.1   Evaluation Experiments

Regarding **non-named entities** and their relations, our "gold standard" consisted of corpora produced from 50 articles annotated by two researchers. We drew from 29 different journals from various research areas (Digital Humanities, Geology, Medicine, Bioinformatics, Biology, Computer Science, Sociology and Anthropology) to try our system with multiple writing styles. The non-named entities extracted belong to the classes *Activity*, *Goal* and *Proposition*, along with their relations *follows(act1, act2)*, *hasPart(act1, act2)*, *hasObjective(act, goal)*, *resultsIn(act, prop)*. The manual annotation process took about 3.5–4 h per article on average. Inter-annotator agreement was 83% (kappa-statistic) based on corpora of 5 articles annotated by both annotators. Annotation produced about 1700 *Activities*, 300 *Goals*, 700 *Propositions*, 1000 *follows ()*, 100 *hasPart()*, 250 *hasObjective()*, 200 *resultsIn()*.

Regarding **named entities** (instances of *Method* class) and the *employs(act, meth)* relation, the dataset was created in the pre-processing phase. A list of 12,000 methods was populated by the NE list creation module. After cleaning, the list was reduced to 7,000 method names and used to retrieve 210 articles, which were automatically annotated and then manually curated by two doctoral students. Inter-annotator agreement was 81% (kappa-statistic) based on corpora of 5 articles annotated by both annotators. Annotation gave about 3,800 *Methods* and 400 *employs( )* relations. For the experiments we used the Stanford NE[4] recognizer, trained/evaluated in the above dataset.

Adopting the framework of [13], we designed two experiments yielding two different confusion matrices in order to conduct one token-based evaluation – where token is defined as any non-empty sequence of characters -, and one entity-based evaluation – where entity is defined as any non-empty sequence of tokens. To be correct, the prediction of the system must exactly equal the answer produced by humans in entity-based evaluation, whereas in token-based evaluation it should only overlap to at least a certain extent. The purpose of the second evaluation is to measure the performance of boundary detection of the system. In our experiment, after testing, the threshold for overlap was set to 86%, a difference of 1–5 tokens in most cases. In order to avoid promoting large entities in token-based evaluation, scores were calculated based on the relative distance to the perfect match, while penalties for remaining extra and missing tokens were assigned proportionally. The micro- and macro-averaged precision, recall and F1 scores based on confusion matrices from entity and token-based evaluation experiments, and individual scores for each type of entities and relations, are displayed in Tables 3, 4 and 5 respectively.

Table 3. Macro & micro averaging scores

	Macro-averaging			Micro-averaging		
	Precision	Recall	F1	Precision	Recall	F1
Entity-based	0.67	0.68	0.68	0.70	0.74	0.72
Token-based	0.87	0.77	0.81	0.84	0.83	0.83

Table 4. Entity extraction

Entity type	Entity-based			Token-based		
	P	R	F1	P	R	F1
Activity	0.70	0.75	0.72	0.79	0.85	0.81
Goal	0.74	0.78	0.76	0.86	0.74	0.80
Proposition	0.76	0.78	0.76	0.82	0.84	0.82
Method	0.80	0.69	0.74	0.75	0.72	0.73

Table 5. Relation extraction

Relation type	P	R	FI
follows	0.69	0.72	0.71
hasPart	0.57	0.54	0.55
hasObjective	0.79	0.78	0.78
resultsIn	0.54	0.58	0.56
employs	0.87	0.92	0.90

---

[4] https://nlp.stanford.edu/software/CRF-NER.html.

6 Discussion

The system performs adequately with F1 scores between 0.68 (lowest overall performance) and 0.83 (highest overall performance). Differences between the micro- and macro-averaged values are expected due to the way these measures are calculated. Because the F1 measure is mostly determined by the number of true positives, micro-averaging exposes effects related to large classes *(Activity, Method, Proposition, follows, employs)*, while macro-averaging does so for small ones *(Goals, hasPart, hasObjective, resultsIn)*. Token-based evaluation gives better scores since it is based on per-token comparison of system-extracted entities with human-extracted ones, thus constitutes a lenient but "closer to the real case" comparison.

The high increase in score values of the *Activity* class (increase of 11%) suggests average boundary detection. Analysis showed this to depend strongly on the complexity of the sentence. Regarding the *Proposition* class, a major source of errors was the rendering of a proposition as a statement in a separate phrase without introduction from an adverbial modifier (e.g. using "that..."). Regarding the NER module for the recognition of *Methods*, analysis showed that the majority of errors were caused by NEs with surface form that contains more than two words or punctuation marks. Regarding the extracted relations, the big difference in performance between the *employs()* relation and the rest can be attributed to the fact that the first involves mainly entities in the same sentence. When the domain and range of a relation were in different sentences or even paragraphs (e.g. *hasPart()*) relation extraction did poorly.

Regarding the entire RS workflow of KB creation, based on our measurements, information extracted from 50 articles –according to the semantics of SO, described in this paper- translates roughly to 100.000 triples, this of course being highly dependent on the writing style and the discipline. Indicative running times (intel i7, 16 GB RAM) for the entire process are approx. 120 secs/paper.

7 Conclusion

RS leverages an ontology of research practices and deep syntactic analysis to extract information from articles and populate a knowledge base published as linked data. RS acquires information from the Web in several ways (API integration, scrapping). Classifiers are automatically trained to recognize named entities of "non-common" type (e.g. research methods) not supported by current serialized models. Using these together with the knowledge captured in the Scholarly Ontology and deep syntactic text analysis, the system achieves extracting entities and relations representing research processes at a level of detail and complexity not addressed before. Future work includes improving recall addressing other types of entities (e.g. tools used in activities), and other types of rhetorical arguments stated in the text, thus improving overall coverage.

References

1. Jurafsky, D., Martin, J.H.: Speech and language processing - an introduction to natural language processing, computational linguistics, and speech recognition (2017)
2. Pertsas, V., Constantopoulos, P.: Scholarly ontology: modelling scholarly practices. Int. J. Digit. Libr. **18**, 173–190 (2017). https://doi.org/10.1007/s00799-016-0169-3
3. Gerber, D., Hellmann, S., Bühmann, L., Soru, T., Usbeck, R., Ngonga Ngomo, A.-C.: Real-time RDF extraction from unstructured data streams. In: Alani, H., et al. (eds.) ISWC 2013. LNCS, vol. 8218, pp. 135–150. Springer, Heidelberg (2013). https://doi.org/10.1007/978-3-642-41335-3_9
4. Lehmann, J., et al.: DBpedia - a large-scale, multilingual knowledge base extracted from Wikipedia. Semant. Web **6**, 167–195 (2015). https://doi.org/10.3233/SW-140134
5. Zimmermann, A., Gravier, C., Subercaze, J., Cruzille, Q.: Nell2RDF: read the web, and turn it into RDF. In: CEUR Workshop Proceedings, pp. 1–7 (2013)
6. Stern, R., Sagot, B.: Population of a knowledge base for news metadata from unstructured text and web data. In: AKBC-WEKEX 2012, pp. 35–40, Montreal, Canada (2012)
7. Alani, H., et al.: Automatic ontology-based knowledge extraction from web documents. IEEE Intell. Syst. **18**, 14–21 (2003)
8. Makki, J., Alquier, A.-M., Prince, V.: Ontology population via NLP techniques in risk management. Int. J. Humanit. Soc. Sci. **3**, 212–217 (2008)
9. Celjuska, D., Vargas-Vera, M.: Ontosophie: a semi-automatic system for ontology population from text. In: ICON 2004 (2004)
10. Buitelaar, P., Cimiano, P., Frank, A., Hartung, M., Racioppa, S.: Ontology-based information extraction and integration from heterogeneous data sources. Int. J. Hum.-Comput. Stud. **66**, 759–788 (2008). https://doi.org/10.1016/j.ijhcs.2008.07.007
11. Pertsas, V.: Modeling and extracting research processes. Athens University of Economics and Business, Athens (2018)
12. Manning, C.D., Raghavan, P., Schutze, H.: Introduction to Information Retrieval. Cambridge University Press, Cambridge (2008)
13. De Sitter, A., Calders, T., Daelemans, W.: A formal framework for evaluation of information extraction, University of Antwerp (2004)
14. Maynard, D., Peters, W., Li, Y.: Metrics for evaluation of ontology based information extraction. In: WWW 2006 Workshop on Evaluation of Ontologies for the Web (2006)

Information Retrieval

Scientific Claims Characterization for Claim-Based Analysis in Digital Libraries

José María González Pinto$^{(\boxtimes)}$ (ID) and Wolf-Tilo Balke (ID)

Institut für Informationssysteme,
Mühlenpfordstrasse 23, 28106 Brunswick, Germany
{pinto, balke}@ifis.cs.tu-bs.de

Abstract. In this paper, we promote the idea of automatic *semantic characterization* of scientific claims to explore entity-entity relationships in Digital collections. Our proposed approach aims at alleviating time-consuming analysis of query results when the information need is not just one document but an *overview* over a set of documents. With the semantic characterization, we propose to find what we called "dominant" claims and rely on two core properties: the consensual support of a claim in the light of the collection's previous knowledge as well as the authors' assertiveness of the language used when expressing it. We will discuss useful features to efficiently capture these two core properties and formalize the idea of finding "dominant" claims by relying on Pareto dominance. We demonstrate the effectiveness of our method regarding quality by a practical evaluation using a real-world document collection from the medical domain to show the potential of our approach.

Keywords: Pareto semantics · Scientific claims · Skyline

1 Introduction

With the exponential growth in the number of publications, information needs are not always easy to satisfy. Consider Julia who wants to write an overview of research findings regarding 'Ibuprofen' and 'headaches'. This type of query is widespread in scientific digital libraries. For instance, Islamaj Dogan et al. [14] analyzed one month of log data from PubMed, consisting of more than 23 million user sessions and more than 58 million user queries. Indeed, the authors found that most PubMed user's type in very few terms (3.54) and more than 30% are token pairs. In fact, more than 55% of the queries are *entity-based* queries such as disease, gene, drug, chemical substance, protein and medical procedure. Moreover, the size of the returned result sets for those queries can be rather difficult to manage: over 10K. Thus, today, Julia would have a time-consuming analysis of the results of the query to build an information space with possible relationships between the entities, find representative papers for each relation and then argue in an informed way to finally accomplish her goal. In other words, Julia is interested in writing a summary of the results of the query and not in a particular document. How can we help? In this work, we promote the idea of distinguishing between 'dominant' and 'dominated' *scientific claims* to help users that share Julia's information needs.

© Springer Nature Switzerland AG 2018
E. Méndez et al. (Eds.): TPDL 2018, LNCS 11057, pp. 257–269, 2018.
https://doi.org/10.1007/978-3-030-00066-0_22

In a nutshell, scientific claims [9–11] are sentences that contain any association between entities, where one entity has some influencing, manipulating, or even causal relationship to another entity with the additional constraint that they represent the primary contribution(s) of a scientific paper. Indeed, the use of claims as metadata can help to build the information space needed for Julia: each claim contains the relation between the entities! However, little attention is paid to enabling a *semantic characterization* of claims to ease Julia's journey.

In this paper we focus on the discovery of properties to *characterize* claims. To do so, we argue that we should focus on the following properties; (a) commitment of the authors: the certainty in the results; (b) overall agreement and disagreement between authors concerning current knowledge. Consider the following example to clarify what we mean: *"[...] we performed a preliminary non-randomized clinical trial with 10 participants and our limited data suggests that Ibuprofen can alleviate headaches [...]"*. In this example the phrase 'our limited data suggests' directly expresses a rather weak assertiveness of the sentence and – generally speaking- the sentence's structure correspond to specific stylistic and linguistic features casting doubt on the information contained. The second aspect in our running example is the expert assessment of the perceived strength of the relationship between the pair of entities judging the context for the claim given in the document (e.g., slightly weak characteristics of some clinical trial).

Our proposed methodology aims to identify what we have called "dominant" and "dominated" claims. The identification of these two types of claims can ease the creation of a summary in two complementary ways. Firstly, "dominated" claims can help to discover specific aspects of associations between entities that need the development of new hypothesis and the design of new experiments. Secondly, "dominant" claims can help to identify documents which represent aspects that are more certain regarding the association between entities.

We approach the problem of automatically annotating claims as "dominant" and "dominated" in a data-driven fashion. Therefore, we proceed as follows with four necessary steps: first, for a given pair of entities $<e_1, e_2>$ we rely on high-quality content from a digital library and retrieve a set of documents relevant to the query. Second, from this set of candidate documents, we extract all scientific claims linking the entities. Thirdly, we then proceed to operationalize the properties above to characterize each scientific claim. Then, to formalize the idea of identifying "dominant" and "dominated" claims, we found that the notion of Pareto *dominance* fits naturally. That means: given a choice between two scientific claims, with one claim being better concerning at least one property but at least equal concerning all other properties, the first claim should always be preferred over the second one (the first claim is said to 'dominate' the second claim). Thus, finally, we use this simple but powerful intuitive concept to annotate "dominant" and "dominated" claims. Our contributions are as follows:

- We introduce a novel approach to annotate scientific claims as "dominant" and "dominated" to serve as high-quality building blocks for claim-based summary analysis in Digital Libraries.

- We provide an entirely data-driven approach to operationalize the concept of "dominant" claims.
- We investigate, with evaluation in the context of a real-world digital library, the scope and limitations of our proposed approach.

2 Related Work

Our work draws motivation from the following two areas: the field of argumentation mining and from the research efforts on credibility analysis in social media.

Argumentation mining has a clear focus on modeling and extracting argument structures for different purposes. Currently, efforts to identify argument components, to find evidence for claims, and to predict arguments structures exist. In particular, the work of Habernal et al. [13] is related to our approach. Here, the convincingness of Web arguments is analyzed and the authors show that it is indeed possible to predict the convincingness for a given argument pair concerning some given topic. For this task researchers annotated a large dataset of pairs of arguments over different topics using crowdsourcing. Then the authors used different features and machine learning algorithms to perform two tasks: to determine from a given pair of arguments which one is more convincing (a classification task) and to rank them by convincingness (a regression task). Our work differs from Habernal et al. [13] in three aspects. Firstly, we target scientific digital libraries with clear-cut claims as first-class citizens. Secondly, we use machine learning algorithms only as part of our pipeline but then rely on the semantics of the skyline operator that captures the intuition of "dominance" more naturally. Moreover, we aim at using all the claims to annotate them as "dominant" or "dominated" and let the user use this semantic filter instead of discarding more convincing claims for argumentation. Also related to our work is the ongoing effort to realize argumentation machines of IBM Watson's Debater [29]. Currently, IBM's system relies on handcrafted argumentation structures created by expert users, see Lippi et al. [19].

Credibility Analysis. Efforts that account for the credibility of online communities are also relevant to our work. For example, the work of Mukherjee et al. [23] uses a probabilistic graphical model to account for user trustworthiness, language objectivity and credibility of postings in online communities. In fact, we build on the idea of using lexicons that are related to bias in language from their work. In a similar line of thought in Mukherjee et al. [22], researchers studied the credibility of news articles jointly modeled with expert-level users judgments and the trustworthiness of the sources. The work of Castillo et al. [5] focuses on analyzing the credibility of news propagated on Twitter. The authors model credibility as a binary classification task (credible vs. not credible) based on features extracted from user behavior (re-tweeting), the presence of URLs and citations in the tweets to external sources. Kumar et al. [15] studied the credibility of Wikipedia articles to detect false information. Here, researchers addressed the automatic discovery of articles that have fabricated (hoax) entities and events as a classification task with the goal of stopping false articles to remain in Wikipedia. The authors also investigated for how long such articles usually survive, then discussing their impact.

3 Problem Definition

In this section, we formalize our goal to annotate scientific claims as "dominant" and "dominated" in a Digital library.

Definition 1. A **scientific claim** is a natural language sentence in a scientific paper that expresses a specific *relationship* between entities. In particular, how one of them affects, manipulates, or causes the other entity.

An example of a claim is the following: *"Smoking cigarettes has the potential to increase the risk of lung cancer"*. In this example, *'cigarettes'* and *'lung cancer'* are the entities, and the relationship between them is *'increase the risk'*.

A scientific claim involves a specific relationship between entities. The set of relationships considered as relevant is domain-dependent. Let $\mathbb{R} = \{r_1, \ldots, r_n\}$ be the set of relevant relations of a given domain of study. For instance: alleviates, causes, and treats.

Pareto Semantics. Here we follow the terminology used by Lofi et al. in [20]. Borzsony et al. in [3] proposed Skyline queries to fill the gap between set-based SQL queries and rank-aware database retrieval. Skyline queries rely on the notion of Pareto semantics from the field of Economics discussed by Gabbay et al. in [7]: some object o_1 dominates an object o_2, if and only if o_1 is preferred over o_2 with respect to some attribute and o_1 is preferred over or equivalent to o_2 with respect to all other attributes. Formally, the dominance relationship is denoted as $o_1 \succ o_2$. This simple concept has been used in the data base community to implement an intuitive, personalized data filter as dominated objects can be safely excluded, resulting in the so-called *skyline set* of the query.

More formally, let us define dominance relationships following Pareto semantics for every database relation $R \subseteq D_1 \times \ldots \times D_m$ over m attributes as follows:

$$o_1 \succ o_2 \Leftrightarrow \exists i \in \{1, .., m\}: o_{1i} \succ o_{2i} \land \forall i \in \{1, .., m\}: o_{1i} \succcurlyeq o_{2i} \qquad (1)$$

Where $o_{j,i}$ denotes the i-th component of the database tuple o_j.

The skyline set is the set of all non-dominated objects of the database instance R. Let $A = \{a_1, \ldots, a_m\}$ be the set of attributes used to characterize claims. Let $\Sigma Claims$ be the set of claims found in a collection of documents D for a given pair of entities $<e_1, e_2>$. In summary, note that we use the Pareto semantics to identify "dominant" and "dominated" claims to achieve our final goal in this paper: semantic annotation of the claims.

We assume that the claims in each document of a given collection D are part of the metadata available or were found using the adaptation of the TextRank algorithm using embedding representations of sentences by González Pinto et al. in [11].

Definition 2 Claim Skyline Set. Let $Claims_{e_1,e_2}$ be an m-dimensional dataset that represents the $\Sigma Claims$ of the entity pair $<e_1,e_2>$, where greater values are preferred. Then, a claim p in $Claims_{e_1,e_2}$ dominates claim q iff claim p is better than or equal to claim q in all attributes A and is strictly better than claim q in at least one of the dimensions. Now we are ready to define our problem:

Definition 3 Finding Dominant Claims. Given $\Sigma Claims$ of the entities $<e_1,e_2>$, we attempt to find the Claim Skyline Set under a set of attributes A with respect to an explicit set of relations $\mathbb{R}$.

Once we have found the dominant claims, we can proceed to the semantic annotation of the claims $\Sigma Claims$ for a given entity pair $<e_1,e_2>$ (Fig. 1).

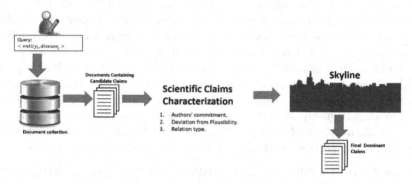

Fig. 1. Approach overview to find dominant claims for a given entity pair

4 Scientific Claim Characterization

In this section we present our approach for determining the attributes A to characterize the claims $\Sigma Claims$ of a given entity pair $<e_1,e_2>$. We propose to model three core content-based properties to build the following set of attributes: author's commitment; deviation from or similarity to a given plausible claim, and the relation type expressed in the claim. In short, a dominant claim is one that (a) expresses high commitment from the authors of the papers: - it is highly objective, and it is unbiased, (b) shares similarity concerning a claim that fits the current body of knowledge, and (c) expresses a relation type that is mutually exclusive with respect to other relations between the given pair of entities. To this end, we consider a set of attributes to capture these properties. Let's provide the details of how we do it.

Authors' Commitment. To capture commitment from the authors, we rely on a set of lexicons from the Natural Language community that proved to model two different stylistic features in written language: concreteness introduced by Brysbaert et al. in [4] and bias introduced by Recasens et al. in [26]. The concreteness lexicon contains ratings of more than 35,000 English words and more than 2,500 bigrams. We hypothesize that this dataset can help our approach to detect the degree of concreteness in the language used by the authors. The lexicon of Recasens et al. in [26] is a set of words that includes "factive verbs", "implicative", "hedges", and "subjective intensifiers". Recasens et al. argue that the identification of unbiased language is a requirement for reliable sources such as scientific articles or encyclopedia. This lexicon was developed and used to remove bias from Wikipedia articles. For our task, we found hedges – word unigrams and bigrams- useful to model the "strength" of the context of scientific claims. Recasens et al. found that the use of hedges reduces one's commitment to the truth of a proposition. Some instances of this lexicon are: might, likely, may, and perhaps.

Deviation from Plausibility. We build on the notion of plausibility based on the knowledge-fit theory from cognitive sciences by Connell et al. in [6] that we proposed for high-quality content preservation in Digital Libraries in González Pinto et al. in [9].

For the sake of completeness, we provide here the intuition of plausibility. The theory from cognitive sciences states that human plausibility judgments consist of two steps: firstly, a mental representation of current knowledge is built and secondly, an assessment examines how a new piece of information fits all prior knowledge. The operationalization of plausibility in an information system is extremely hard in general settings. However, we show in González Pinto et al. in [9] that for some specific type of collections where scientific claims are first-class citizens, the approach proved to be useful to support quality content in Digital Libraries. Given that we share the same domain in this work, we rely on the approach to find a document with a plausible claim as detailed in our previous work for each entity pair used in our experiments.

Relation Types. To define the relation types worth modeling, we first perform a manual exploration of our dataset. Manual exploration led us to distinguish between relations relevant to the domain, yet incomparable. In particular, we focus on the following 'semantic types': *beneficial, non-beneficial, no-effect* and *unknown*. The following examples will clarify the meaning of each type:

- Example 1: *"Recent studies suggest that occasional drinking of coffee might offer protection from pancreatic cancer."* Example 1 will be mapped to the 'beneficial' semantic type, because - as stated in the claim- drinking coffee offers protection from pancreatic cancer.
- Example 2: *"Coffee consumption may weakly increase the risk of pancreatic cancer."* Example 2 will be mapped to the 'non-beneficial' semantic type because according to the claim, coffee consumption increases the risk of cancer.
- Example 3: *"After adjustment for demographic and dietary characteristics, there was no association between pancreatic cancer risk and the intake of coffee, beer, red wine, hard liquor or all alcohol combined."* This example corresponds to the 'no-effect' semantic type.
- Example 4: *"Recent observations of association of risk with coffee consumption and with use of decaffeinated coffee require further evaluation."* This example corresponds to the 'unknown' semantic type.

In summary, given a set of claims $\Sigma Claims$ of a given pair of entities $<e_1, e_2>$, we aim at automatically identifying the semantic type of the relationship between the entities. In addition to the relation type, the following are the specific attributes A that we used to characterize scientific claims $\Sigma Claims$ to obtain $Claims_{e_1, e_2}$ for a given entity pair $<e_1, e_2>$. In the features 1-3, $context_i$ of a claim $claim_i$ refers to the abstract of the paper that contains the corresponding claim.

1. Concreteness score: for each claim $claim_i$, we consider the sum of the ratings of each word in the concreteness lexicon occurring in its context $context_i$.
2. Bias score: for each claim $claim_i$, we consider its context $context_i$ to compute the relative frequency of words in the bias lexicon occurring in $context_i$, thus obtaining the bias score:

$$bias = num(words\ in\ lexicon\ present\ in\ context_i)/length(context_i).$$

3. The similarity concerning a plausible claim: cosine similarity of a Topic Model [2] representation of $context_i$, with respect to the plausible claim found. We rely on this popular latent Dirichlet allocation algorithm that assigns probabilistically $context_i$ to different topics in an unsupervised manner. Let $T = \{t_1, \ldots t_n\}$ be the set of different topics. Then $context_i$ is represented as a vector representation of these n topics. This representation is used to compute cosine similarity with respect to the *context* representation of a plausible claim.
4. Distance concerning a plausible claim: word mover's distance of $claim_i$ with respect to the plausible claim found. In other words, given a plausible claim, we compute for each claim $claim_i$ a semantic distance using word mover's distance that has been shown to outperform other approaches using the semantics of word embeddings, see the details of the work of Kusner et al. in [16].

To find a plausible claim and thus compute features 3 and 4, we proceed as follows: let $contradict(claim_i, D)$ and $support(claim_i, D)$ be functions that compute a cumulative sum of similarities concerning the documents whose corresponding claims contradict and support $claim_i$ respectively. We select as plausible the document with the claim that has the highest similarity score difference between supported and contradicted documents.

We chose to work only with these four attributes for two reasons: firstly, to avoid one of the drawbacks of the Skyline operator: the skyline size. Researchers have shown that with datasets with five up to 10 attributes, the skyline set can contain 30% or more of the entire dataset [1, 3, 8]. Secondly, to interpret the results and evaluate the potential of our proposed approach. However, the approach can be applied considering some other aspects of documents, such as citations counts, the prestige of authors, or altmetrics see for instance the work of Priem in [24]. Regardless of the attributes used, the approach can be applied keeping in mind a manageable size of the skyline set for the task at hand.

Finally, to find the Claim Skyline Set within the set of claims and being able to annotate all the claims semantically, we performed two steps: firstly we generated a dataset where each claim is a data point with the features outlined above. Secondly, we applied the skyline operator on the dataset.

5 Evaluation

In this section, we report the evaluation of two aspects of our approach. Firstly, the semantic type detection of scientific claims. In other words, whether a given claim corresponds to one of the four semantic types, we defined in Sect. 4. Secondly, we evaluate to determine the degree of success of our proposed approach to distinguish between scientific claims that are "dominant" or "dominated". Thus, in the following, we detail the document collection, the algorithms, and metrics used in our evaluation.

Document Collection. Firstly, to find documents with claims relevant to a pair of entities, we relied on PubMed as our primary data source and used the following query pattern [9] for the entity pair: $<entity, disease>$:

(help AND prevent) OR (lower AND risk) OR (increase OR increment AND risk) OR (decrease OR diminish AND risk) OR (factor AND risk) OR (associated AND risk) AND (entity AND disease).

To evaluate our proposed approach on a real-world dataset, we used twenty entity pairs from work in nutritional sciences of [9, 27] linking entities investigated in researchers papers concerning their impact on cancer. Therefore, we used the following entities: coffee, tea, salt, lycopene, wine, milk, sugar, potato, pork, onion, olive, lemon egg, corn, cheese, carrot, butter, bread, beef, and bacon. We used the new "Best matches" algorithm from PubMed to retrieve relevant documents to our queries. As stated on PubMed's website, the new algorithm uses machine learning to combine over 150 signals that are helpful to find matching results. In summary, it automatically expands our query pattern to account for synonyms, MeSH terms, and medical terms. The final collection size consisted of 12,616 research papers. The text mining techniques used only the abstracts of the documents retrieved. There were two reasons behind our decision: first, not all retrieved research papers featured full-text access. Thus, to be fair and avoid bias, we decided to rely only on the abstracts. Second, we assume that the abstracts convey enough information and the relevant context to summarize research papers accurately.

(a) **Semantic relation-type detection.** To detect the semantic relation-type automatically, we annotated a set of claims to build a supervised learning system that can predict the semantic type in unseen claims. As basic machine learning algorithms we used logistic regression and support vector machines with the following features: word n-grams (unigrams, bigrams, and trigrams). We call this set of features 'Bag of Words' features. We compare this approach to the respective word embedding representation of the claims. In particular, we use word embedding based on the models presented in [17, 21]. These vector representations are obtained in an unsupervised fashion from a large textual corpus. They have been used in many applications in text classification systems and serve to support more advanced deep learning models, see [18, 28]. Due to their potential usefulness in similar tasks, we used them to compare to more traditional approaches such as 'Bag of Words'.

In our experiments, we relied on the word2vec vectors trained by Pyysalo et al. [25] on a combination of all publication abstracts from PubMed and all full-text documents from the PubMed Central Open Access subset. Word2vec was run using the skip-gram model with a windows size of 5, hierarchical softmax training, and a frequent word subsampling threshold of 0.001 to create 200-dimensional vectors. Given that these vectors come from a representative collection of the domain that we study, we decided to use them to represent each sentence in our dataset. We refer to the sum representation of the word vectors of each claim using word2vec as 'Embedding' features. This representation of sentences corresponds to the one used in the experiments reported by Lev et al. in [18] that achieved comparable results with more time-consuming deep learning models in different classification tasks. We evaluate the performance of our learned model regarding overall accuracy as well as per-class precision, recall, and F-measure. The results are calculated using 10-fold cross-validation. In Table 1 we summarize the results; 'Logi BoW' and 'SVM BoW' are the models using logistic regression and support vector machines with Bag of Words features respectively; 'Logi Embed' and 'SVM Embed' correspond to the logistic regression and support vector machines models trained with Embedding features. We employ SVM with RBF kernel with one-vs-rest decision function. In Table 2 we show the detailed per-class results of the best model: a support vector machine trained with Bag of Words features. Finally, Table 3 shows statistics of the data used for this first task.

Table 1. Results of semantic relation types

	P	R	F
Logi BoW	0.84	0.85	0.84
SVM BoW	0.85	0.85	0.84
Logi Embed	0.72	0.72	0.71
SVM Embed	0.66	0.64	0.64

Table 2. Per-class results of the best model

	P	R	F
Beneficial	0.89	0.91	0.85
Non-beneficial	0.82	0.88	0.78
No-effect	0.78	0.78	0.78
Unknown	0.80	0.57	0.67

Table 3. Statistics of the data used to assess performance of the different models

#sentences per category	Number of samples
Beneficial	260
Non-beneficial	161
No-effect	84
Unknown	71

Discussion of the Results: To our surprise the Embedding features models were outperformed by its counterparts using simple Bag of Words features. That may due the size of the data that is rather small compared to the number of examples needed according to Goodfellow et al. [12] required to bring significant improvements over more shallow traditional approaches. Nevertheless, the data confirms that traditional machine learning approaches can be used to model relations to guarantee a certain degree of success.

We also observed one limitation of our current approach: we assumed that each claim must fit in one out of the four classes that we have previously defined. This assumption is the reason behind our single-label multi-class problem approach. However, some claims require considering a multi-class multi-label classification approach. For instance, consider the following claim: *"Risk of pancreatic cancer decreased with increasing tea consumption but was unrelated to coffee consumption."* We leave a richer feature engineering approach to solve this issue for future work.

(b) **Finding dominant and dominated claims.** To evaluate our overall approach we performed a Crowdsourcing task on CrowdFlower[1]. We randomly selected 22 claim-pairs. Each pair consisted of a claim in the skyline set, in other words, a dominant claim. The other member of each pair was a claim not in the skyline set. For each pair we asked workers to decide which claim was more convincing. Five workers evaluated each pair of claims; we took majority vote as the final judgment and we consider that as our ground truth. To further control quality of the workers we set a minimum of 70% of correct answer concerning the gold standard questions we provided.

The idea of evaluating a pair of argumentative units – in our case scientific claims– in a crowdsourcing environment have been shown to deliver high-quality results. In particular, in the work of Habernal et al. [13], workers evaluated pairs of arguments to decide "which one is more convincing." Motivated by the results of the authors, we considered a similar approach to evaluation. However, two fundamental challenges in our setting are the use of scientific collections instead of web collections and the fact that 'convincingness' is just an approximation of what we are trying to accomplish. Remember that in our setting we hypothesize that a dominant scientific claim has properties that include but are not limited by stylistic features.

When we considered the crowdsourcing results as ground truth, our approach achieved 86% of accuracy. The results of the experiment look indeed promising.

Discussion of the Results: We manually examined some of the pairs that were difficult to assess for the workers. Indeed, we found that it was very challenging for the workers because of two reasons: firstly, they were asked not to use any external source to assess which of the pair of claims was more convincing. Secondly, our scientific claims characterization goes beyond the claim itself and these properties, as well as the reasoning behind them, was not available to the workers. This latter observation explains that examples such as the following pair were difficult to evaluate:

• Claim 1: *"In a prospective study of coffee intake with the largest number of pancreatic cancer cases to date, we did not observe an association between total, caffeinated, or decaffeinated coffee intake and pancreatic cancer."*
• Claim 2: *"Based on an analysis of data from the European Prospective Investigation into Nutrition and Cancer cohort, total coffee, decaffeinated coffee, and tea consumption are not related to the risk of pancreatic cancer."*

Nevertheless, the data shows that the attributes we proposed can approximate human interpretation up to a certain degree of accuracy.

---

[1] https://www.crowdflower.com/

6   Conclusions and Future Work

In this work, we promote the idea of finding "dominant" and "dominated" scientific claims to annotate them semantically. We devised a set of features that focused on the claim and its context (as given by the respective abstract of the underlying research paper). We then relied on the intuitive semantics of Pareto dominance and thus applied the skyline operator on the real-world datasets used in our experiments to subsequently derive the annotation for scientific claims. We evaluated our data-driven approach using a set of crowdsourcing tasks and achieved an accuracy above 80% that proves the usefulness of our proposed approach.

For the near future, we foresee work to use our current approach in an application setting to facilitate users the discovery of topics of research in need of more experiments and new hypothesis given our semantic annotation of claims.

References

1. Balke, W.-T., Zheng, J.X., Güntzer, U.: Approaching the efficient frontier: cooperative database retrieval using high-dimensional skylines. In: Zhou, L., Ooi, B.C., Meng, X. (eds.) DASFAA 2005. LNCS, vol. 3453, pp. 410–421. Springer, Heidelberg (2005). https://doi.org/10.1007/11408079_37

2. Blei, D.M., Ng, A.Y., Jordan, M.I.: Latent Dirichlet allocation. J. Mach. Learn. Res. **3**, 993–1022 (2003). https://doi.org/10.1162/jmlr.2003.3.4-5.993

3. Borzsony, S., Kossmann, D., Stocker, K.: The skyline operator. In: Proceedings of the 17th International Conference on Data Engineering, pp. 1–20 (2001). https://doi.org/10.1109/icde.2001.914855

4. Brysbaert, M., Warriner, A.B., Kuperman, V.: Concreteness ratings for 40 thousand generally known English word lemmas. Behav. Res. Methods **46**, 904–911 (2014). https://doi.org/10.3758/s13428-013-0403-5

5. Castillo, C., Mendoza, M., Poblete, B.: Information credibility on Twitter. In: Proceedings of the 20th International Conference on World Wide Web - WWW 2011, p. 675 (2011). https://doi.org/10.1145/1963405.1963500

6. Connell, L., Keane, M.T.: A model of plausibility. Cogn. Sci. **30**, 95–120 (2006). https://doi.org/10.1207/s15516709cog0000_53

7. Gabbay, D.M., Guenthner, F.: Handbook of Philosophical Logic. Springer, Dordrecht (2002). https://doi.org/10.1007/978-94-017-0462-5

8. Godfrey, P.: Skyline cardinality for relational processing. In: Seipel, D., Turull-Torres, J.M. (eds.) FoIKS 2004. LNCS, vol. 2942, pp. 78–97. Springer, Heidelberg (2004). https://doi.org/10.1007/978-3-540-24627-5_7

9. González Pinto, J.M., Balke, W.-T.: Can plausibility help to support high quality content in digital libraries? In: Kamps, J., Tsakonas, G., Manolopoulos, Y., Iliadis, L., Karydis, I. (eds.) TPDL 2017. LNCS, vol. 10450, pp. 169–180. Springer, Cham (2017). https://doi.org/10.1007/978-3-319-67008-9_14

10. González Pinto, J.M., Balke, W.-T.: Result set diversification in digital libraries through the use of paper's claims. In: Choemprayong, S., Crestani, F., Cunningham, S.J. (eds.) ICADL 2017. LNCS, vol. 10647, pp. 225–236. Springer, Cham (2017). https://doi.org/10.1007/978-3-319-70232-2_19

11. González Pinto, J.M., Balke, W.-T.: Offering answers for claim-based queries: a new challenge for digital libraries. In: Choemprayong, S., Crestani, F., Cunningham, S.J. (eds.) ICADL 2017. LNCS, vol. 10647, pp. 3–13. Springer, Cham (2017). https://doi.org/10.1007/978-3-319-70232-2_1

12. Goodfellow, I., Bengio, Y., Courville, A.: Deep Learning, vol. 521, p. 800. MIT Press, Cambridge (2016). https://doi.org/10.1038/nmeth.3707

13. Habernal, I., Gurevych, I.: Which argument is more convincing? Analyzing and predicting convincingness of web arguments using bidirectional LSTM. In: ACL, pp. 1589–1599 (2016)

14. Islamaj Dogan, R., Murray, G.C., Névéol, A., Lu, Z.: Understanding PubMed® user search behavior through log analysis. Database (2009). https://doi.org/10.1093/database/bap018

15. Kumar, S., West, R., Leskovec, J.: Disinformation on the web: impact, characteristics, and detection of wikipedia hoaxes. In: WWW, pp. 591–602 (2016). https://doi.org/10.1145/2872427.2883085

16. Kusner, M.J., Sun, Y., Kolkin, N.I., Weinberger, K.Q.: From word embeddings to document distances. In: Proceedings of the 32nd International Conference on Machine Learning, vol. 37, pp. 957–966 (2015)

17. Le, Q., Mikolov, T.: Distributed representations of sentences and documents. In: International Conference on Machine Learning - ICML 2014, vol. 32, pp. 1188–1196 (2014). https://doi.org/10.1145/2740908.2742760

18. Lev, G., Klein, B., Wolf, L.: In defense of word embedding for generic text representation. In: Biemann, C., Handschuh, S., Freitas, A., Meziane, F., Métais, E. (eds.) NLDB 2015. LNCS, vol. 9103, pp. 35–50. Springer, Cham (2015). https://doi.org/10.1007/978-3-319-19581-0_3

19. Lippi, M., Torroni, P.: Argumentation mining: state of the art and emerging trends. ACM Trans. Internet Technol. **16**, 10 (2016). https://doi.org/10.1145/2850417

20. Lofi, C., Balke, W.-T.: On skyline queries and how to choose from pareto sets. In: Catania, B., Jain, L.C. (eds.) Advanced Query Processing, vol. 36, pp. 15–36. Springer, Heidelberg (2013). https://doi.org/10.1007/978-3-642-28323-9_2

21. Mikolov, T., Corrado, G., Chen, K., Dean, J.: Efficient estimation of word representations in vector space. In: Proceedings of the International Conference on Learning Representations (ICLR 2013), pp. 1–12 (2013). https://doi.org/10.1162/153244303322533223

22. Mukherjee, S., Weikum, G.: Leveraging joint interactions for credibility analysis in news communities. In: Proceedings of the 24th ACM International on Conference on Information and Knowledge Management, pp. 353–362 (2015)

23. Mukherjee, S., Weikum, G., Danescu-Niculescu-Mizil, C.: People on drugs: credibility of user statements in health communities. In: KDD 2014 Proceedings of the 20th ACM SIGKDD International Conference on Knowledge Discovery and Data Mining, pp. 65–74 (2014). https://doi.org/10.1145/2623330.2623714

24. Priem, J.: Altmetrics. In: Beyond Bibliometrics: Harnessing Multidimensional Indicators of Scholarly Impact, pp. 263–287 (2014)

25. Pyysalo, S., Ginter, F., Moen, H., Salakoski, T., Ananiadou, S.: Distributional semantics resources for biomedical text processing. In: Proceedings of LBM 2013 (2013)

26. Recasens, M., Danescu-Niculescu-Mizil, C., Jurafsky, D.: Linguistic models for analyzing and detecting biased language. In: Proceedings of the 51st Annual Meeting of the Association for Computational Linguistics, pp. 1650–1659 (2013)

27. Schoenfeld, J.D.: Is everything we eat associated with cancer? A systematic. Am. J. Clinincal Nutr. **97**, 127–134 (2013). https://doi.org/10.3945/ajcn.112.047142.1

28. Zhang, Y., Wallace, B.: A sensitivity analysis of (and practitioners' guide to) convolutional neural networks for sentence classification, pp. 253–263 (2015)

29. IBM Debating Technologies. http://researcher.watson.ibm.com/researcher/view_group.php?id=5443. Accessed 11 Oct 2017

Automatic Segmentation and Semantic Annotation of Verbose Queries in Digital Library

Susmita Sadhu and Plaban Kumar Bhowmick(✉)

Indian Institute of Technology, Kharagpur 721302, India
susmita90iitkgp@gmail.com, plaban@gmail.com

Abstract. In this paper, we propose a system for automatic segmentation and semantic annotation of verbose queries with predefined metadata fields. The problem of generating optimal segmentation has been modeled as a simulated annealing problem with proposed solution cost function and neighborhood function. The annotation problem has been modeled as a sequence labeling problem and has been implemented with Hidden Markov Model (HMM). Component-wise and holistic evaluation of the system have been performed using gold standard annotation developed over query log collected from National Digital Library (NDLI) (National Digital Library of India: https://ndl.iitkgp.ac.in). In component-wise evaluation, the segmentation module yields 82% F1 and the annotation module performs with 56% accuracy. In holistic evaluation, the F1 of the system has been obtained to be 33%.

Keywords: Query segmentation · Semantic annotation
Semantic search · Simulated annealing · Hidden Markov model

1 Introduction

Though facets have become integral part of digital library interfaces, usage of the facets varies greatly in controlled and uncontrolled environment [1]. Low adoption of faceted search in user communities has been attributed to cognitive burden in form of choice overload, visual complexity and information overload [1] specifically for novice users. Without facets, complex information needs in many cases are expressed through *verbose queries* that are considerably longer and the query language is more closer to natural language than simple bag of keywords. Though the verbose queries are convenient to express, they pose processing difficulty to the traditional search engines. For better retrieval, longer queries require *semantic treatment* to map different meaningful segments of the queries to appropriate facets. For example, the query "graph traversal algorithm book in computer science authored by kurt mehlhorn springer" conveys information need that can be semantically represented as

© Springer Nature Switzerland AG 2018
E. Méndez et al. (Eds.): TPDL 2018, LNCS 11057, pp. 270–276, 2018.
https://doi.org/10.1007/978-3-030-00066-0_23

[graph traversal algorithm]<Keyword [book]<Resource Type> in [computer science]<Subject Domain> authored by [kurt mehlhorn]<Author> [springer]<Publisher>

The semantic analysis will be helpful in achieving desired retrieval effect by performing a faceted search subsequently taking cues from the analysis. The task of semantic query analysis can be decomposed into two subtasks: query segmentation and semantic annotation. Most of the existing works focus only on query segmentation task to improve web information retrieval. The existing works on semantic analysis of query make use of comprehensive structured background knowledge base for query segment annotation or disambiguation. For example, Shekarpour et al. [2] links segments in a query into different DBpedia resources. The model parameters of HMM-based disambiguation strategy are computed using the link structure of DBpedia and other knowledge graphs. The existing approaches in semantic analysis of queries have been very effective in Question Answering over Linked Data (QALD) task [3]. However, their scope becomes limited in domain like digital libraries where faceted search is still dominant.

In this work, we propose a system for automatic semantic analysis of English verbose queries in digital library.

2   Proposed Query Analysis System

The query analysis pipeline has four distinct stages, namely, *Generate, Selection, Annotation and Ranking*. The input query is fed to the *Segmentation Generator* module which generates different possible segmentations. The *Selection* module interacts with the *Generator* module to keep track of N-best solutions based on solution cost and feeds N fittest solutions to the next stage of pipeline, i.e., *Segment Annotator*. The segment annotation module takes each candidate segmentation generated in the previous stage and annotates each segment with a tag that represents one of the bibliographic metadata fields. It then generates N annotated segmentations each associated with a probability score. The *Ranker* module orders the annotated segmentations based on the probability scores and chooses the top ranked one as predicted annotation.

3   Query Segmentation Model

3.1   Segmentation Problem

We pose the segmentation generation task as an optimization problem. We make use of the following definitions to present segmentation generation problem formally.

Definition 1 (Segment). *For a given query $Q = (k_1 k_2 \ldots k_n)$ of length n, a segment $S_{i \to j}$ is a sequence of keywords having start and end indices to be i and j respectively. A segment is denoted as $S_{i \to j} = (k_i, k_{i+1}, \ldots, k_j)$.*

Definition 2 (Segmentation). *A segmentation* $\mathbb{S}$ *is formally defined as* $\mathbb{S} = [S_{1 \to i}, S_{i+1 \to j}, ..., S_{m \to n}]$ *and consists of non-overlapping segments arranged in a continuous order. For two consecutive segments* S_x, S_{x+1}, $Start(S_{x+1}) = End(S_x) + 1$.

A partial order can be defined over all possible segments with respect to goodness of segmentation. The quantitative measure of goodness is formulated with following properties of a good segmentation:

Property 1 (High Probable Segment). Each segment $S_x \in \mathbb{S}$ occurs with high probability in different usage scenarios, e.g., corpora, query log etc.

Property 2 (Separateness). Two adjacent segments $S_x, S_{x_1} \in \mathbb{S}$ are well separated. Thus the probability of the string generated as $End(S_x).Start(S_{x+1})$ ('.' implies string concatenation) should be significantly low.

The probability (P_c) of a string (R) is measured by taking contribution of probabilities estimated from Google N-gram language model and N-gram language model trained on NDLI metadata for 12 million resources.

$$P_c(R) = \alpha P_{Google}(R) + (1 - \alpha)P_{NDLI}(R), \qquad 0 \le \alpha \le 1 \qquad (1)$$

The *Goodness Score Function* $(\mathcal{G} : \mathbb{S} \to \mathbb{R})$ for a given segmentation $(\mathbb{S})$ is computed based on Property 1 and Property 2 and is given below

$$\mathcal{G}(\mathbb{S}) = \sum_{S_x \in \mathbb{S}} P_c(S_x) \times (1 - \Delta(S_x)) \qquad (2)$$

The factor $\Delta(S_x)$ for a segment $S_x \in \mathbb{S}$ measures the drop in language model probability between two adjacent segments S_x and S_{x+1} and is given by

$$\Delta(S_x) = \begin{cases} 0, & \text{if } x = |\mathbb{S}| \\ P_c(End(S_x).Start(S_{x+1})) & otherwise \end{cases} \qquad (3)$$

Definition 3 (Optimal Segmentation Problem). *Let* $Q = (k_1 k_2 \ldots k_n)$ *be a query of length n and* ξ *is the set of all possible segmentations that can be obtained from Q. The optimal segmentation is defined as*

$$\xi^* = \underset{i}{\operatorname{argmax}} \; \mathcal{G}(\xi_i) \qquad (4)$$

3.2   Simulated Annealing Approach

Optimal query segmentation problem belongs to combinatorial optimization problem[1]. In this work, we intend to explore the effectiveness of simulated annealing [4] technique in solving optimal query segmentation problem. Important modeling parameters are as follows:

---

[1] For a query of length n, the number of valid segmentations is 2^{n-1}.

Initial Solution: To design initial configuration or solution, we make use of knowledge about the query domain that allows us to mark some of the segments to be *unbreakable*. We use a set of simple rules to mark unbreakable segments. Any random segmentation over 'unbreakable' marked segments is treated as an initial solution in simulated annealing method.

Solution Cost: The solution cost function is taken to be Goodness Score Function ($\mathcal{G}$) in Eq. 2.

Neighborhood Generation Strategy: For an input solution, the proposed neighborhood generation strategy randomly selects a segment and generates new segment by mixing words of the neighboring segments and the current segment if they are breakable; otherwise the selected segment is broken into two segment. The generated segments replace the participating segments.

Cooling Schedule and Acceptance Probability: The new temperature (T_{k+1}) is computed to be $T_{k+1} = \beta \times T_k$ where T_k is the temperature at previous iteration and β is the cooling rate. At each iteration, the changed segmentation having better $\mathcal{G}$-score is accepted. At k^{th} iteration, a changed segmentation ($\mathbb{S}_{prop}$) having inferior $\mathcal{G}$-score is accepted with probability $exp^{\frac{\mathcal{G}(\mathbb{S}_{prop}) - \mathcal{G}(\mathbb{S}_{curr})}{\mathcal{G}(\mathbb{S}_{curr}) \times T_k}}$ where $\mathbb{S}_{curr}$ is the current solution.

Maintaining N-Best Solution: A fixed size priority queue is used to maintain top N segmentation obtained during the course of execution of simulated annealing.

N-best segmentations generated by simulated annealing stage are fed to the segmentation annotation stage.

4   Model for Semantic Annotation of Query

We model segmentation annotation task as a sequence labeling task that has been implemented with Hidden Markov Model (HMM). The *Hidden State Space* consists of 26 states comprising of: (a) 16 states representing different metadata fields like *Contributor, SourceOrganization, LearningResourceType, Language, EducationalLevel etc.* and (b) 10 states represent connectives that appear in context of the metadata fields. For example, 'written by', 'by', 'author' etc. represent *AuthorConnective*. Set of all the segments in the training data acts as the observation symbols space. The observation, state transition and initial state probabilities are estimated following usual notion of HMM. The HMM is trained using the Baum-Welch Algorithm and the decoding problem, i.e., annotating segments with tags, has been addressed with Viterbi algorithm.

5   System Evaluation

5.1   Semantic Query Dataset

The semantic query dataset consists of 1100 queries that have been picked up from queries collected from user query log in NDLI portal for a period of

February'16 - December'17[2]. The selected queries have been segmented and annotated with the set of 26 tags representing different metadata fields and their corresponding connectives. Different evaluation metrics have been used to evaluate different components of the system. We present below performance of each module with respect to relevant evaluations metrics.

5.2 Segmentation Task

Evaluation Metrics: Evaluation of segmentation task has been performed with WindowDiff as well as recently developed *Boundary Similarity*-based measures (B) [5]. We use the measures B-measure Precision (B-P), B-measure recall (B-R) and B-measure F1 (B-F1) along with WindowDiff to measure performance of this task.

Parameter Tuning: Through grid search technique applied over development dataset the β (cooling rate) and α (relative contribution of Google n-gram) have been set to be 0.9 and 0.2 respectively.

Performance Results: The performance measures of the segmentation module are presented for best segmentation and N-best segmentation as output. For N-best segmentation, all the measures have been computed based on micro and macro average. Table 1 presents values of different performance metrics computed over the predicted segmentations for the entire dataset.

Table 1. Individual performance of query segmentation model (for $N = 2$)

Measure	N-Best		Best
	Macro average	Micro average	Average
WindowDiff	0.420	0.411	0.384
B-Precision	0.676	0.683	0.708
B-Recall	0.919	0.919	0.926
B-F1	0.779	0.784	0.802

5.3 Segment Annotation

In evaluation of *segment annotation* module, we have taken the manually segmented queries for testing. Each segment in an input segmentation is classified into either of the metadata fields. Consequently, the performance of the annotation system has been measured with accuracy metric. We have performed 10-fold cross validation. We used majority tag system as the baseline where each segment is tagged with most frequent metadata field (i.e., Learning Resource

---

[2] The queries having between 3 to 15 words in the raw query log are included in the dataset.

Type). Accuracies of the baseline system and HMM-based annotator are 15.26% and 55.88% respectively. It is observed that the proposed HMM-based annotator performs much superior to the majority tag based baseline system.

5.4 Holistic Evaluation

The holistic evaluation of the system has been performed by putting the segmentation and annotation module together where N-best segmentations are passed down to the annotator module which subsequently annotates each segment for a given segmentation into one metadata field.

We have used standard Precision, Recall and F1 measures. Here, we have also used 10-fold cross validation to compute the performance measures. The only parameter that is to be fixed is the value of N, the number of top segmentation that are passed down to the next phase. It is observed that feeding the single best segmentation to the next stage of the pipeline provides the best result. It may be conjectured that for most of the queries the best segmentation is always retrieved in the top position.

Best performance obtained for holistic evaluation is with Top-1 segmentations where Precision = 0.337, Recall = 0.333 and F1 = 0.33. The presented results are encouraging given the complexity of the task. The primary impediment to attain high performance measure is limited training dataset.

6   Conclusion

In this work, we presented a multi-stage system for automatic semantic processing of verbose queries. The first stage of the pipeline, namely, segmentation, has been modeled as an optimization problem and has been implemented with simulated annealing technique. The segment annotation stage has been formulated as a sequence labeling problem and HMM has been used in implementation. The segmentation module yields a reasonable performance while tested with state-of-art segmentation evaluation metrics. However, there are several scopes for improvement. Firstly, the training dataset has to be grown to a significant volume for improvement in segment annotation performance. Secondly, a better strategy to combine predictions for multiple best segmentations has to be explored.

Acknowledgement. This work is partially supported by National Digital Library of India project, sponsored by MHRD, Govt of India and IBM Research.

References

1. Niu, X., Hemminger, B.: Analyzing the interaction patterns in a faceted search interface. J. Assoc. Inf. Sci. Technol. **66**(5), 1030–1047 (2015)
2. Shekarpour, S., Marx, E., Ngomo, A.-C.N., Auer, S.: SINA: semantic interpretation of user queries for question answering on interlinked data. Web Semant. Sci. Serv. Agents World Wide Web **30**, 39–51 (2015). Semantic Search

3. Usbeck, R., Ngomo, A.-C.N., Haarmann, B., Krithara, A., Röder, M., Napolitano, G.: 7th open challenge on question answering over linked data (QALD-7). In: Dragoni, M., Solanki, M., Blomqvist, E. (eds.) SemWebEval 2017. CCIS, vol. 769, pp. 59–69. Springer, Cham (2017). https://doi.org/10.1007/978-3-319-69146-6_6
4. Aarts, E., Korst, J., Michiels, W.: Search methodologies. In: Burke, E.K., Kendall, G. (eds.) Simulated Annealing, pp. 187–210. Springer, Boston (2005). https://doi.org/10.1007/0-387-28356-0_7
5. Fournier, C.: Evaluating text segmentation using boundary edit distance. In: Proceedings of the 51st Annual Meeting of the Association for Computational Linguistics (Volume 1: Long Papers), Sofia, Bulgaria, pp. 1702–1712. Association for Computational Linguistics, August 2013

Recommendation

Open Source Software Recommendations Using Github

Miika Koskela, Inka Simola, and Kostas Stefanidis[✉]

University of Tampere, Tampere, Finland
miika.s.koskela@gmail.com, inkariina.simola@gmail.com,
kostas.stefanidis@uta.fi

Abstract. The focus of this work is on providing an open source software recommendations using the Github API. Specifically, we propose a hybrid method that considers the programming languages, topics and README documents that appear in the users' repositories. To demonstrate our approach, we implement a proof of concept that provides recommendations.

1 Introduction

Recommender systems have become indispensable for several systems and Web sites, such as Amazon, Netflix, Yelp and Google News, helping users navigate through the abundance of available data items (e.g., [1–4,6]. In this paper[1], we introduce a recommender system for suggesting open source software using the Github API. Getting these recommendations requires users to have, in addition to a Github account, at least one of the following: (i) starred repositories, (ii) repositories followed, or (iii) users own repositories. As this information is public and available through Github's public API, recommendations can be generated to *any* user by *any* user. For generating recommendations, we exploit a hybrid method that combines three different similarity measures on three different feature sets. Specifically, we consider separately the languages found in a user's repositories, the topics present in the user's repositories, and the README documents of the user's repositories. To demonstrate our approach, we implement a proof of concept prototype for providing software recommendations.

2 Dataset

The dataset contains information on approximately 1000 software repositories and consists of languages, topics and README-files retrieved via Github's API. In order to recommend repositories some information on the users skills and interests is needed. We call this user information user profile. Specifically, the user profile consists of a combination of information from repositories the user

---

[1] The work was partially supported by the TEKES Finnish project Virpa D.

E. Méndez et al. (Eds.): TPDL 2018, LNCS 11057, pp. 279–285, 2018.
https://doi.org/10.1007/978-3-030-00066-0_24

has somehow been associated with in the past. We consider the following as proof of an association between the user and a repository: (i) the user has starred the repository, (ii) the user has followed the repository, or (iii) the user has forked[2] the repository. For simplicity, we refer to these collectively as users repositories. For example, assume that user `mkoske` has starred the repository `php-ai/php-ml`, and has `mkoske/scatter-r` as his own repository. To construct a profile for this user, we collect the following pieces of information on these two repositories: (i) topics assigned to a repository by its owner, (ii) programming languages used in a repository, and (iii) README document of a repository. Next we will take a look at these pieces of information separately.

Topics: As an example, consider the `php-ai/php-ml`-repository, which is starred by user `mkoske`, and contains, among other topics, the following: php, machine-learning, classification, and data-science. Topics are specified manually by the repository owner or someone with proper permissions and is therefore a sparse source of information. In our sample dataset of 1000 items, almost half of the repositories (47.30%) are missing topics entirely.

Programming Languages: In the previous example, the `php-ai/php-ml`-repository contains two different languages: PHP and Shell. The latter seems to be used, e.g., to generate PHPUnit tests coverage reports and is therefore listed among the languages of the repository. The repository programming language is detected automatically[3] and no user interaction is required for that. In contrast, topics have to be input manually by the repository owner and are not detected automatically. In our sample dataset, over 95% of the repositories have a language or languages specified.

README: README-files are the third source of information on a repository. When accessing the repository at Github via web browser, the README is one of the first things that user encounters. It contains a free-form description of the repository in plain-text format and can be considered a front page for the repository. Table 1 shows some statistics on the lengths of the README files in our data set. The standard deviation is quite high, indicating that there is much

Table 1. Statistics on the lengths of the README files in our data set.

Statistic	Value
Min	115.00
Max	505 402.00
Mean	13 257.88
Median	6 891.00
Std	25 287.51

---

[2] https://help.github.com/articles/fork-a-repo/.

[3] https://help.github.com/articles/about-repository-languages/.

variation in the lengths of the descriptions repository owners are assigning to their projects.

3 Method

We follow a hybrid method to produce open source software recommendations. Specifically, we combine three different similarity measures on three different feature sets to make a single list of recommendations. The final similarity score is a linear combination of three similarity scores: (i) all the languages found in a user's repositories are compared to the languages present in a given non-user repository, (ii) all the topics present in the user's repositories are compared to the topics present in a given non-user repository, and (iii) an averaged vector representation of the README documents of the user's repositories is compared to the vector representation of the README document of a given non-user repository.

We construct a language vector, a topic vector and a README vector for each repository. For the sake of practicality, all user repository vectors are collapsed together per type to form 3 different user vectors per user: a user language vector, a user topic vector and a user README-vector. Each of these vectors is then compared against the respective language, topic and README vector of each repository not associated with the user for whom a recommendation is to be generated. A linear combination of the resulting language, topic and README similarity scores is then calculated to obtain the final repository ranking.

Languages: The repository metadata contains information on the programming languages[4] used to write the software in question. Continuing with the example we had earlier, repository `php-ai/php-ml` is mainly written in PHP, but also contains some shell scripts. This is a small number of languages compared to, e.g., Visual Studio Code by Microsoft, which contains numerous languages.

In our dataset, languages detected in a repository were first transformed into binary vectors. If the repository contained a language, e.g., aforementioned PHP, the feature was assigned a value of 1, and 0 otherwise. The length of a repository language vector is the number of all programming languages found in the entire dataset, including the user repositories. Below is an example of a transformed binary language vector.

Assembly	awk	c	...	Typescript	vim_script	vb
1	0	1	...	0	1	0

The union of languages present in a user's repositories was used to form the user language vector. The mean of the user language vector was then subtracted from the user language vector, ensuring that cases where the user language

---

[4] Languages are automatically detected.

vector lacks a language present in a repository language vector get a lower score than cases where neither user nor repository language vector contain a given language. This reflects the author's assumption that a user is most versed in languages present in his or her own repositories and not much else.

Ordinary cosine similarity does not differentiate between the aforementioned cases, and would give them equal language rankings. Adjusted cosine similarity, on the other hand, would needlessly penalize cases where the repository language vector lacks a language present in the user language vector. However, a repository does not have to contain all the user vector languages in order to be recommendable to the user. Hence, the mean was only subtracted from the user language vector and not the repository language vectors. For calculating the similarity between a user language vector and each repository language vector, we used a hybrid cosine similarity measure defined as: $sim_{cos}(a,b) = \frac{(a-\bar{a})\cdot b}{||(a-\bar{a})||\times||b||}$.

After subtracting the mean from the user language vector, our hybrid cosine similarity was calculated between the user language vector and each non-user repository language vector.

Topics: Repository topics resemble the commonly used *tags*: they contain at most a few words of free-form text that the author of the repository has chosen to describe the repository. Topics can also contain names of the programming languages used in writing the software. We did not filter out these potential overlaps and used the topics as they were. The length of a topic vector is the number of topics found in the entire dataset. Below is an example of a binary topic vector.

d	Ad-blocker	Admin	...	Youtube	zeit	zsh
1	0	0	...	1	0	0

The union of topics present in a user's repositories was used to form the user topic vector. Jaccard similarity was then calculated between the user topic vector and each repository's topic vector. The Jaccard similarity measure, sometimes also called the *intersection over union* similarity, was used to compute similarities related to the topic component of the dataset. If a user vector contains (i.e., has values of 1 for) topics $a_0, a_1, \ldots, a_i$ and a repository vector contains topics $b_0, b_1, \ldots, b_j$, the Jaccard similarity between vectors $\mathbf{a}$ and $\mathbf{b}$ becomes: $sim_{jac}(\mathbf{a}, \mathbf{b}) = \frac{|\{a_0,a_1,\ldots a_i\}\cap\{b_0,b_1,\ldots b_j\}|}{|\{a_0,a_1,\ldots a_i\}\cup\{b_0,b_1,\ldots b_j\}|}$.

README: In our current dataset, all repositories have a README document. The READMEs were also retrieved using the Github API. Each repository README document was subjected to the following preprocessing operations: (i) tokenization, i.e., splitting the long string of text to tokens, (ii) removal of words with less than 3 characters, (iii) removal of content between any kind of brackets, (iv) removal of content matching certain frequently observed patterns (e.g., url, email address), (v) removal of English stopwords (e.g., 'and', 'when'), and (vi) part-of-speech tagging and removal of words that are not nouns in singular form.

A vector representation was then generated for each preprocessed README using TF-IDF as implemented in the TF-IDF-Vectorizer function of *Scikit-learn*:

- For each word appearing in the corpus, the documents containing the word are counted.
- Words appearing in just one or more than 95% of the documents of the corpus are removed.
- The inverse document frequency for each preserved word (~3000 words) is calculated.
- $IDF_i = \log \frac{\text{total \# documents}}{\text{\# of documents containing word i}}$.
- For each README, the normalized term frequency for each preserved word is calculated.
- The TF-IDF score (term_frequency x IDF score) is calculated for each preserved word in each README document, yielding a vectorized representation for each README document.

Below is an example of README-vector.

Abilities	Abort	...	Zeros	Zones	Zookeeper
0.0	0.001	...	0.0	0.002	0.0

README vectors from the user's repositories were averaged to obtain a user README vector. Oridnary cosine similarity between the user README vector and each non-user-owned repository README vector was then calculated. It is possible for a repository README to not contain any of the words preserved by the TF-IDF transformation. In these cases, the cosine similarity between the user and repository README vectors is zero.

4 Recommendations

After feature extraction was completed, a linear combination of the language, topic and README similarities was calculated using weights w_l, w_t and w_r for language, topic and README respectively. For the time being and in the absence of user feedback, all weights were initialized to 0.33. The maximum and minimum scores thus assigned to the most and least recommended repositories respectively, were denoted *max_score* and *min_score*. The final scores were then obtained by normalizing the results using the following formula:

$$final\_score_{a,b} = \frac{w_l * lang\_sim_{a,b} + w_t * topic\_sim_{a,b} + w_r * rcadme\_sim_{a,b}}{max\_score - min\_score}.$$

Figure 1 shows the first five recommendations given to user `inkasimola` based on the public repositories on her Github account and 1000 retrieved repositories.

The project is public and located at: https://github.com/mkoske/recom mender/. The software implementation was written in Python, while the *requests*

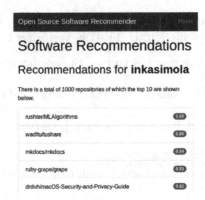

Fig. 1. Recommendations for user `inkasimola`.

and *Pandas* libraries were used for data retrieval from Github. As preprocessing - and especially part-of-speech tagging - entire README documents is time-consuming, it was done beforehand using Python scripts. Finally, the web application was written using *Flask*, and the Bootstrap CSS-framework.

5   Summary

In this paper, we propose a hybrid method that considers the programming languages, topics and README documents present in Github users' repositories, to generate open source software recommendations. The future work could evaluate the system by performing experiments with real users. Moreover, we could implement a model that takes into account the popularity of repositories and involves a feedback loop to allow for learning of user-specific feature weighting and further context-aware personalization of recommendations [5,7].

References

1. Adomavicius, G., Kwon, Y.O.: Multi-criteria recommender systems. In: Ricci, F., Rokach, L., Shapira, B. (eds.) Recommender Systems Handbook, pp. 847–880. Springer, Boston, MA (2015). https://doi.org/10.1007/978-1-4899-7637-6_25
2. Kyriakidi, M., Stefanidis, K., Ioannidis, Y.E.: On achieving diversity in recommender systems. In: ExploreDB (2017)
3. Ntoutsi, E., Stefanidis, K., Rausch, K., Kriegel, H.: Strength lies in differences: diversifying friends for recommendations through subspace clustering. In: CIKM (2014)
4. Sandvig, J.J., Mobasher, B., Burke, R.D.: A survey of collaborative recommendation and the robustness of model-based algorithms. IEEE Data Eng. Bull. **31**(2), 3–13 (2008)
5. Stefanidis, K., Koutrika, G., Pitoura, E.: A survey on representation, composition and application of preferences in database systems. ACM Trans. Database Syst. **36**(3), 19:1–19:45 (2011)

6. Stefanidis, K., Ntoutsi, E.: Cluster-based contextual recommendations. In: EDBT (2016)
7. Stefanidis, K., Pitoura, E., Vassiliadis, P.: Managing contextual preferences. Inf. Syst. **36**(8), 1158–1180 (2011)

Recommending Scientific Videos Based on Metadata Enrichment Using Linked Open Data

Justyna Medrek[2], Christian Otto[1,3](✉) [iD], and Ralph Ewerth[1,3] [iD]

[1] Leibniz Information Centre for Science and Technology (TIB), Hannover, Germany
{christian.otto,ralph.ewerth}@tib.eu
[2] Leibniz Universität Hannover, Hannover, Germany
justa@mail.de
[3] L3S Research Center, Leibniz Universität Hannover, Hannover, Germany

Abstract. The amount of available videos in the Web has significantly increased not only for entertainment etc., but also to convey educational or scientific information in an effective way. There are several web portals that offer access to the latter kind of video material. One of them is the TIB AV-Portal of the Leibniz Information Centre for Science and Technology (TIB), which hosts scientific and educational video content. In contrast to other video portals, automatic audiovisual analysis (visual concept classification, optical character recognition, speech recognition) is utilized to enhance metadata information and semantic search. In this paper, we propose to further exploit and enrich this automatically generated information by linking it to the Integrated Authority File (GND) of the German National Library. This information is used to derive a measure to compare the similarity of two videos which serves as a basis for recommending semantically similar videos. A user study demonstrates the feasibility of the proposed approach.

Keywords: Video recommendation · Semantic enrichment
Linked data

1 Introduction

Videos hold a great potential to communicate educational and scientific information. This is, for instance, reflected by e-Learning platforms such as Udacity (https://udacity.com) or Coursera (http://www.coursera.org). Another type of Web portals also offers access to scientific videos, one of them is the TIB AV-Portal (https://av.tib.eu) of the Leibniz Information Centre for Science and Technology (TIB). Researchers can provide, search, and access scientific and educational audiovisual material, while benefiting from a number of advantages compared to other portals. First, submitted videos are reviewed to check whether they contain of scientific or educational content. Second, videos are represented in a persistent way using DOIs (digital object identifier), potentially even at

E. Méndez et al. (Eds.): TPDL 2018, LNCS 11057, pp. 286–292, 2018.
https://doi.org/10.1007/978-3-030-00066-0_25

the segment and frame level, making it easy and reliable to reference them. Finally, audiovisual content analysis is applied in order to allow the user to not only search for terms in descriptive metadata (e.g., title, manually annotated keywords), but also in the audiovisual content, i.e., in the speech transcript, in the recognized overlaid or scene text through video OCR (optical character recognition), and keywords derived from visual concept and scene classification.

Usually, recommender systems in online shopping platforms or video portals mainly rely on user-based information such as the viewing history [2] or current trends [1]. In this paper, we investigate the question how relevant videos can be recommended based on their metadata, especially by also making use of automatically extracted metadata from audiovisual content analysis. This is relevant, for example, when users do not agree to track their search behavior or sufficient amount of user data is not available. In particular, we propose to further exploit and enrich the entire set of available metadata, be it created manually or extracted automatically, in order to improve recommendations of semantically similar videos. In a first step, we utilize a Word2Vec approach [3] to make the semantic content of two videos comparable based on title, tags, and abstract. Then, the automatically extracted metadata about the audiovisual content is enriched by linking it to the Integrated Authority File (GND: Gemeinsame Normdatei) of the German National Library (DNB: Deutsche Nationalbibliothek). These two kinds of information are used to derive a measure to compare the content of two videos which serves as a basis for recommending similar videos.

The paper is structured as follows. First, we give a brief overview of related work in Sect. 2. The framework to generate video recommendations is presented in Sect. 3. Section 4 describes the conducted user study to evaluate the proposed approach, while Sect. 5 concludes the paper.

2   Related Work

Scientific Video Portals: Yovisto is a scientific video portal that allows the user to search for information via text-based metadata [9,10]. The users can reduce the number of search results by refining their query via additional criteria and grouping videos by language, organization, or category. On the contrary, to increase the scope of possible results, a tool for explorative search reveals interrelations between different types of videos in order to present a broader spectrum of results to the user. This is achieved by exploiting an ontology structure, which is part of each video element and Linked Open Data (LOD) resources, namely DBpedia (http://wiki.dbpedia.org). Another similar portal is described by Marchionini [4], where the uploaded content is automatically fed into an automatic data analysis chain similar to the TIB AV-Portal. However, the semantic entities are then assigned to each video segment resulting in a storyboard comprising the video content. In contrast to the AV-Portal, these metadata are hidden from the user. Marchionini's approach focuses on providing a good explorative search tool, i.e., a user should be able to find what s/he is looking for even when being unsure about the correct phrasing.

Recommender Systems for Scientific Videos: Clustering semantically similar videos is a possible approach to provide video recommendations based on a given, currently watched video. A fundamental problem of this research is the semantic gap between low-level features and high-level semantics portrayed in visual content [6]. To circumvent this problem, textual cues can be used in addition to visual content. These can be manually added tags by the author of the video or automatically extracted keywords by machine learning algorithms. Either way, they are often superficial, noisy, incomplete or ambiguous which makes the process of clustering a challenge. Vahdat et al. [8] enrich the set of tags by modelling them from visual features and correct the existing ones by checking their agreement with the visual content. They show that the proposed method outperforms existing ones that use either modality and even the naive combination. Wang et al. [11] discover that by incorporating hierarchical information the semantics of a video can be described even better. Despite only using two levels of abstraction in their hierarchical multi-label random forest model, strong correlations between ambiguous visual features and sparse, incomplete tags could be found. Our approach will also make use of this idea.

3   Enriching Video Metadata Through Linked Open Data

In this section, we present our approach to enrich metadata with open data sources. First, the set of available metadata is described before the acquisition of additional information from an open data source is explained in Sect. 3.1. Second, a similaritiy measure to compare videos based on a Word2Vec representation and enriched metadata is derived in Sect. 3.2. The overall workflow is displayed in Fig. 1. The input of our system consists of manually generated and automatically extracted information, where the former comprises abstract and title. Additional inputs are the following automatically extracted **Tags** (see Fig. 1) derived from: (1) Transcript based on speech recognition, (2) Results of video OCR, and (3) results of visual concept and scene classification. All of them have a representation in the German National Library, which is the key requirement for the enrichment process.

3.1   Acquiring Additional Information from Open Data Source

Automatically generated tags usually contain a certain amount of errors and noise. Although state-of-the-art algorithms can achieve human performance [7] in specific tasks and settings, issues with audio quality in lecture rooms or hardly legible handwritings can cause errors. We try to circumvent this problem by evaluating additional information provided by the German National Library. Besides information such as synonyms and related scientific publications, the *Dewey Decimal Classification (DDC)* for every tag is provided. The DDC is a library classification system, which categorizes technical terms into ten classes via three-digit arabic numerals [5]. These main classes are then further divided into subcategories denoted by the decimals after these three digits, where additional decimals

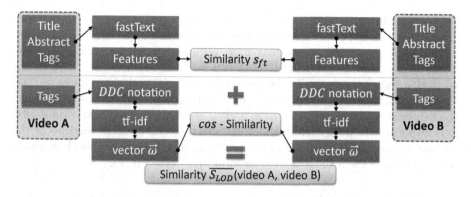

Fig. 1. The general workflow of the approach combining the *method without LOD* (upper half) with the features from the DDC notation (lower half).

depict a more specific subject. For instance, *SPARQL* is contained in *006.74 - Markup Language, 005.74 - Data files and Databases* and *005.133 - Individual Programming Languages*, which yields valuable contextual information.

3.2   Defining a Similarity Measure for Scientific Videos

Simply comparing two videos for mutual tags is not sufficient to determine semantic similarity. Even if two sets of tags have little to no overlap they might be highly correlated when their context is considered. We address this issue by utilizing fastText [3] to generate word embeddings, which has several advantages for this task. First, semantically similar words are modeled closer to one another so that a simple distance measure indicates the correlation of two words. Second, since fastText works on substrings rather than whole words it is able to produce valuable features even for misspelled or words unknown to the word embedding. Finally, a pre-trained model is available for a large number of languages. Title, tags, and abstract are taken from the metadata and processed via fastText. It generates a 300-dimensional feature vector for every word in the metadata. The average of these vectors is our representation for a particular video. This approach is our baseline and denoted as *method without LOD* in the sequel.

The improvement of this already powerful feature extraction method is the main contribution of this paper. It is achieved by incorporating the information provided by the DDC notation in addition to the fastText embeddings. As a preprocessing step we need to create a vector ω, which consists of all DDC tags that occur in our dataset and which will be assigned to every video entry v. Since the upper level classes of the notation are also encoded in the codes of the classes at lower levels, we divide them accordingly. Therefore, the length of ω equals the total number of these tag fragments. For instance, if the video corpus would only contain the tags 005.74 and 005.133, we would split them into $5_1, 57_2, 51_2, 574_3, 513_3, 5133_4$ (indices mark the level in the hierarchy) resulting in a vector ω of length 6. If a particular tag fragment occurs in a video, we set the corresponding bin in ω to the *term frequency - inverse document frequency (tf-idf)*, or zero otherwise. This assures that the more specific, and therefore more informative, DDC classes have more influence on the result. For example, if two tags share the main DDC class *Science and Mathematics*, it does not mean that they are necessarily closely correlated, but if both share the class *Data Compression* they most likely cover a similar topic. For the "method with LOD" the two vectors ω_i and ω_j of video v_i and v_j are compared via cosine similarity. It is important to note that this method also uses the fastText features of the *method without LOD*. In order to compute the overall similarity, both methods are applied and the average is used to form s_{LOD} (see Fig. 1).

4   Experimental Results

Videos of the TIB AV-Portal were used in the experiment. The complete stock of metadata that falls under the Creative Commons License CC0 1.0 Universal is made available by the TIB (https://av.tib.eu/opendata) as Resource Description Framework (RDF) triples. To extract the necessary annotations we utilized SPARQL. In a first step, it was necessary to keep only videos that allowed *"derivate works"* in addition to the CC0 1.0 license, since content analysis is applied. 2066 samples satisfied these conditions[1]. Unfortunately, word embeddings of two different languages cannot be directly compared forcing us to use a subset of videos with the same language (German in this case, 1430 videos). Annotations are represented in JSON format to make them easily accessible for future tasks without rebuilding the RDF graph. After gathering all tags of an entry, we employed another SPARQL query assigning a GND (German: *Gemeinsame Normdatei*, English: Integrated Authority File) link to each tag, which is the key part of linking it to the data of the German National Library (DNB) and retrieving the corresponding DDC notations.

We evaluated the quality of our similarity measure by conducting a user study with eight participants, five men and three women. A random selection of 50 videos was presented to every participant along with ten video recommendations, randomly either completely provided by the *method without LOD* or the *method with LOD*. The results were integrated by a Greasemonkey script in the Firefox browser. Every participant had to rate each of the ten recommendations from 0–3, i.e., 0: not relevant; 1: low relevance; 2: medium relevance; 3: highly relevant. The results are displayed in Fig. 2.

The results show that the *method with LOD* increases the number of video recommendations with medium (4.56%) and low relevance (11.29%), while the effect is small (0.97%) for the highly relevant recommendations. However, the *method with LOD* significantly decreases the number of irrelevant recommendations (by 18.17%). This indicates that this method is superior to the text-based method, most likely due to the hierarchical nature of the DDC notation. We assume that the rather small improvement for the very relevant recommendations is a result of the restrictions we had to oblige to (license and language), i.e.,

---

[1] As of June 16, 2017.

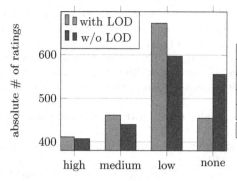

Relevance ↓	with LOD	w/o LOD
High	411	407
Medium	461	440
Low	673	597
None	455	556
SUM	2000	2000

Fig. 2. Absolute number of votings for each relevance level in the user study.

the relatively small set of remaining videos (1 430) does not contain more highly relevant samples. A chi-square test shows that the *method with LOD* is significantly better than our baseline (Chi-Square = 15.1471, p-value = 0.001695).

5   Conclusions

In this paper, we have proposed a method to generate recommendations for scientific videos based on automatically extracted (and partially noisy) tags by utilizing linked open data to weave in hierarchical semantic metadata. This enables users to find relevant information more quickly, which improves their overall learning experience. In future work, we plan to incorporate recommendations for scientific papers or definitions of technical terms through linked open data.

References

1. Covington, P., Adams, J., Sargin, E.: Deep neural networks for Youtube recommendations. In: Proceedings of the 10th ACM Conference on Recommender Systems, pp. 191–198. ACM (2016)
2. Davidson, J., et al.: The Youtube video recommendation system. In: Proceedings of the Fourth ACM Conference on Recommender Systems, pp. 293–296. ACM (2010)
3. Joulin, A., Grave, E., Bojanowski, P., Mikolov, T.: Bag of tricks for efficient text classification. arXiv preprint arXiv:1607.01759 (2016)
4. Marchionini, G.: Exploratory search: from finding to understanding. Commun. ACM **49**(4), 41–46 (2006)
5. Reiner, U.: Automatic analysis of dewey decimal classification notations. In: Preisach, C., Burkhardt, H., Schmidt-Thieme, L., Decker, R. (eds.) Data Analysis, Machine Learning and Applications, pp. 697–704. Springer, Heidelberg (2008). https://doi.org/10.1007/978-3-540-78246-9_82
6. Beyer, K., Goldstein, J., Ramakrishnan, R., Shaft, U.: When is "nearest neighbor" meaningful? In: Beeri, C., Buneman, P. (eds.) ICDT 1999. LNCS, vol. 1540, pp. 217–235. Springer, Heidelberg (1999). https://doi.org/10.1007/3-540-49257-7_15

7. Shi, B., Bai, X., Yao, C.: An end-to-end trainable neural network for image-based sequence recognition and its application to scene text recognition. IEEE Trans. Pattern Anal. Mach. Intell. **39**(11), 2298–2304 (2017)

8. Vahdat, A., Zhou, G.-T., Mori, G.: Discovering video clusters from visual features and noisy tags. In: Fleet, D., Pajdla, T., Schiele, B., Tuytelaars, T. (eds.) ECCV 2014. LNCS, vol. 8694, pp. 526–539. Springer, Cham (2014). https://doi.org/10.1007/978-3-319-10599-4_34

9. Waitelonis, J., Sack, H.: Augmenting video search with linked open data. In: I-SEMANTICS, pp. 550–558 (2009)

10. Waitelonis, J., Sack, H.: Towards exploratory video search using linked data. Multimedia Tools Appl. **59**(2), 645–672 (2012)

11. Wang, J., Zhu, X., Gong, S.: Video semantic clustering with sparse and incomplete tags. In: AAAI, pp. 3618–3624 (2016)

Posters

TIB-arXiv: An Alternative Search Portal for the arXiv Pre-print Server

Matthias Springstein[1](✉)(iD), Huu Hung Nguyen[1](iD), Anett Hoppe[1](iD), and Ralph Ewerth[1,2](iD)

[1] Leibniz Information Centre for Science and Technology (TIB), Hannover, Germany
{matthias.springstein,hung.nguyen,anett.hoppe,ralph.ewerth}@tib.eu
[2] L3S Research Center, Leibniz Universität Hannover, Hannover, Germany

Abstract. *arXiv* is a popular pre-print server focusing on natural science disciplines (e.g., physics, computer science, quantitative biology). As a platform with an emphasis on easy publishing services it does not provide enhanced search functionality – but offers programming interfaces which allow external parties to add these services. This paper presents extensions of the open source framework *arXiv* Sanity Preserver (SP). With respect to the original framework, it derestricts SP's topical focus and allows for text-based search and visualisation of all papers in *arXiv*. To this end, all papers are stored in a unified back-end; the extension provides enhanced search and ranking facilities and allows the exploration of *arXiv* papers by a novel user interface.

Keywords: arxiv · Academic search · Web user interface
Social networks

1 Introduction

For scientists it is crucial to keep up to date in their field of research – but this task is becoming increasingly difficult. One reason for this problem is that the number of new publications per day increases dramatically. Another reason is that today's scientists have several ways to publish their article and the publication system is becoming more heterogeneous. As a result, scientists have to devote more and more time to find articles that are relevant for their research.

A good indicator of this trend is the *arXiv* pre-print server. The number of articles in the repository has increased linearly over the last 25 years, with more than 10,000 articles per month in 2017 [1]. The platform's objective is to provide a good and easy-to-use publishing service, whereas it does not particularly focus on user interface and search. Thus, functionalities for enhanced search and sorting are missing. Anyhow, its maintainers are open to external partners developing novel services on top of *arXiv*'s existing infrastructure [6].

This paper presents TIB-arXiv[1], a web-based tool for enhanced search and exploration of publications in *arXiv*. A comfortable interface and specifically developed retrieval and ranking methods enable scientists to easily keep track

---

[1] https://labs.tib.eu/arxiv.

E. Méndez et al. (Eds.): TPDL 2018, LNCS 11057, pp. 295–298, 2018.
https://doi.org/10.1007/978-3-030-00066-0_26

of the current development in their research field, while social and collaborative functionalities facilitate interactive research processes.

2   Related Work

arXiv is a very popular source for academic literature search. Thus, web-based tools for its (more) efficient use already exist (see [4] for a survey). Most of the applications create a topic or user-specific news feed, but they differ in their presentation of the retrieved articles: *Arxiv Sanity Preserver* [3] generates thumbnails, *Arxivsorter* [7] shows figures extracted from the papers. Some of them explore re-ranking techniques, considering for instance the number of tweets referring to the article [3] or the authors' names [5]. A more global approach is presented by *PaperScape* [2]: The tool visualises the *arXiv* dataset in form of a map. In concrete terms, each paper in *arXiv* is visualized by a circle whose size represents the number of citations and the position depends on the cited paper.

Overall, all tools have one big disadvantage in common: They usually limit their topical scope to a single domain of interest (e.g., *Arxiv Sanity Preserver* focuses on computer vision research). There is thus no real support for the cross-sectional search necessary for interdisciplinary search. Instead, the user would have to refer to different, domain-specific toolsets. Furthermore, the applications limit their scope to the re-ranking and representation of result lists. In contrast, we aim to provide an integrated platform for individualised re-ranking and search, and exploration of the document collection.

3   TIB-arXiv: Focusing on Search and User Interface

The web-based tool TIB-arXiv bundles the existing features of *arXiv* and *Arxiv Sanity Preserver*, and extends and improves them. It enables access to all *arXiv* papers, offers efficient search and individualised ranking functionalities, provides an easy-to-use user interface, which features an integrated PDF reader and additional visualisations.

Data Sources: The TIB-arXiv project is based on the entire *arXiv* data set and is synchronised daily. The website currently manages 1.3 m articles and preview images have been generated for more than half of them. The collected meta information contains the versions, title, authors and abstract of an article and is further extended by version metadata and a categorisation chosen by the author, based on the *arXiv*-supported category set (e.g., cs.AI for computer science and artificial intelligence).

Ranking: TIB-arXiv offers several ways to rank the result list based on different criteria:

- **Date:** Release date of the latest version
- **Twitter:** Number of tweets that mention a certain paper
- **Collection:** Number of copies in the individual users' collections
- **Relevance:** Ranking based on the full-text search engine

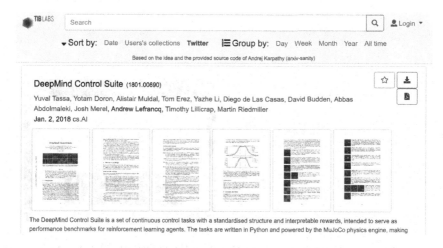

Fig. 1. Desktop version of TIB-arXiv's user interface

Additionally, TIB-arXiv allows to restrict the list to results from a certain time span – thus, the user can for instance explore papers which have been most popular on Twitter on a specific day or during the last month.

Searching: The search is realised using Elasticsearch[2] – a full-text search engine based on the popular Lucene indexing and searching framework[3]. The index of TIB-arXiv relies on the metadata provided by the *arXiv* data set. Search can be limited to data fields selected by the user, for example, displaying only papers written by a certain author.

User Interface: TIB-arXiv's user interface aims for clarity and responsiveness, and adapts to different display sizes, see Fig. 1. Research articles are presented by their title, author list, research domains, thumbnails of their page content and their abstract. Below the summary of the article, the user can find more detailed information about the selected document, including links to tweets that have mentioned this article. Furthermore, the tool allows for direct interaction with relevant research articles – they can be marked and stored for later usage, downloaded or read directly on the platform. To handle different user preferences, TIB-arXiv allows to change the arrangement between the list view and the embedded PDF viewer. Finally, the interface can be displayed on mobile devices as well. The entire user interface is shown in Fig. 2.

4   Conclusions

This paper describes the current state of development of TIB-arXiv. It resolves some of the shortcomings of related tools and (a) enables access to all *arXiv*

---

[2] https://www.elastic.co/products/elasticsearch.
[3] https://lucene.apache.org/.

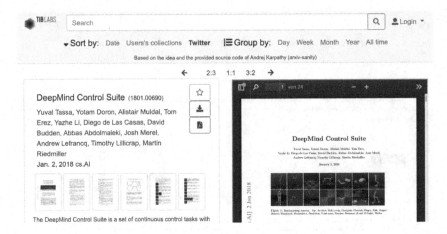

Fig. 2. Search results side by side with the embedded PDF viewer

papers, (b) offers efficient text-based search, (c) contains individualised ranking functionalities, and (d) provides a user interface with an integrated PDF reader, mobile access and additional visualisations.

The tool is work in progress and several developments are planned for the near future: The PDF viewer is to be more interactive – interfacing with reference managers such as Mendeley could allow direct transfer and storage of citations and comments. Mass download services for individualised article packages (e.g., papers of a certain group, appeared in a certain time span, on a specific topic) could streamline scientific inquiry further.

Current focus is on the development of enhanced retrieval methods: deep learning-based approaches for text analysis might allow further improvement of ranking and personalisation. Furthermore, we plan to explore specialised visualisation techniques to enable an engaging exploration of interdisciplinary research papers in *arXiv*.

References

1. arxiv.org::stats. https://arxiv.org/help/stats/2017_by_area/index
2. George, D.P., Knegjens, R.J.: Paperscape. http://paperscape.org. Accessed 11 Apr 2018
3. Karpathy, A.: Arxiv sanity preserver. http://www.arxiv-sanity.com. Accessed 11 Apr 2018
4. Marra, M.: Astrophysicists and physicists as creators of arxiv-based commenting resources for their research communities. An initial survey. Inf. Serv. Use **37**(4), 371–387 (2017)
5. Ménard, B., Magué, J.P.: Arxivsorter. http://www.arxivsorter.org. Accessed 12 Apr 2018
6. Rieger, O.Y., Steinhart, G., Cooper, D.: arxiv@ 25: key findings of a user survey. arXiv preprint arXiv:1607.08212 (2016)
7. Robert Simpson, V.M., Hotan, A.: arxiver. http://arxiver.net. Accessed 11 Apr 2018

An Analytics Tool for Exploring Scientific Software and Related Publications

Anett Hoppe[1]([✉]) [iD], Jascha Hagen[2], Helge Holzmann[3] [iD], Günter Kniesel[4], and Ralph Ewerth[1,3] [iD]

[1] Leibniz Information Centre for Science and Technology (TIB), Hannover, Germany
{anett.hoppe,ralph.ewerth}@tib.eu
[2] Leibniz Universität Hannover, Hannover, Germany
jascha_hagen@yahoo.de
[3] L3S Research Center, Leibniz Universität Hannover, Hannover, Germany
holzmann@l3s.de
[4] University of Bonn, Bonn, Germany
gk@iai.uni-bonn.de

Abstract. Scientific software is one of the key elements for reproducible research. However, classic publications and related scientific software are typically not (sufficiently) linked, and tools are missing to jointly explore these artefacts. In this paper, we report on our work on developing the analytics tool SciSoftX (https://labs.tib.eu/info/projekt/scisoftx/) for jointly exploring software and publications. The presented prototype, a concept for automatic code discovery, and two use cases demonstrate the feasibility and usefulness of the proposal.

Keywords: Software reproducibility · Source code exploration
Cross-modal relations

1 Introduction

The open science movement works towards the general availability of scientific insight and is considered one answer to the so-called "reproducibility crisis" [2]. Science results are often generated by a combination of software, data, and parameters, all of which contribute to the final result (and its interpretation). The complexity of all these elements is hardly describable in a single article – and often the publication does not allow the full reproduction of the achieved results. In the line of work towards consequent reproducibility of scientific results, there are three main tasks to be tackled: (a) motivate researchers to reproduce past results; (b) develop novel ways for the integrated presentation of scientific results; (c) develop tools which allow for exploration of existing scientific works.

The work at hand focuses on the two latter objectives. It presents a tool which facilitates the examination of existing research involving software by joint exploration of a scientific article and the respective source code. The prototype allows the exploration of both in one interface, and the semi-automatic creation

© Springer Nature Switzerland AG 2018
E. Méndez et al. (Eds.): TPDL 2018, LNCS 11057, pp. 299–303, 2018.
https://doi.org/10.1007/978-3-030-00066-0_27

of semantic relations between them. The software is extended by basic visualisations. This kind of work is related to research areas, which have been active for decades: (a) automatic code analysis, and (b) automatic analysis of scientific publications. Solutions for automatic code analysis aim at generating textual documentation [7], summarising code [8], or at generating visualisations [4]. Also common is the generation of formal code models using semantic technologies [1] or logical constructs as realised in tools such as JTransformer[1]. While there is much work on linking code to other (textual) resources (e.g. traceability [3]), to documentation [4], on the automatic understanding of scientific publications [5], or on linking publications with software and archiving them [6], there has been little work on *joint* analytics of scientific software and publications, yet [9].

2   SciSoftX: Scientific Software Explorer

The Scientific Software Explorer provides researchers with functionalities for the exploration of external article-software ensembles and/or annotation of own works for better comprehensibility. Its final version will provide functionalities such as (a) manual annotation of article-software relations, (b) semi-automatic discovery of relations, and (c) visualisations for relation exploration.

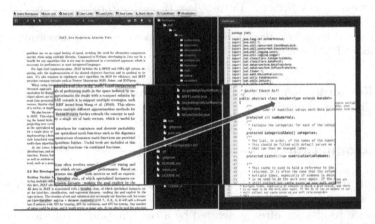

Fig. 1. Main window of the GUI: linked code references are highlighted in colour. (Colour figure online)

2.1   Functionality

SciSoftX allows the user to open and simultaneously view a software project and a publication (Fig. 1). Parsing and processing of source code is realised

---

[1] http://sewiki.iai.uni-bonn.de/research/jtransformer/start.

Fig. 2. Graph-based view on connections between software and publication. Red nodes: mentions in publication; blue nodes: source code packages. (Colour figure online)

using ANTLR[2] (Another tool for Language Recognition) that supports most of the relevant programming languages, while publications are processed via PDF.js[3]. The user can manually link code identifiers to relevant locations in the publication. When the user moves the mouse over a linked identifier in the publications, a tool tip shows the relevant source code positions.

Automatic Discovery of Code Identifiers and Snippets: At the current stage, the tool contains a basic method for the detection of code-relevant text snippets: It relies on the common convention of setting code elements in monospace fonts. The found identifiers are used to search the code model produced by ANTLR, multiple finds are disambiguated based on vicinity. In a random sample of 24 articles from computer science, the monospace-based linker was able to correctly detect 89.9% of the links annotated by a human expert.

Manual Annotation of Links: As a facilitator of exchange between scientists the tool also allows for the manual annotation of resources. In a step-wise process, the user marks article snippets, code elements and annotates the established link with one of the pre-defined labels. The created set of links can be exported to an XML format and imported by an interested reader.

Visualisations: Graph-based visualisations illustrate relations between software and publication on different levels of abstraction. Figure 2 shows an example displaying the connections at the package (software) and page (publication) level.

---

[2] http://www.antlr.org/.
[3] https://mozilla.github.io/pdf.js/.

2.2   Use Cases

Use Case 1 – Reader-side: A researcher reads a publication that refers to a blob of software and then tries to understand the structure and rationale of the software. This time-consuming task can be supported by the automatic creation of links between textual description and actual source code, and the visualisations provided by SciSoftX. The user can click on nodes in the visualisation or on text elements that are highlighted in the publication and explore the implementation details, discover additional parameters, and understand the relevant code part step by step. Furthermore, it is possible to manually add and save useful information and metadata, which can help future users to explore the software.

Use Case 2 – Author-side: Paper authors can use SciSoftX to ensure their software is easily understood, e.g. in a reviewing process or for re-use. Therefore, they make use of the manual and automatic methods to annotate the semantic relations between their paper and the underlying software and publish the annotations. The visualisation of cross-modal relations can aid the authors (and the reviewers) to decide whether all relevant code parts and parameters are covered by the publication. In this way, the tool helps to evaluate the quality of the software description in a paper.

3   Conclusion

Reproducibility is one of the major issues of today's scientific landscape. In this paper, we have reported on work in progress for an analytics tool that allows users to explore relations between scientific software and publications. To this date, the tool features simple mechanisms for detecting links between software and publications which serve as a proof of concept. Future work will explore (a) more powerful infrastructures for code analysis, (b) more sophisticated means for text/image analysis, e.g. mapping diagrams and formulas to source code.

References

1. Atzeni, M., Atzori, M.: Codeontology: RDF-ization of source code. In: d'Amato, C. (ed.) ISWC 2017. LNCS, vol. 10588, pp. 20–28. Springer, Cham (2017). https://doi.org/10.1007/978-3-319-68204-4_2
2. Baker, M.: 1,500 scientists lift the lid on reproducibility. Nat. News **533**(7604), 452 (2016)
3. Borg, M., Runeson, P., Ardö, A.: Recovering from a decade: a systematic mapping of information retrieval approaches to software traceability. Empir. Softw. Eng. **19**(6), 1565–1616 (2014). https://doi.org/10.1007/s10664-013-9255-y
4. Chen, X., Hosking, J.G., Grundy, J.: Visualizing traceability links between source code and documentation. In: IEEE Symposium on Visual Languages and Human-Centric Computing, Innsbruck, Austria, pp. 119–126 (2012). https://doi.org/10.1109/VLHCC.2012.6344496

5. Constantin, A.: Automatic structure and keyphrase analysis of scientific publications. Ph.D. thesis, University of Manchester, UK (2014). http://www.manchester.ac.uk/escholar/uk-ac-man-scw:230124

6. Holzmann, H., Sperber, W., Runnwerth, M.: Archiving software surrogates on the web for future reference. In: Fuhr, N., Kovács, L., Risse, T., Nejdl, W. (eds.) TPDL 2016. LNCS, vol. 9819, pp. 215–226. Springer, Cham (2016). https://doi.org/10.1007/978-3-319-43997-6_17

7. Moser, M., Pichler, J.: Documentation generation from annotated source code of scientific software: position paper. In: Proceedings of the International Workshop on Software Engineering for Science, SE4Science@ICSE 2016, 14 May 2016–22 May 2016, Austin, Texas, USA, pp. 12–15. ACM (2016). https://doi.org/10.1145/2897676.2897679

8. Nazar, N., Hu, Y., Jiang, H.: Summarizing software artifacts: a literature review. J. Comput. Sci. Technol. **31**(5), 883–909 (2016). https://doi.org/10.1007/s11390-016-1671-1

9. Witte, R., Li, Q., Zhang, Y., Rilling, J.: Text mining and software engineering: an integrated source code and document analysis approach. IET Softw. **2**(1), 3–16 (2008). https://doi.org/10.1049/iet-sen:20070110

Digital Museum Map

Mark Michael Hall[(✉)] [iD]

Martin-Luther-Universität Halle-Wittenberg,
Von-Seckendorff-Platz 1, 06120 Halle, Germany
mark.hall@informatik.uni-halle.de

Abstract. The digitisation of cultural heritage has created large digital collections that have the potential to open up our cultural heritage. However, the search box, which for non-expert users presents a significant obstacle, remains the primary interface for accessing these. This demo presents a fully automated, data-driven system for generating a generous interface for exploring digital cultural heritage (DCH) collections.

Keywords: Digital cultural heritage · Generous interfaces · Browsing

1 Introduction

The ongoing digitisation of cultural heritage artefacts has made large swathes of our cultural heritage, which previously were hidden in storage, available to the general public [7]. However, available does not mean accessible, particularly not for non-expert users, as the primary interaction method remains the search box. For non-expert users this often represents a significant hurdle, as they struggle to construct appropriate search queries [3,4,11].

Generous interfaces aim to address this by allowing the non-expert to explore the collection without requiring them to immediately enter query terms. Using a variety of approaches, they are more generous in offering up what the user might want to and can look at. What current generous interfaces generally lack is the ability to give the user an initial overview over what is available, before letting them freely explore that overview. The Digital Museum Map attempts to address this by providing an automatically generated physical-museum-style overview map, which the users can then use to explore the collection.

2 Generous Interfaces

"Generous interfaces" [10] aim to address the limitations of the search box by providing a generous overview or sample of the kind of things the user can find in the collection, which can then be explored through browsing. This primacy of browsing has the potential to enable rich interactions with the collection [1,2,8], particularly for non-expert users who have a strong preference for browsing [6,9]. Whitelaw overviews generous interfaces [10] and the Tyne & Wear Collection

© Springer Nature Switzerland AG 2018
E. Méndez et al. (Eds.): TPDL 2018, LNCS 11057, pp. 304–307, 2018.
https://doi.org/10.1007/978-3-030-00066-0_28

Dive [5] is also often highlighted as a good generous interface. These examples highlight one of the main limitations of current generous interfaces; they either do not provide a complete overview, or, where there is an overview, it is very shallow and does not allow the user to zoom into the data.

3   Digital Museum Map

The Digital Museum Map[1] (DMM) system addresses these limitations by providing an initial overview visualisation that provides the user with an overview over what is available in the collection and which the user can then interactively zoom into to explore the collection.

Data Loading: The DMM is a fully automated and data-agnostic system, thus to apply it to any DCH collection, all that is needed is a custom loading component. In this demo the source data is a collection of 1962 items from the National Museums Liverpool's World Museum Egyptology collection. These cover a wide range of objects from personal jewellery to architectural elements drawn from all periods of ancient Egypt. The items in this collection are annotated with the following meta-data attributes: category, culture, materials, measurements, date made, place made, measurements, collector, and date collected. Not all items are annotated with values for all attributes and the map generation takes this into account when generating the overview.

Structure Generation: To generate the structure that underlies the DMM the system first generates a fully data-driven hierarchical organisational structure and then annotates each node in the structure with context information.

The basic principle of the hierarchical structure is to recursively split the data-set into individual "rooms" of around 50–70 items that belong together conceptually (as defined through their meta-data). The target room size is driven by the number of items that can easily be displayed in the interface without requiring too much scrolling. The splitting algorithm uses a recursive divide-and-conquer approach together with a series of heuristics to automatically determine which attribute to use to split the data and how to group attribute values together to create cohesive nodes higher in the structure.

At each recursive step the algorithm analyses the distribution of attribute values for the current data sub-set to split. For each attribute it calculates how much of the data sub-set the attribute values cover and how many unique values that attribute has in the sub-set. It then chooses the attribute with the highest coverage and if there is a tie the one with the smallest number of attribute values as the attribute to use for partitioning.

The initial partitioning generally creates more than 10 nodes, many of which have few items, which would create a poor structure. Thus in the next step partition values are aggregated, either to create "rooms" with 50–70 items or at

---

[1] Available at https://museum-map.uzi.uni-halle.de.

most 10 higher-level organisational nodes. Where the attribute has an intrinsic ordering (numeric, temporal, spatial) the values are first sorted using that ordering. Then consecutive values are grouped to achieve either the target room size or target number of nodes. Where there is no intrinsic ordering the values are sorted by the number of items assigned to each value. Then a round-robin approach is used to evenly distribute the values into the target number of nodes.

After the organisational structure is generated, a title is assigned to each node. For ordered attributes, the first and last value are used, while for all other attributes all values are joined using commas. Each nodes is also annotated with context information drawn from Wikipedia.

Map Display: The resulting organisational structure forms the basis for the browser-based digital museum, which consists of three aspects: the museum map, the main item viewer, and the contextual information display (see Fig. 1). For the main item viewer the aim was to have a display that lets the user view a large number of items at the same time. Thus initially only the item images are shown. The user has to click on an image to see the item's detailed meta-data. This is driven by the experience of visiting a physical museum, where upon entering a room, the visitor can see all items from a distance, and can then choose which ones they want to look at in more detail.

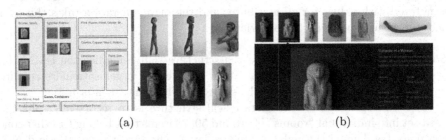

(a)   (b)

Fig. 1. (a) One "floor" of the museum map with the selected "room" highlighted and items shown on the right. (b) The item viewer showing details for one item.

The museum map provides an overview over the top two levels of the organisational structure, where each group in the top level becomes a "floor" of the digital museum, while groups from level two form "rooms" within the "floor". The visual size of the rooms represents the approximate number of items in that room. To place a room, the ratio between the number of items in the room to place and the number of items in the other rooms of the floor is calculated. The space in the "floor" is then split using these ratios. It is always the longer side of the free space that is split, to ensure a relatively even combination of vertical and horizontal splits. For each room, a sample of items are shown as small thumbnails to give an idea of what the room contains.

When the user selects a room on the map, the main display shows the items in that room. Additionally, in the bottom-left corner the contextual information is updated to list the Wikipedia articles that the room was annotated with.

4   Conclusion

The Digital Museum Map automatically generates a data-driven, explorable overview for any DCH collection. It uses a series of heuristics to recursively split the data until the leaf nodes contain 50–70 items for display. This structure is used to provide the non-expert user with an overview over what the digital museum contains and allows them to then freely explore that overview.

At the same time there are a number of planned improvements to the system. Currently attribute semantics are not taken into account when grouping. Additionally, while browsing is a powerful interaction method, users will also want to search the data-set and we plan to integrate search into the current interface. Finally, while the DMM's heuristics are tuned to provide a sensible split of the data, we are also investigating how to let users specify how they would like to see the data, to provide a fully user-driven and personalised DMM interface.

Acknowledgements. Thank go to National Museums Liverpool for providing their data.

References

1. Bates, M.J.: What is browsing—really? A model drawing from behavioural science research. Inf. Res. **12**(4) (2007)
2. Chang, S.J., Rice, R.E.: Browsing: a multidimensional framework. Ann. Rev. Inf. Sci. Technol. (ARIST) **28**, 231–276 (1993)
3. Geser, G.: Resource discovery – position paper: putting the users first. Resource Discovery Technologies for the Heritage Sector **6**, 7–12 (2004)
4. Hall, M.M., de Lacalle, O.L., Soroa, A., Clough, P.D., Agirre, E.: Enabling the discovery of digital cultural heritage objects through Wikipedia. In: Proceedings of the LaTeCH Workshop Held at EACL 2012 (2012)
5. John, C.: I don't know what i'm looking for: better understanding public usage and behaviours with Tyne & Wear Archives & Museums online collections, 29 January 2016 (2016)
6. Lopatovska, I., Bierlein, I., Lember, H., Meyer, E.: Exploring requirements for online art collections. In: Proceedings of the Association for Information Science and Technology, vol. 50, no. 1, pp. 1–4 (2013)
7. Petras, V., Hill, T., Stiller, J., Gäde, M.: Europeana - a search engine for digitised cultural heritage material. Datenbank-Spektrum **17**(1), 41–46 (2017). https://doi.org/10.1007/s13222-016-0238-1
8. Toms, E.G.: Understanding and facilitating the browsing of electronic text. Int. J. Hum.-Comput. Stud. **52**(3), 423–452 (2000)
9. Walsh, D., Hall, M., Clough, P., Foster, J.: The ghost in the museum website: investigating the general public's interactions with museum websites. In: Kamps, J., Tsakonas, G., Manolopoulos, Y., Iliadis, L., Karydis, I. (eds.) Research and Advanced Technology for Digital Libraries, pp. 434–445. Springer International Publishing, Cham (2017). https://doi.org/10.1007/978-3-319-67008-9_34
10. Whitelaw, M.: Generous interfaces for digital cultural collections **9**(1) (2015). http://www.digitalhumanities.org/dhq/vol/9/1/000205/000205.html
11. Wilson, M.L., Elsweiler, D.: Casual-leisure searching: the exploratory search scenarios that break our current models. In: Proceedings of HCIR, pp. 28–31 (2010)

ORCID iDs in the Open Knowledge Era

Marina Morgan[1](✉)(iD) and Naomi Eichenlaub[2](iD)

[1] Florida Southern College, Lakeland, FL 33801, USA
mmorgan@flsouthern.edu
[2] Ryerson University, Toronto, ON M5B 2K3, Canada
neichenl@ryerson.ca

Abstract. The focus of this poster is to highlight the importance of sufficient metadata in ORCID records for the purpose of name disambiguation. In 2017 the authors counted ORCID iDs containing minimal information. They invoked RESTful API calls using Postman software and searched ORCID records created between 2012–2017 that did not include affiliation or organization name, Ringgold ID, and any work titles. A year later, they reproduced the same API calls and compared with the results achieved the year before. The results reveal that a high number of records are still minimal or orphan, thus making the name disambiguation process difficult. The authors recognize the benefit of a unique identifier that facilitates name disambiguation and remain confident that with continued work in the areas of system interoperability and technical integration, alongside continued advocacy and outreach, ORCID will grow and develop not only in number of iDs but also in metadata robustness.

Keywords: ORCID · Open data · Public data · Persistent identifiers
Researcher identifiers · Interoperability · Name disambiguation
Metadata assessment · API

1 ORCID, ORCID iDs, and Name Disambiguation

1.1 ORCID and ORCID iDs

ORCID is a non-profit organization that has facilitated an open registry of persistent digital identifiers since 2012. ORCID is backed by a vast community comprised of research organizations and institutions, publishers, professional associations, individual research and organizational members.

ORCID iD, the Open Researcher and Contributor Identifier, is a persistent digital identifier provided by ORCID to researchers across disciplines. The purpose of the ORCID iD is to distinguish researchers by accrediting their research and activities, and improving recognition and discoverability. The only required metadata fields for creating and registering an ORCID iD are first name and primary email address [1]. The researcher has the flexibility of controlling what data is entered in their ORCID iDs, which can be made private, public, or shared with a limited group. It is the researcher's responsibility to create and maintain their ORCID record as ORCID is not in a position to modify any incorrect data [2].

© Springer Nature Switzerland AG 2018
E. Méndez et al. (Eds.): TPDL 2018, LNCS 11057, pp. 308–311, 2018.
https://doi.org/10.1007/978-3-030-00066-0_29

1.2   Name Disambiguation

Author name remains an ongoing challenge in the field of scholarly communication. Extensive changes to authorship and collaboration in recent decades have resulted in an acceleration in multi-authorship. These changes can be attributed in large part to the present day facility for online collaboration but also to an increase in interdisciplinary collaboration and a heightened focus on research and publication outputs. As such, today it is not uncommon for science articles to be authored by hundreds of authors [3], but the trend of increasing co-authorship applies in the social sciences and humanities as well [4, 5]. A further challenge has been the growth in research output from countries where authors have recurrent similar names [6]. All of these authorship issues impact researcher identity and increase the urgency for author name disambiguation.

2   Methodology and Results

2.1   Background

Approaching from the perspective of exploring the quality and utility of author name metadata, in 2017 the authors investigated and queried the ORCID records and metadata fields using the public ORCID API to evaluate the completeness of metadata in ORCID records [7]. ORCID iDs are useful in the author disambiguation process only if they provide enough information to distinguish one author from another. Thus, the focus of the investigation was to count the records with only the minimal required information, specifically name and email. The search focused on records created from 2012 to May 2017 that did not include affiliation name, any work titles, and Ringgold ID (a numerical identifier assigned to scholarly institutions). The results revealed that approximately 65% of the ORCID records were minimal, or "orphan". Moreover, it was observed that some records included false or misleading metadata such as random names (John Doe or Jane Doe), misleading funding (Awards-R-Us or Grant-R-Us), and filler text in the biographies (lorem ipsum), further hampering disambiguation.

2.2   Methodology and Results

To get a better sense of how many ORCID records have only minimal information and in order to generate the results needed for the comparison with the original results, we reran the same public ORCID API calls for records created between 2012 and 2017. The 2017 count includes records created between January and May. The same authentication token and Postman software were used to invoke RESTful API calls.

As seen in Table 1, even though there is a slight improvement in the number of records that do not include affiliation name, Ringgold ID, and any work titles, orphan or empty records are still prevalent. Additionally, the results reveal that on average there has been a 6.5% improvement, an unexpectedly low percentage. However, this does not come as a surprise as ORCID "does not absolutely prevent multiple iDs/records from being intentionally created and maintained by an individual" [8]. Despite ORCID's policies in place to avoid duplicate records, the high number of orphan records may be because of duplicates since individuals who have multiple email addresses can in theory create as many records as email addresses.

Table 1. Comparison of public API calls for minimal ORCID records for 2017 and 2018.

Year	2017 count	2018 count	Improvement percentage
2012	25,351	23,724	6.42%
2013	258,182	239,582	7.20%
2014	370,074	344,213	6.99%
2015	479,144	448,453	6.41%
2016	709,046	666,447	6.01%
2017*	372,709	344,405	7.59%
2012–2017*	2,216,944	2,070,491	6.61%

* January–May 2017

Table 2 illustrates a total count of ORCID records starting with 2012. The 2017 count revealed that approximately 65% of the total records were minimal. When the count was repeated in April 2018, the results showed that nearly 45% are minimal. This decrease from 65% to 45% represents a significant reduction. However, with just under half of ORCID records remaining without enough metadata for proper disambiguation, there is still room for improvement.

Table 2. Total ORCID records

Date	Total ORCID records	Minimal ORCID records	Percentage of minimal ORCID records
2012–May 17, 2017	3,391,358	2,216,944	65.37%
2012–April 03, 2018	4,625,545	2,070,491	44.76%

The total number of ORCID records continues to grow at an impressive rate of 36.4% from May 2017 to April 2018. Therefore, the rapid uptake of ORCID by researchers shows that while many records do not have adequate metadata, there is the potential for both comprehensive researcher uptake as well as more robust and complete associated metadata in the future [9]. ORCID initiatives such as Collect & Connect, which aim to validate research affiliations through authentication, have likely contributed to the reduction in orphan records shown in Table 2 [10].

3   Conclusions

This project involved a follow-up investigation of work done in 2017 to determine the completeness of metadata in ORCID records. Specifically, the authors used API calls to conduct searches to count records created between 2012 and May 2017 that did not include affiliation name, any work titles, and Ringgold ID. API calls were repeated in April 2018 in order to determine changes in counts. The 2017 count revealed that more than 65% of the total records were minimal. When the count was repeated in April 2018, the results showed that approximately 45% are minimal, a substantial decrease.

Since ORCID does not modify incorrect data, but they may correct invalid data such as empty or wrongly formatted fields, this improvement is likely due to the researchers' recognition of the importance of accurate data, or perhaps adding additional metadata to a record that was previously created.

Lastly, the authors recognize the benefit of a unique identifier that facilitates name disambiguation and the ongoing ORCID initiatives to increase metadata robustness and record quality through initiatives such as Collect & Connect.

References

1. Building your ORCID record & connecting your iD. https://support.orcid.org/knowledge base/articles/142948-name. Accessed 18 June 2018
2. ORCID Trust. https://orcid.org/about/trust/integrity. Accessed 18 June 2018
3. King, C.: Multiauthor papers: onward and upward. ScienceWatch Newsletter (2012). http://archive.sciencewatch.com/newsletter/2012/201207/multiauthor_papers/. Accessed 05 Apr 2018
4. Taylor & Francis Group, Co-authorship in the Humanities and Social Sciences: A global view. A white paper from Taylor & Francis (2017). http://authorservices.taylorandfrancis.com/wp-content/uploads/2017/09/Coauthorship-white-paper.pdf. Accessed 05 Apr 2018
5. Kuld, L., O'Hagan, J.: The proportion of co-authored research articles has risen markedly in recent decades. LSE Impact Blog (2018). http://blogs.lse.ac.uk/impactofsocialsciences/2018/04/04/the-proportion-of-co-authored-research-articles-has-risen-markedly-in-recent-decades/. Accessed 05 Apr 2018
6. Youtie, J., Carley, S., Porter, A.L., Shapira, P.: Tracking researchers and their outputs: new insights from ORCIDs. Scientometrics 113, 437 (2017)
7. Eichenlaub, N., Morgan, M.: ORCID: using API calls to assess metadata completeness. In: Proceedings of the International Conference on Dublin Core and Metadata Applications, pp. 104–107. Dublin Core Metadata Initiative, Washington D.C. (2017)
8. Managing Duplicate ORCID iDs. https://orcid.org/blog/2014/01/09/managing-duplicate-iDs. Accessed 05 Apr 2018
9. Klein, M., Van de Sompel, H.: Discovering scholarly orphans using ORCID. In: 2017 ACM/IEEE Joint Conference on Digital Libraries (JCDL) (2017)
10. ORCID: Welcome to Collect & Connect: ORCID's integration and engagement program. https://orcid.org/content/collect-connect. Accessed 10 Apr 2018

Revealing Historical Events Out of Web Archives

Quentin Lobbé[✉] [iD]

LTCI, Télécom ParisTech, Université Paris Saclay & Inria, Paris, France
quentin.lobbe@telecom-paristech.fr

Abstract. As the living Web expands, worldwide volumes of Web archives constantly increase, making difficult to identify relevant archived contents. Here we propose an application for detecting historical events out of a corpus of Web archives and based on an entity called *Web Fragment*: a semantic and syntactic subset of a given Web page. The Web fragment has the particularity to be indexed by its edition date instead of its archiving date. We apply our framework on an archived Moroccan forum and witness how it reacted to the Arab Spring at the end of 2010.

Keywords: Web archives · Event detection
Online migrants collectives

1 Introduction

Since the creation of the Web in the early 90's [2], the loss of the digital content that constitutes the Web itself has been considered a major issue [6]. Whereas related works mainly focus on upstream Web archive acquisition [8], we choose here to perform the exploration of an existing corpus. But apart from the online portal of the WayBack Machine[1], the majority of corpora of Web archives only allows local consultation points, with no remote access or API. In this paper, we first introduce the usage of a new entity called *Web fragment* to guide researchers through Web archives at retrieval time (Sect. 2). As we think that most explorers of Web archives pursue the discovery of events of some sort, we describe an application called *Web Archive Explorer* (WAE) for detecting historical events (Sect. 3). We finally use the WAE to understand how the Moroccan forum *yabiladi.com* reacted to the Arab Spring at the end of 2010 (Sect. 4).

2 Setup

An Online Migrants Collective. As input data for WAE we use the Moroccan section of the e-Diasporas Atlas [3]. The Atlas revealed diasporic communities

---

[1] https://archive.org/Web/.

© Springer Nature Switzerland AG 2018
E. Méndez et al. (Eds.): TPDL 2018, LNCS 11057, pp. 312–316, 2018.
https://doi.org/10.1007/978-3-030-00066-0_30

organized as networks of migrant Web sites connected to each other through hypertext links. But facing the partial or total disappearance of some of the observed Web sites, it was decided to start archiving them. Thus, the corpus was archived from March 2010 to September 2014, covering 254 Web sites[2]. In Sect. 4, we will focus on the Moroccan forum *yabiladi.com*: an old established Web site, representing a set of 2.8 million archived Web pages.

Web Fragment. A Web fragment is a coherent set of textual, audiovisual or animated contents extracted from a Web page and understandable on its own. It can be a meaningful object like a post inside a forum, a news article, or a comment, and it has the particularity to be indexed by its edition date (the time when it was written). We assume that an original *edition date* will always be more historically accurate than the *download date* of the parent archived Web page. In practice, a Web fragment is the result of the agglomerative clustering of some of the HTML nodes that constitute a given Web page. To extract them, we extend the boilerplate method from [7] and use a combination of vision-based [1] and tag-based scraping strategies [5].

Event Detection Model. Following our logic of archive exploration, we don't want to detect specific events with expert knowledge, so we avoid patterns and clustering methods. We instead use a threshold-based heuristic [4] within a sliding time frame of one week. We define an event as a detected outlier in the temporal distribution of a set of Web fragments that matches a given keyword. We try explaining the events by finding semantic correlation between the text content of the Web fragments (splitted in bigrams) and a set of Morrocan news titles. As *yabiladi.com* is a combination between a news provider and a forum, we choose to construct an index of potential events using the titles and the edition dates extracted from its news section. To sum up, a historical event is the semantic encounter between a well-dated news title and a burst of web fragments.

3   Architecture

We now introduce the components of WAE[3]. We refer to Fig. 1 as an illustration of its architecture: (1) Our data set is recorded under the Digital Archive File Format (DAFF) formalism that separates the metadata (URL, download date, etc.) from the archived data contents (original HTML content). Our Moroccan corpus results in a 30 GB metadata DAFF file and a 300GB data DAFF file. (2) The ArchiveMiner component grabs the files using a Java extractor which uploads them into a Hadoop Distributed File System (HDFS). Then a distributed Spark[4] pipeline ingests the HDFS and groups the metadata by time-stable versions and joins them to the data contents. A set of filters based on

---

[2] Publicly available at http://maps.e-diasporas.fr/index.php?focus=map&map=5& section=5.

[3] Open source and available at https://github.com/lobbeque/archive-miner and https://github.com/lobbeque/peastee.

[4] See http://hadoop.apache.org/, http://spark.apache.org/ and http://lucene.apache. org/solr/.

download dates or domain names are then applied. (3) We enrich the original corpus by adding qualitative informations such as the main language (French, Moroccan, Spanish, etc.) or category (forum, blog, media, etc.) of each Web site. (4) The FragmentsExtractor component divides each archived Web page into Web fragments (Sect. 2). Every edition date is translated from a natural language format into a normalized date format. Additionally, the component extracts the text content, author and title of each Web fragment and joins them with the information inherited from the parent Web pages (URL, download date, etc.). (5) The ArchiveSearch component indexes the Web fragments into a Solr search engine. A lemmatizer is then applied to increase the accuracy of the full-text facilities. Custom requestHandlers are built to allow different time query strategies. (6) The WAE provides two different inverted indexes (Sect. 2): first the Web fragments extracted from the forum section of *yabiladi.com* and then the events extracted from its news section. (7) The ArchiveViz component provides an interface to request the archives by writing a set of keywords and choosing the granularity of the query: Web pages or Web fragments. The results are displayed as a list of documents, illustrated with histograms and a bigrams viewer linked to the events detection system.

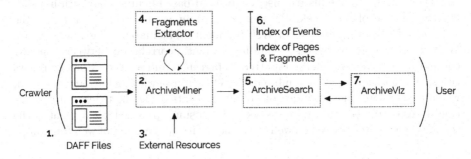

Fig. 1. Architecture of the Web Archive Explorer (WAE)

4   Demonstration Scenario

We now describe[5] a set of use cases where WAE helps to reveal historical events: (1) The user first tries to query the Wayback Machine. But, as it is built on top of a search-by-URL system, the keywords *morroco king* do not match the real content of the archived Web pages. They can only match strict URLs or HTML titles. (2) With our system, the user request for *roi* (meaning king in french) and selects *pages* for granularity. The user has to pick up a range of dates to filter the archives. The top ten resulting archived pages are ordered by default Lucene similarity. (3) The user now specializes his query by focusing on one of the main author contributions. The 5 most prolific authors are displayed

---

[5] See the accompanying video https://youtu.be/snW4O-usyTM for a peek at the GUI.

in the facets section of the GUI. (4) One can use the first histogram to see the number of matching Web pages by week. Below, there is a line chart displaying the ratio of matching bigrams by weeks. There, the user can follow the evolution of the word *king* in the corpus. The event detection system does not find any matching event because the user chose to use Web pages as a scale. But pages are timestamped by download date without regard for any historical accuracy. (5) Now the user switches to the Web fragment level, enters the same query and witnesses that she has the possibility to study Web fragments written up to 2003. The event detection system now understands that around the late 2004 an event concerning the king may have focused the conversations on *yabiladi.com*. The system identifies it as an official visit of the Morrocan king to Mexico in November 2004. (6) The WAE supports multiple queries (using comma as a separator) for comparison purpose. It displays a coloured line in the n-gram viewer for each query and a union of the resulting fragments in the list below. The user can clearly see a growing percentage of the phrase *Ben Ali* (the former Tunisian president) during the late 2010 that we may correlate to the beginning of the Arab Spring in Tunisia. This assumption is reinforced by a triggered event about the destitution of Ben Ali in January 2011. (7) Finally, more seasonal keywords can be entered in the search box such as the muslim month of fasting *ramadan*. Here the user observes a temporal pattern in the archives that can be explained by the cultural specificity of our Moroccan corpus.

5   Conclusion

In this paper, we proposed an application to reveal historical events. We introduced a new entity called *Web fragment* to guide researchers through an exploration of Web archives at retrieval time. We described the architecture of our application and, as a demonstration, we witness how the online community of the Moroccan forum *yabiladi.com* reacted to the Arab Spring at the end of 2010. In the future, we will feed our application with more diverse sets of Web archives (social media streams, blogging platforms, etc.) and work in close collaboration with sociologists and historians to investigate multidisciplinary research questions based on Web archive analysis.

References

1. Cai, D., Yu, S., Wen, J.R., Ma, W.Y.: VIPS: a vision-based page segmentation algorithm (2003)
2. CERN: The document that officially put the world wide web into the public domain (1993). http://cds.cern.ch/record/1164399
3. Diminescu, D.: e-Diasporas Atlas. Explorations and Cartography of Diasporas on Digital Networks. Ed. de la Maison des Sciences de l'Homme, Paris (2012)
4. Fung, G.P.C., Yu, J.X., Yu, P.S., Lu, H.: Parameter free bursty events detection in text streams. In: Proceedings of the 31st International Conference on Very Large Data Bases, pp. 181–192. VLDB Endowment (2005)

5. Jatowt, A., Kawai, Y., Tanaka, K.: Detecting age of page content. In: Proceedings of the 9th annual ACM International Workshop on Web Information and Data Management, pp. 137–144. ACM (2007)
6. Kahle, B.: Preserving the internet. Sci. Am. **276**(276), 82–83 (1997)
7. Kohlschütter, C., Fankhauser, P., Nejdl, W.: Boilerplate detection using shallow text features. In: Proceedings of the Third ACM International Conference on Web Search and Data Mining, WSDM 2010, pp. 441–450. ACM, New York (2010)
8. Masanès, J.: Web Archiving. Springer, New York (2006). https://doi.org/10.1007/978-3-540-46332-0

The FAIR Accessor as a Tool to Reinforce the Authenticity of Digital Archival Information

André Pacheco^(✉) (iD)

University of Coimbra, Coimbra, Portugal
andrez.pacheco@gmail.com

Abstract. The constant changeability of the digital environment raises a complex series of issues regarding the preservation of authentic, accessible, intelligible and reusable digital information. An implementation of the FAIR Accessor, a technology developed with the goal of delivering findable, accessible, interoperable and reusable research data, is discussed as a means of supporting archival description with the goal of ensuring its authenticity. A qualitative literature review focused on some of the main tenets of digital preservation in the fields of Information Science, Diplomatics and research data is followed by a discussion on how the core criteria of each area overlap and complement each other. It is concluded that the FAIR Accessor can assist in providing a rich archival description, ultimately helping to determine the authenticity of records.

Keywords: Information Science · FAIR data · Diplomatics
Archival description · Digital preservation

1 Introduction

The still existing lack of inadequate practices in addressing the issues of preservation of the authenticity of records on the long term, which frequently leads to the inability to constitute evidence [1] demands further research. This study attempts to contribute with an unprecedented theoretical approach that analyzes the concept and structure of a FAIR Accessor, a resource architecture based on the core principles that research data needs to be findable, accessible, interoperable and re-usable (FAIR), and discusses its reproducibility within the framework of Information Science, as a means of bolstering archival description, specifically for ascertaining a digital record's authenticity.

2 Methodology

In research data management, the focus has been on how to make research findings openly available, whereas archival science shares the same purpose with the need to add an authenticity layer. This paper provides a literature review that unravels points of contact between these methodologies. The principles of Diplomatics in respect to the

© Springer Nature Switzerland AG 2018
E. Méndez et al. (Eds.): TPDL 2018, LNCS 11057, pp. 317–320, 2018.
https://doi.org/10.1007/978-3-030-00066-0_31

preservation of authentic records are also considered in the literature review in order to strengthen their connection. The findings from the literature review are summarized into a novel exploratory research that bridges the methodologies of these fields by demonstrating how the structure of a FAIR Accessor can prove useful for the description of authentic archival information, within the framework of Information Science.

3 Theoretical Exploration

3.1 The Object of Information Science

Today it is increasingly common to acknowledge that the object is social information [2], which gains form and existence through a document, its carrier. As a result, the management of digital archival information has to look beyond the scope of the individual record and also consider the information technological system that provides the connective tissue for its creation, handling and use [3]. Consequently, the presumption of authenticity shifts from the sheer appreciation of the medium to the understanding of a record's associated metadata, which is expected to accurately provide a description of its context and nature, and to ensure the continuous access for as long as necessary [4].

3.2 Diplomatic Authenticity

From a diplomatic perspective, a record is a conceptual embodiment of internal and external elements that must be analyzed in order to assess its authenticity [5]. In order to assess a record's authenticity, one must be able to establish its identity and demonstrate its integrity [6]. Records that result from the same function exhibit the same documentary form, and are therefore linked by an archival bond, which represents the organic relationship that a record shares with other records of the same system created in the course of the same activity [7].

In analog records, to ensure the conservation of its medium is equivalent to safeguard its integrity, identity and its authenticity, since it embodies both form and content. However, in the case of digital records, these components exist separately, scattered across the information system. For this reason, authenticity is no longer observable *in the documentt itself, but in the procedures of creation, maintenance and preservation* [8]. As a result, *the best method of ensuring ongoing authenticity of electronic records is external to the records themselves and involves a tight control on record-making and record keeping procedures* [6], by the inclusion of preservation metadata.

3.3 Structure of a FAIR Accessor

The FAIR principles are a theoretical guideline that define characteristics that contemporary data resources, tools, vocabularies and infrastructures should exhibit to assist discovery and reuse by third-parties. Data can be considered findable when it has a unique persistent identifier; accessible when there is a clear protocol with clear access rules; interoperable when data is machine-actionable using shared vocabularies and/or ontologies; and reusable with it is findable, accessible, and sufficiently well-described with contextual information, allowing discoverability and use by both humans and machines [9].

In order to comply to these criteria, Wilkinson and his colleagues started by developing the FAIR Accessor, a lightweight HTTP Interface that provides unique identifiers for all data entities with a machine-readable metadata. It uses the Linked Data Platform architecture to describe the function or purpose of each Uniform Resource Identifier (URI), and the nature of that resource. It is structured in two layers that can be managed independently, the "Container Resource" which provides metadata about any research object (either a repository, a database, a workflow, a record), and the "MetaRecord", which describes the individual resources [9]. Each description level possesses a unique identifier that helps discoverability of other records. In addition, since it does not require the development of additional technology, as it uses web technologies, it becomes a low-cost implementation that allows for any web-crawler agent to discover the data, acting as a web page.

In short, the FAIR Accessor is characterized by the inclusion of rich metadata that facilitates discovery and interoperability of both repository and record-level information, described by widely accepted vocabularies.

4 Discussion

A discussion will follow on how this illustrative implementation can serve the interests of records management in respect to ensuring the authenticity of digital information. Based on the notions revealed by the literature review, it is possible to formulate three essential characteristics that digital records should exhibit regarding their authentic preservation, access and description: (i) the ability to demonstrate their identity and integrity; (ii) a clear archival bond; (iii) permanent access. The first criteria relies heavily on descriptive metadata for content, structure and context, whereas for the archival bond it is necessary to make explicit the organic relations that a record has with other records in the system created in the course of the same action or procedure. Lastly, records are permanently accessible when its data is continuously available.

The FAIR Accessor can assist in boosting archival description according to an infrastructure that allows to fulfill these objectives. In regard to identity, both layers of the Accessor provide a proper and predictable place to include every metadata considered relevant, whether at a repository or funds level (*e.g.*, information about custody or administrative history), or even at a record level (*e.g.*, author, title, creator, medium). The choice of these fields could and should comply to international archival description standards, such as ISAD(G) and to metadata standards (*e.g.* Dublin Core, PREMIS, METS), promoting interoperability of descriptions. Integrity can be demonstrated by including metadata on changes made to the record.

The archival bond is guaranteed, on one hand, by the fact that each resource level possesses an indication of its place in the hierarchy, *i.e.*, each layer of the accessor exhibits information on the resources it contains/is contained in, therefore connecting a record to others that are functionally related to it. On the other hand, the rich descriptive metadata also helps identifying the bond (*e.g.*, who the fonds belongs to).

Finally, permanent access is promoted by web-based nature of the structure, not requiring any specific API for data exploration. This means that resources are permanently available and discoverable in the web, both by human and machines. Each resource possesses a unique identifier that can be used as its access point. The interoperability of archival description, necessary to interpret resources in different technological contexts, can be ensured by using the .xml format to store any pertinent metadata.

5   Conclusions

This study revealed that the contribute of the FAIR Accessor to describe digital archival information in a way that promotes its authentic preservation, and ensures that it is findable, accessible and reusable, is not only possible, but also potentially significant. It was shown to provide an infrastructure that includes rich and comprehensive metadata; preserves and makes explicit the archival bond; allows for both machine and human discoverability; has universal access; and provides unique identifiers.

Acknowledgments. This study was fully supported by the Portuguese Foundation for Science and Technology, under the PhD research grant SFRH/BD/131004/2017.

References

1. Duranti, L.: Theoretical Elaborations into Archival Management in Canada (TEAM Canada): Implementing the theory of preservation of authentic records in digital systems in small and medium-sized archival organizations. SSHRC CURA InterPARES 3 Project Proposal, sections 1–7, pp. 11–29 (2007). https://doi.org/10.1016/j.lcats.2008.02.004
2. Cook, T.: Electronic records, paper minds: the revolution in information management and archives in the post-custodial and post-modernist era. Arch. Soc. Stud.: J. Interdisc. Res. 1(0), 399–443 (2007)
3. da Silva, A.M.: A Informação: da compreensão do fenómeno e construção do objeto científico. Edições Afrontamento, Porto (2006)
4. ISO 15489-1: Information and documentation — Records management — Part 1: General. ISO, Geneva (2001)
5. Ribeiro, F.: O perfil profissional do arquivista na sociedade da informação. Trabalhos de Antropologia E Etnologia 45(1), 49–57 (2005)
6. Duranti, L., Blanchette, J.: The authenticity of electronic records: the InterPARES approach. In: The Society for Imaging Science and Technology, pp. 215–220. IS&T, Virginia (2004)
7. Macneil, H.: Archival theory and practice: between two paradigms. Arch. Soc. Stud.: J. Interdisc. Res. 1(1), 6–20 (2007)
8. InterPARES 1: Authenticity task force report. In: The Long-term Preservation of Authentic Electronic Records: Findings of the InterPARES Project. Archilab, Michigan (2005)
9. Wilkinson, M. et al.: Interoperability and FAIRness through a novel combination of Web technologies. PeerJ Comput. Sci. 3(e110) (2017). https://doi.org/10.7717/peerj-cs.110

Who Cites What in Computer Science? - Analysing Citation Patterns Across Conference Rank and Gender

Tobias Milz[1](✉) [iD] and Christin Seifert[2] [iD]

[1] University of Passau, 94030 Passau, Germany
tobias.milz@uni-passau.de
[2] University of Twente, PO BOX 217, 7500 Enschede, The Netherlands
c.seifert@utwente.nl

Abstract. Citations are a means to refer to previous, relevant scientific bodies of work. However, little is known about how citations behave with respect to venue reputation. Do A* papers get more often cited by C papers or vice versa? What is the source and sink of a citation in terms of venue reputation? In this work, we investigate this issue by analysing the DBLP database of computer science publications, utilizing rank information from the CORE database. Our analysis shows that authors tend to cite publications from the same or higher ranked venues more often than from lower tier venues. Self-citations, on the contrary, are especially focused on same-tier venues. The gender of the first author does not seem to have any impact on the citations from and to differently ranked mediums.

Keywords: Citations · Self-citations · Analysis · DBLP · CORE

1 Introduction

Citations are a means to refer to previous scientific bodies of work, and are also used to calculate impact factors for journals [1,4] and performance measures for scientists [3] and thus have become a valuable commodity in science. Research has been concerned with finding influencing factors for citations (e.g. [11]), and most prominently to identify the influence of self-citations on citations and subsequently on indicators of scientific performance, e.g. [2,5]. Multi-authored, as well as papers with male first author have been found to have a higher self-citation rate [2,5,9], while self-citation rates generally vary over fields and countries [14]. To the best of our knowledge, the only study that investigated the relation of self-citations and the scientific reputation of the publication venue is in the economics domain [8]. The authors found that the proportion of self-citations increased with the impact factor of ecology journals.

This paper contributes to the knowledge of citation and self-citation by analysing the domain of computer science. Specifically, we investigate the DBLP computer science bibliography [10] w.r.t. ranking of the conferences/journals and gender of the first author.

© Springer Nature Switzerland AG 2018
E. Méndez et al. (Eds.): TPDL 2018, LNCS 11057, pp. 321–325, 2018.
https://doi.org/10.1007/978-3-030-00066-0_32

2   Problem Statement

Citations can either be synchronous (outgoing) or diachronous (incoming) [7], the former refers to the number of publications a paper cites and the latter how often a publication gets cited. Analogously, outgoing and incoming self-citations are citations from and to publications of the same author. The self-citation rate is defined as the ratio of the self-citations normalized by the total number of citations and can be calculated for both, incoming and outgoing self-citations. In this paper, we analyse incoming and outgoing citations and self-citation rates with respect to the conference/journal rank. For instance, if paper P cites paper Q, and P was published at an A* conference while paper Q was published at a C conference, the citation counts as an outgoing citation for A* and incoming citation for C.

3   Method

For our analysis, we use the DBLP citation graph [13], supplemented with the paper's ranking information and a gender attribute for the authors. The rankings are extracted from the Computing Research and Education Association of Australasia (CORE) database[1] using a rule-based string matching method of the venue name. The focus of this method is to find the most likely match, but without introducing any false-positives in favour of Recall. The publication year of the papers is also considered in order to take rank changes of venues into account. We follow previously suggested methods to determine an author's gender by matching their first name (given name) to country-specific name lists [6]. For author identity, we rely on the quality of the DBLP citation graph, which already employs author name disambiguation approaches [12]. Out of all 3,079,007 papers in DBLP covering the publication and citation period from 1946–2018, 55.66% (1,744,449) were assigned a binary (female/male) gender based on the first author's inferred gender. A CORE rank was assigned to 14.15% (435,823), while both information could be assigned to 7.86% (242,096) of all papers.

4   Results

The heatmaps in Fig. 1 show the fraction of outgoing and incoming citations and self-citations for publications from each conference/journal rank. The initial theory is, that publications will more often cite highly ranked papers, as they have more visibility. According to the results, this hypothesis seems to hold true. For example, 93.6% of all outgoing citations from publications with a B rating, cite other publications with the same or higher rating (top left). Furthermore, A, B and C ranked papers receive more than half of all their incoming citations from publications of the same rank (top right). For self-citations, this effect is

---

[1]   http://www.core.edu.au, accessed 2018-03-02.

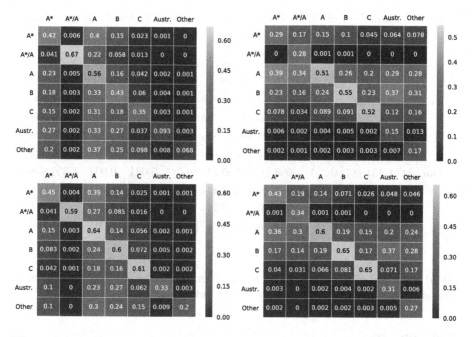

Fig. 1. Ratio of citations (top) and self-citations (bottom) from venues with specific rank. Rows indicate the source and columns the target of citations. Left: normalized by the total number of outgoing citations per rank; right: normalized by total incoming citations per rank.

even more prominent especially for the categories C and Australasian, which have much lower citation rates (35.1% and 9.3% respectively) than self-citation rates (61.2% and 33.2% respectively) towards same-tier publications (bottom). In other words, authors prefer to cite higher ranked publications, but self-citations are more commonly towards publications of the same conference/journal rank. Please note, that although a difference is observable in values for categories Australasia and Other, we abstain from an interpretation, since both categories only contain 4318 (0.9%) of the papers with an assigned rank.

Table 1 shows the statistics w.r.t. venue rank and gender of the first author. For example, out of all 1,957,108 outgoing citations towards papers with a male lead author, 13.8% are cited in publications from conferences/journals with an A* rating. This citation-rate indicates how citations from/to differently rated mediums are affected by the first author's gender of the cited/citing paper. The results show that despite the lower number of papers with female leading authors (410,262 papers with female and 1,334,187 with male lead author), the distribution of the incoming and outgoing citation rate stays the same. In other words, the gender of the leading author has no significant effect on the citations of papers when considering their identified rating.

Further studies are required to shed light on the reason for the difference in citation/self-citations behaviour w.r.t. rank. An interesting future question

Table 1. Comparison of citations by gender (M - male, F- female, X - unisex, ? - unknown) and conference/journal rank

		Papers	Conference/journal rank							$\sum$
			A*	A*/A	A	B	C	Austr.	Other	Citations
In	M	1,334,187	0.138	0.003	0.387	0.319	0.143	0.006	0.003	1,957,108
	F	410,262	0.126	0.001	0.398	0.325	0.143	0.005	0.003	417,655
	X	609,101	0.134	0.002	0.371	0.343	0.144	0.005	0.003	748,836
	?	725,453	0.117	0.001	0.355	0.346	0.174	0.005	0.003	676,809
Out	M	1,334,187	0.234	0.008	0.427	0.239	0.087	0.004	0.002	1,355,908
	F	410,262	0.226	0.003	0.433	0.248	0.084	0.004	0.001	430,910
	X	609,101	0.231	0.004	0.432	0.244	0.084	0.004	0.001	733,721
	?	725,453	0.237	0.002	0.427	0.235	0.094	0.004	0.001	761,150

would be, whether a homophily property in citation behaviour can be observed, i.e., whether a specific gender tends to cite authors of the same gender.

References

1. Time to remodel the journal impact factor. Nature **535**(7613), 466–466 (2016)
2. Aksnes, D.W.: A macro study of self-citation. Scientometrics **56**(2), 235–246 (2003)
3. Alonso, S., Cabrerizo, F., Herrera-Viedma, E., Herrera, F.: h-index: a review focused in its variants, computation and standardization for different scientific fields. J. Inform. **3**(4), 273–289 (2009)
4. Callaway, E.: Beat it, impact factor! Publishing elite turns against controversial metric. Nature **535**(7611), 210–211 (2016)
5. King, M.M., Bergstrom, C.T., Correll, S.J., Jacquet, J., West, J.D.: Men set their own cites high: gender and self-citation across fields and over time. Socius **3**, 2378023117738903 (2017)
6. Larivière, V., Ni, C., Gingras, Y., Cronin, B., Sugimoto, C.: Bibliometrics: global gender disparities in science. Nature News **504**, 211–213 (2013)
7. Lawani, S.M.: On the heterogeneity and classification of author self-citations. J. Am. Soc. Inf. Sci. **33**(5), 281–284 (1982)
8. Leblond, M.: Author self-citations in the field of ecology. Scientometrics **91**, 943–953 (2012)
9. Milz, T., Seifert, C.: Analysing author self-citations in computer science publications. In: Elloumi, M. (ed.) Database and Expert Systems Applications. Springer, Cham (2018). https://doi.org/10.1007/978-3-319-99133-7_24
10. Ley, M.: The DBLP computer science bibliography: evolution, research issues, perspectives. In: Laender, A.H.F., Oliveira, A.L. (eds.) String Processing and Information Retrieval, pp. 1–10. Springer, Heidelberg (2002). https://doi.org/10.1007/3-540-45735-6_1
11. Medoff, H.M.: The efficiency of self-citations in economics. Scientometrics **69**, 69–84 (2006)

12. Tang, J., Fong, A.C.M., Wang, B., Zhang, J.: A unified probabilistic framework for name disambiguation in digital library. IEEE Trans. Knowl. Data Eng. **24**(6), 975–987 (2012)
13. Tang, J., Zhang, J., Yao, L., Li, J., Zhang, L., Su, Z.: Arnetminer: extraction and mining of academic social networks. In: Proceedings of SIGKDD International Conference on Knowledge Discovery and Data Mining, pp. 990–998. ACM, New York (2008)
14. Thijs, B., Glänzel, W.: The influence of author self-citations on bibliometric meso-indicators. The case of European universities. Scientometrics **66**, 71–80 (2006)

Back to the Source: Recovering Original (Hebrew) Script from Transcribed Metadata

Aaron Christianson, Rachel Heuberger, and Thomas Risse[✉]

University Library J. C. Senckenberg, Goethe-University Frankfurt,
Frankfurt, Germany
{a.christianson,r.heuberger,t.risse}@ub.uni-frankfurt.de

Abstract. Due to technical constrains of the past, metadata in languages written with non-Latin scripts have frequently been entered using various systems of transcription. While this transcription is essential for data curators who may not be familiar with the source script, it is often an encumbrance for researchers in discovery and retrieval. Until 2011 the Judaica collection in Hebrew and Yiddish of the University Library J. C. Senckenberg were catalogued with transcription only. The aim of this work is to develop an open-source system to aid in the automatic conversion of Hebrew transcription back into Hebrew script, using a multi-faceted approach.

1 Introduction and Problem

There is a long history of collecting, preserving and cataloging Jewish literature. One challenge for catalogers and users is the use of non-Latin scripts in the original publication and Latin script that is supported by the library cataloging system. This was especially important in the early days of digital library systems, when it was not possible to mix different scripts. Even today, though modern catalogs are capable of supporting any script, the use of Romanization continues to have a role buy facilitating access for librarians and others who may not be familiar with the original script. However, the absence of the original script in many older records is a problem. For users familiar with the original script, the use of Romanization adds an extra layer of cognative indirection and often involves guesswork, since there is no single standard for conversion.

In Germany, the proof of the materials in a central catalogue system is based on the conversion of the original script into Latin as the dominant script. Until 2006, Hebrew in German catalogs was transcribed according to tables of the "Instructions for the Alphabetical Catalogues of the Prussian Libraries (PI)" published in 1899 [5]. These rules, originally binding only on Prussian libraries, including the Frankfurt University Library, were later adopted by the wider German-speaking world and remained in effect until they were replaced by the rules for alphabetical cataloguing (RAK) in 1977 and the introduction of the rules for alphabetical cataloguing (RAK-WB) in 1983. The transcription rules

© Springer Nature Switzerland AG 2018
E. Méndez et al. (Eds.): TPDL 2018, LNCS 11057, pp. 326–329, 2018.
https://doi.org/10.1007/978-3-030-00066-0_33

developed within the framework of the PI were the basis of the DIN 31636 [1] transcription of the Hebrew alphabet issued in April 1982, with minor modifications introduced for Modern Hebrew [4]. The new DIN standard 31636 of February 2006 is based closely on the rules of the American Library Association/Library of Congress (ALA-LC Romanization Rules for Hebrew) [3].

Catalogers have always used whichever transcription method was standard at the time. As a result, in long-term maintained catalogs, entries with different rules applied appear. Most users are not fully aware of the rules and their changes over time. Therefore, they are not able to find all relevant literature.

The University Library J. C. Senckenberg has a very large Judaica collection with many works in Hebrew and Yiddish. For more than 100 years, until 2011, all these works were catalogued with transcription only. To enable retrieval and exploration with different scripts the transcriptions need to be reversed. Due to the volume of entries, a manual re-cataloguing is not a feasible solution. Therefore an automated process has to be developed.

The different Romanization methods which have been developed for Hebrew typically fall into the categories of transliteration and phonetic transcription for Hebrew [6]. Transliteration is a precise representation of the source writing system in the target writing system, whereas transcription is focused on representing the phonological properties of the source language using the writing system of the target language. However, most modern systems are not pure representations of either transliteration or transcription but represent a compromise between the two points, as well as frequent compromises made for simplicity of representation for the target writing system.

The ALA-LC Romanization tables which our institution has used since 2006 are closer to a transcription than transliteration, with a stronger emphasis on sound [3]. While this seems to be an improvement for users, it also means there is significant loss of detail about the writing system, and there is no simple deterministic way to reverse the standard.

By contrast, the older systems represent a much deeper level of detail about the writing system – as well as many details which are interesting to philologists, but not relevant for reconstructing the Hebrew. This detail is represented through the extensive use of diacritical markings. For example, the Hebrew word שלום will be represented according to the 1982 version of DIN 31636 [2] as šālôm, whereas it will simply be *shalom* according the ALA-LC rules.

Indeed, the 1982 DIN [2] claims to be automatically reversable. However, it is clear that its authors did not attempt to reverse it. There are several small ambiguities remaining which make it impossible to determine the original Hebrew with certainty. Nonetheless, the additional details should make it easier to reason about the source text.

Unfortunately, the emphasis on detail in this standard appears to have made it more difficult for catalogers to generate Romanization. Metadata cataloged using this and similar standards contains a high rate of technical errors. In the end, the regularity with which these errors appear means we have to anticipate them, which, in turn, means it is not much less ambigious than the newer transcription for the purposes of automatic reversal. In addition, there are also ambiguities related to Hebrew itself, about whether the current official spellings are used, or a different historical orthography will be favored.

2   Approach and Implementation

The fist step is generating a list of every Hebrew form that could be naively conjectured from the given Romanized form. The second step is narrowing down these possibilities to what we would consider the most likely alternative, for which we have tried several methods and where we continue to experiment. The methods tried so far including simple spell checking, human audit, and matching against existing Hebrew metadata from other catalogs, which will be discussed in the following section.

We address the naive generation of all theoretically possible Hebrew forms for a given Romanized word with a Python library called *deromanize*. This library is language agnostic and simply allows the programmer to define a set of tokens in the source script (Romanized Hebrew, in our case), paired with a list of possible conversions. It also allows the use of different sets of tokens for different parts of the word and includes a simple pattern language for the generation of additional composite tokens, which can be used to address different types of syllables.

These conversion tables can be expressed as serialized data which is consumed by the library. We currently use YAML, but the system will work with any format which supports JSON-like data types. Once the appropriate data structures are generated, the programmer can use some of the built-in decoder functions, or construct their own in Python. The decoder will essentially fetch all the conversions for each token and hypothesize Hebrew words as a Cartesian product of the conversions. For example, *shalom* will be tokenized as sh|a|l|o|m, and the given conversions for each token will be, respectively, 'שׁ' | '', 'א' | 'ל' | 'ו', '', 'א' | 'ם', which combine to produce these hypothetical reconstructions:

שלום, שלם, שלאם, שאלום, שאלם, שאלאם

The first two hypotheses (starting from the right) are real Hebrew words, the first being correct. The order is guessed at this point by the order in which the users supplies the possible conversions. The user may optionally add additional weight if a particular conversion is deemed to be rare. Weights start at 0 and may go up indefinitely. There are also methods for converting these weights into numbers between 1 and 0 if such a representation is suited to their use.

3   Evaluation of Selection Methods

The most basic solution tried so far for selection is spell checking – simply ensuring that generated words are real Hebrew. In an audit of 408 titles containing 2036 words, the top suggestion for words was correct in 92.9% of occurances. However the entire title was correct in only 73.8% of cases.

We also have used matching with pre-existing Hebrew data from the National Library of Israel's catalog. By indexing this data in Apache Solr, using our generated Hebrew forms to build queries, and ensuring both the title and author fields have a Levenshtein distance within 10% of the overall length of the field, we have recovered Hebrew for 7,586 titles. In the audit of 200 titles, all matches were accurate. One error was found outside of the formal audit due the small degree of fuzziness we allowed. The correct work was matched, but a misspelling in the other catalog would have proliferated to ours.

4 Conclusion and Outlook

In this paper we addressed the challenge of reversing the Romanization applied on the catalog entries in Hebrew in the past 100 years in German libraries. The major challenge was to handle ambiguties introduced by different methods applied for the transliteration. We presented an approach that generates (1) all orthographically possible version of a term and (2) uses various methods to decide which of the results are most probable. The benefit of this approach is that it is independent from the original applied transliteration method. The library itself and the documentation are available as Open Source[1].

We continue to refine our approach to discover more matches to our catalog. Additionally, using the data we have verified so far, we have accumulated a set of 11,597 distinct Hebrew forms correlated with their Romanized forms and reductions based on their phonology which we will use to provide more accurate suggestions. Looking even further ahead, we are evaluating the use of machine learning to provide a level of context sensitivity over our simple word-based approach.

Acknowledgments. This work was partially funded by the DFG Project "Specialised Information Service Jewish Studies".

References

1. Deutsches Institut für Normung: DIN 31636:2011–01: Information and documentation - Romanization of the Hebrew alphabet. Beuth (2011)
2. Deutsches Institut für Normung: Publikation und Dokumentation: 2. Erschließung von Dokumenten, DV-Anwendungen in Information und Dokumentation, Reprographie, Bibliotheksverwaltung (1989)
3. Library of Congress: ALA-LC Romanization Tables. https://www.loc.gov/catdir/cpso/roman.html. Accessed 29 Mar 2018
4. Maher, P.E.: Hebraica cataloging: a guide to ALA/LC romanization and descriptive cataloging. Cataloging Distribution Service, Library of Congress (1987)
5. Marquardt, S.: Transliteration und Retrieval. Zur Problematik des Auffindens hebräischsprachiger Medien in Online-Katalogen. Berliner Handreichungen zur Bibliothekswissenschaft 157 (2005)
6. Ornan, U., Leket-Mor, R.: Phonemic conversion as the ideal romanization scheme for hebrew: implications for hebrew cataloging. Jud. Librariansh. **19**, 43–72 (2016)

---

[1] https://github.com/FID-Judaica/deromanize.

From Handwritten Manuscripts
to Linked Data

Lise Stork[1,2(✉)], Andreas Weber[3], Jaap van den Herik[1,2], Aske Plaat[1,2],
Fons Verbeek[1,2], and Katherine Wolstencroft[1,2]

[1] Leiden Institute of Advanced Computer Science, Leiden, The Netherlands
{l.stork,k.j.wolstencroft,f.j.verbeek,a.plaat}@liacs.leidenuniv.nl
[2] The Leiden Centre of Data Science, Leiden, The Netherlands
h.j.vandenherik@law.leidenuniv.nl
[3] University of Twente, Enschede, The Netherlands
a.weber@utwente.nl

Abstract. Museums, archives and digital libraries make increasing use
of Semantic Web technologies to enrich and publish their collection items.
The contents of those items, however, are not often enriched in the same
way. Extracting named entities within historical manuscripts and disclosing the relationships between them would facilitate cultural heritage
research, but it is a labour-intensive and time-consuming process, particularly for handwritten documents.

It requires either automated handwriting recognition techniques, or
manual annotation by domain experts before the content can be semantically structured. Different workflows have been proposed to address
this problem, involving full-text transcription and named entity extraction, with results ranging from unstructured files to semantically annotated knowledge bases. Here, we detail these workflows and describe the
approach we have taken to disclose historical biodiversity data, which
enables the direct labelling and semantic annotation of document images
in hand-written archives.

Keywords: Linked data · Cultural heritage
Handwriting recognition · Semantic annotation
Named entity recognition

1 Introduction

Digital libraries often provide web-accessible, digitised images of handwritten
manuscripts from various domains. However, the challenge remains to elucidate the handwritten content in a way that will enable exploration and further
research. This involves the transformation of the content into a searchable knowledge base. Historical documents are especially difficult due to the hard-to-read
handwriting, often in multiple languages, and the historical context of the text,
which makes them difficult to interpret. To enrich and elucidate the content of
manuscripts, different workflow methods have been developed. Digitised images

© Springer Nature Switzerland AG 2018
E. Méndez et al. (Eds.): TPDL 2018, LNCS 11057, pp. 330–334, 2018.
https://doi.org/10.1007/978-3-030-00066-0_34

of content can be annotated by human domain experts through nichesourcing [3], or computationally using automated handwriting recognition or word spotting techniques [2,7]. Most workflows, however, produce flat files or semi-structured output. This is useful for further searching and processing (e.g. using text mining techniques), but it does not enable content to be interlinked, semantically queried, or compared to other collections. We argue that this can be facilitated by labelling *and* semantically annotating word-zones - single word segments extracted from document images - using a domain ontology, resulting in a rich knowledge base that can be queried and interlinked with external resources.

2   Workflows for Elucidating Contents

Figure 1 roughly presents common workflows for the enrichment of handwritten documents.

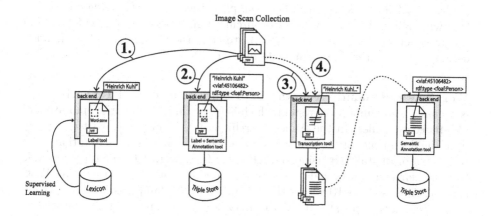

Fig. 1. Manuscript enrichment workflows

Workflow 1. The Automated Labelling of Word-Zones in Images. The *HistDoc* project is an example of a *Handwritten Text Recognition* (HTR) system that uses experts to harvest labels as input to a learning system [2]. Another example is *Transkribus*, where users can label sentences which are then used for training using HTR [5]. This project implements a form of semantic enrichment: labellers can flag certain named entities, e.g., locations or persons, with a user created tag set. Lastly, the aim of the *MONK* handwriting recognition system [7] is not full-transcription per se, but rather searchability of the informative content. The system does not rely on a language model and is therefore adaptive to its input. Labelling is targeted: the system retrieves and labels words that are visually similar to word-zones that are labelled by users.

Workflow 2. Semantic Annotation of Manuscript Images. *Accurator*[1] uses an expert crowd to annotate digital images, in specific digitised items from cultural heritage collections, such as paintings. Web users can help museums describe their collection items by providing expert knowledge. Users are prompted to annotate cultural heritage items with carefully selected controlled vocabularies. Annotations are stored in RDF format and linked to the digital images using the Web Annotation Data Model [3]. Another example, the *Semantic Field Book (SFB) Annotator*[2] [8], labels and simultaneously semantically annotates the most informative content of digitised manuscripts from natural history collections using an application ontology and the Web Annotation Data Model.

Workflow 3. Full-Text Transcription. The *Field Book Project*, a collaboration between the Smithsonian Institution Archives and the National Museum of Natural History, uses the crowd to harvest full-text transcriptions from historical biodiversity field books [1]. Another example is the *Transcribe Betham* initiative that will digitise and, also via crowdsourcing, fully transcribe 12,500 folios from the jurist Jeremy Bentham (1748–1832), stored in the University College London digital archive, through a media-wiki interface [6].

Workflow 4. Semantic Annotation of Fully Transcribed Text. *Annotea*[3] is a shared web annotation system which is based on the semantic annotation of web-based text files. In the Annotea architecture, annotations exist externally from the documents on *annotation servers*. The system lets an annotation point to a piece of digital text using the XPointer framework.[4] Other users are able to add their own additional annotations. Annotea makes use of existing W3C specifications, such as RDF and HTTP [4]. Another example is the *From Documents To Datasets* project [9]. Biodiversity field books are first fully transcribed and then semantically enriched.

Combining Automated Word-Zone Labelling with Semantic Annotation. In the Making Sense project,[5] methods are being developed for automated semantic annotation of natural history collections [10]. Our use case consists of 8,000 field book pages gathered by the Committee for Natural History of the Netherlands Indies between 1820 and 1850. A field book contains records that report species observations: their anatomy, characteristics, habitat and behaviour. Aiming for targeted, semantic annotation, the Making Sense project currently operates workflow 1, through the *MONK* handwriting recognition system, and workflow 2, through the *SFB-Annotator*. Initial results, an ontology

---

[1] http://www.accurator.nl/.
[2] https://github.com/lisestork/SFB-Annotator/.
[3] https://www.w3.org/2001/Annotea/.
[4] https://www.w3.org/TR/xptr-framework/.
[5] http://www.makingsenseproject.org/.

and a web application (see footnote 2) are available. Our final goal, however, is to combine workflows 1 and 2. Expert curated labels and semantic annotations can be used as input to a supervised learning system, combining handwriting and named entity recognition to perform semi-automated semantic annotation, thereby streamlining the process of elucidating, labelling and interlinking named entities.

3   Conclusion and Future Work

In this study, we enumerated the different workflow approaches that have been used to extract and structure the content of historical manuscripts, illustrated by example projects that utilise them. Although full-text transcription is an effective procedure that is often used, it cannot scale for all archived data and it falls short for further exploration and interpretation. Tools should be developed that reduce the requirement for full-text transcription and facilitate semantic annotation of extracted text to enable richer content descriptions. In our case study, we show that by providing tools to enable the direct semantic annotation of named entities, we can reduce the full-text transcription burden. In future work we will develop automated methods for semantic annotation.

By publishing the results online as Linked Open Data, the contents can be disclosed as a rich, structured resource that can be searched and combined with other cultural heritage collections.

References

1. The Field Book Project. https://siarchives.si.edu/about/field-book-project. Accessed 14 Mar 2018
2. Baechler, M., Fischer, A., Naji, N., Ingold, R., Bunke, H., Savoy, J.: HisDoc: historical document analysis, recognition, and retrieval. In: Proceedings of Digital Humanities, pp. 94–96. University of Hamburg, July 2012
3. Dijkshoorn, C., De Boer, V., Aroyo, L., Schreiber, G.: Accurator: nichesourcing for cultural heritage. Computing Research Repository, abs/1709.09249 (2017)
4. Kahan, J., Koivunen, M.R., Prud'Hommeaux, E., Swick, R.R.: Annotea: an open RDF infrastructure for shared web annotations. Comput. Netw. **39**(5), 589–608 (2002)
5. Kahle, P., Colutto, S., Hackl, G., Mühlberger, G.: Transkribus-a service platform for transcription, recognition and retrieval of historical documents. In: 2017 14th IAPR International Conference on Document Analysis and Recognition (ICDAR), vol. 4, pp. 19–24. IEEE (2017)
6. Moyle, M., Tonra, J., Wallace, V.: Manuscript transcription by crowdsourcing: transcribe bentham. Liber Q. **20**(3–4), 347–356 (2011)
7. Schomaker, L.: Design considerations for a large-scale image-based text search engine in historical manuscript collections. IT - Inf. Technol. **58**(2), 80–88 (2016)
8. Stork, L., et al.: Semantic annotation of natural history collections. Web Semant.: Sci. Serv. Agents World Wide Web. (2018). https://doi.org/10.1016/j.websem.2018.06.002

9. Thomer, A., Vaidya, G., Guralnick, R., Bloom, D., Russell, L.: From documents to datasets: a mediawiki-based metod of annotating and extracting species observations in century-old field notebooks. ZooKeys **209**, 235–253 (2012)

10. Weber, A., Ameryan, M., Wolstencroft, K., Stork, L., Heerlien, M., Schomaker, L.: Towards a digital infrastructure for illustrated handwritten archives. In: Ioannides, M. (ed.) ITN-DCH 2017. LNCS, vol. 10605, pp. 155–166. Springer, Cham (2018). https://doi.org/10.1007/978-3-319-75826-8_13

A Study on the Monetary Value Estimation of User Satisfaction with the Digital Library Service Focused on Construction Technology Information in South Korea

Seong-Yun Jeong[✉]

Department of Future Technology and Convergence Research,
Korea Institute of Civil Engineering and Building Technology,
283 Goyangdae-ro, Ilsanseo-gu, Goyang-Si, Gyeonggi-do, South Korea
syjeong@kict.re.kr

Abstract. Korea Institute of Civil Engineering and Building Technology has been constructing a database by collecting, classifying, and processing the construction technology data required for construction engineers and providing a database information service through the Construction Technology Digital Library portal since 2001. In this study, the monetary value of the user satisfaction with digital library service was estimated by applying the double-bounded dichotomous choice contingent valuation method for the purpose of using the limited information service budget to improve the user satisfaction.

Keywords: Construction Technology Digital Library ·
Contingent valuation method

1 Introduction

The Construction Technology Digital Library (CODIL) is the construction technology portal service system, which is equipped with a comprehensive distribution network of construction-technology-related information and documents (including practical construction materials) as well as construction projects, construction plans, and construction cost-saving cases, and allows the public (including construction engineers) to find the information and documents that they need in one place. An average number of 1.94 million hits and 0.88 million downloads per year ('13–'17). As of December 2017, the CODIL has gained 203,000 cumulative members and metadata for more than about 45,000 construction technology documents have been constructed.

The engineers and practitioners who use the CODIL are requesting for service improvements such as various and up-to-date construction related contents every year, but the information service budget required to oblige with such requests is either the same or lower each year. Therefore, it was necessary to estimate the user satisfaction with the digital library service focused on construction technology information using a quantitative value. It is possible to use the double-bounded dichotomous choice-based contingent valuation method (DBDC-based CVM), which is commonly utilized to estimate the monetary values of non-marketable goods, such as information services. In

© Springer Nature Switzerland AG 2018
E. Méndez et al. (Eds.): TPDL 2018, LNCS 11057, pp. 335–339, 2018.
https://doi.org/10.1007/978-3-030-00066-0_35

this study, a survey model was designed by including the demographic characteristics information of the respondents capable of affecting the monetary value estimation of the construction technology information service, the user satisfaction with the digital library service focused on construction technology information, and the acceptance of the maximum offer price as survey items. Next, after the quantitative and statistical processing of the correlations between the dependent and independent variables using DBDC-based CVM, the influence of the statistically significant variables on the value of satisfaction with the information service was analyzed, and their monetary values were estimated.

2   Theoretical Background

As information services of CODIL have the attributes of non-marketable goods, it is difficult to convert their benefits directly into monetary values. As for such monetary values of information services, the willingness to pay (WTP) prices of the users for information services are collected after a virtual market or situation is set. WTP refers to the maximum cost that users can pay to be satisfied with information services or to obtain intangible goods. Next, the values of information services can be estimated through the probability statistics processing procedure [1]. CVM suggests the maximum WTP price to the users according to the preference of the use of non-marketable goods. As for estimating the values of non-marketable goods using CVM, the utility theory that represents the potential preference for accepting or rejecting the suggested prices of non-marketable goods is applied as shown in Eq. (1).

$$U(I, M, S) = V(i, M, S) + \varepsilon_i, \ i = 0 \, or \, 1 \tag{1}$$

where U is the utility function, M is a variable representing the demographic characteristics of the respondent, S is the income of the respondent, and V is the deterministic part of WTP in the indirect utility function of the individual respondent. ε_i is the random probability variable whose average is zero and which has an independent, and the same distribution for income S, meaning the stochastic part of the uncertainty of the WTP in the indirect utility function of the individual respondent. i means that there is an intention to pay the suggested price or no such intention.

The answer can be "yes" or "no" to the question of whether they accept or reject the suggested price. Therefore, this method is simpler and may yield more reliable results than the other question methods [2]. The DBDC method is most widely used to estimate the values of non-marketable goods in a stable manner. The WTP price for individual respondent (i) can be estimated based on the probability cumulative area of the function. As the calculation is performed using the probability cumulative area that can be estimated up to infinity [3], the monetary value of user satisfaction with the digital library service is calculated using the average and median of the WTP.

3 Value Estimation of the User Satisfaction

3.1 Analysis of Correlations Between the Variables

In this study, a survey model was designed to estimate the value of user satisfaction with the construction technology information service using DBDC-based CVM. For example, the first suggested price may directly affect the WTP price estimation results. Therefore, the first suggested prices of US$5, 10, 20, 30, 40, and 50 were selected considering a case study in which the cost of an original document related to overseas construction technology was estimated to be US$8.2, and that of an original domestic document was estimated to be US$21.9 on average using the direct question method without the statistical analysis process, and another case study, in which the WTP for the construction technology original document service was estimated to be between US$32.8 and $46.5 using the Tobit model [4]. Next, analyzed the results obtained from the survey such as the demographic characteristics of respondents, the respondents' satisfaction with each digital library service area, and the suggested price conducted using the DBDC method. In the value estimation model for satisfaction with the construction technology information service, the WTP price, which is a dependent variable, and the demographic characteristics of the samples and user satisfaction with the digital library service, which are independent variables, have nonlinear mutual correlations. Table 1 shows the quantitative statistics results of the independent variables, which affect the WTP of the respondents for user satisfaction, obtained using the single bounded dichotomous choice (SBDC) and DBDC-based Logit models.

Table 1. Quantitative statistics results of the explanatory variables

Variables		SBDC-based Logit model			DBDC-based Logit model		
		Coefficient	Z-Statistic	Prob.	Coefficient	Z-Statistic	Prob.
C		−6.2037	−4.2835	0.0000	−2.741	−2.604	0.009
Company type		−0.3050	−1.1411	0.2538	−0.366	−1.617	0.106
Business field		0.3356	1.3167	0.1879	0.020	0.097	0.923
Age		0.4281	1.1058	0.2688	0.208	0.595	0.552
Career		0.0628	0.2022	0.8397	−0.338	−1.168	0.243
Construction report	Usage frequency	0.8092	1.4456	0.1483	−0.101	−0.280	0.779
	Volume	0.9035	1.6933	0.0904*	−0.026	−0.063	0.950
	Quality	−0.1350	−0.1962	0.8444	−0.030	−0.062	0.950
Construction working	Usage frequency	−0.1051	−0.2126	0.8317	0.408	1.199	0.231
	Volume	−0.4330	−0.7981	0.4248	0.010	0.025	0.980
	Quality	−0.0440	−0.0789	0.9371	0.212	0.498	0.619
Standard estimation/market unit price	Usage frequency	0.0788	0.1748	0.8612	0.524	1.760	0.078*
	Volume	0.0801	0.1310	0.8958	0.197	0.454	0.650
	Quality	−0.6746	−0.9379	0.3483	−0.506	−1.016	0.310
Bid		0.0006	7.9734	0.0000	0.000	6.457	0.000
McFadde R^2		0.594			0.505		
LR statistic		207.8630			192.3583		
Log likelihood		−71.04617			−94.07444		

where * rejects the null hypothesis that "the average difference between each variable is zero" at the 10% significance level through $p < 0.1$.

Examination using the double-bounded model revealed that the respondents with higher stages in the business field and the age exhibited a higher WTP for user satisfaction. On the other hand, in terms of the companies for which the respondents worked, the small and medium-sized companies exhibited a higher WTP than the conglomerates and public institutions. As for the career, the WTP decreased 0.73-fold, which was $\exp(-0.338)$, as the career increased by one unit. Therefore, it was found that the respondents with shorter careers had higher demands for the information service. Meanwhile, as regards the R^2 value, which is used to assess the suitability of the estimation results, the closer it is to zero, the more the observations that are properly predicted. As the value was 0.505 in this study, the suitability of the estimation results can be assessed at the medium level.

3.2 WTP Price Estimation

The estimation results of the probability function of the WTP price for the information service per person·use were obtained, as shown in Table 2. Because WTP was accepted focused on the small suggested price, there was a large difference between the WTP_{mean} and WTP_{median} values. Meanwhile, approximately 580,000 people visited the CODIL in 2017. The annual monetary value of user satisfaction with the digital library service can be estimated by multiplying the number of visitors and the WTP price per person.

Table 2. WTP-price-based value estimation results (Unit: US$)

Logit Model	WTP_{mean} per person	WTP_{median} per person	Annual WTP_{mean}	Annual WTP_{median}
SBDC	0.073	4.811	42,441	2,790,470
DBDC	0.758	2.032	439,521	1,178,581

4 Conclusion

This study is significant because DBDC-based CVM was used to estimate the value of user satisfaction with the digital library service focused on construction technology information in South Korea. Furthermore, the statistical significance and the correlations between the explanatory variables and the WTP price were examined.

The limitations of this study are as follows. First, the value of user satisfaction estimated in this study was limited to the information service provided by the CODIL. Second, the quality of user satisfaction was not specifically analyzed in this study. To complement this limitation, it is necessary to use the Delphi method to verify the survey items from experts and to apply a method for evaluating the quality of digital library services such as LIBQUAL. Finally, additional research is required on the monetary value estimation of non-use values, such as opportunity costs, which may provide new information services in the future due to continuous database (DB) accumulation.

References

1. Ikeuchi, A., Tsuji, K., Yoshikane, F., Ikeuchi, U.: Double-bounded dichotomous choice CVM for public library services in Japan. Procedia – Soc. Behav. Sci. **73**, 205–208 (2013)
2. Heberlein, T.A.: Measuring resource values: the reliability and validity of dichotomous contingent valuation measures. Paper Presented at the American Sociological Association Meeting, New York (1986)
3. Hanemann, W.M., Kanninen, B.: The statistical analysis of discrete response CV data. Working paper No. 798. Department of Agricultural and Resource Economics, University of California at Berkeley (1996)
4. Jeong, S.Y.: The estimation of domestic construction technology full-text services using Tobit model. J. Korea Acad.-Ind. Coop. Soc. **17**(6), 656–662 (2016). [Written in Korean]

Visual Analysis of Search Results in Scopus Database

Ondrej Klapka[(✉)] and Antonin Slaby

Faculty of Informatics and Management, University of Hradec Kralove,
Hradec Kralove, Czech Republic
{ondrej.klapka, antonin.slaby}@uhk.cz

Abstract. The enormous growth of research and development has been accompanied by a growing number of scientific publications in recent decades. These publications are collected and processed by a number of digital libraries. Although these digital libraries provide basic search tools, more advanced methods such as visualization and a visual analysis can be implemented only by using special software. This article presents the possibilities how to visually analyse the content of digital libraries using the CiteViz tool developed in [6] and shows its implementation using of the Scopus database.

Keywords: Visualization · Data mining · Scopus · CiteViz

1 Introduction

Visualization forms a compact and intuitive way of representing big and complex data of various nature [1–3]. A very often used visualization method is based on networks (graphs), which are quite easily machine-processable at the same time and enables to apply the theory of discrete mathematics, which deals with this data structure and its processing. It also makes it possible to calculate different scales and statistics concerning analysed network and present them to the user. [4] Visualization can be understood as the process of displaying data or information in the form of various objects. [5] Processing data, as well as data mining from an information system, is not always a simple operation. The reason may lie in a large amount of data being processed or varied approaches to data representation which the information system architecture is not prepared for.

2 CiteViz Tool

CiteViz visualization tool invented by authors of the article and presented in [6] allows to visualize the relationships between scientific publications. CiteViz can visualize database records in various ways and provide selected graphic representations of data. The developed solution proposed in [6] has its own REST-based server request interface and the server uses its own, manually managed database that provides data to the client's visualization application. Better user applicability of the CiteViz tool is

E. Méndez et al. (Eds.): TPDL 2018, LNCS 11057, pp. 340–343, 2018.
https://doi.org/10.1007/978-3-030-00066-0_36

achieved by communication based on an appropriate link to some of the large digital libraries that collect scientific publications.

3   CiteViz with Scopus API

Processing of information on scientific publications is linked to the existence of a number of scientific databases that store information on publications as well as their full texts. [7] The number of publications in these libraries is estimated to be several tens of millions. [8] These libraries enable a search based on several criteria, but the possibilities of presenting relationships between publications are limited. [7]

Based on previous research into possibilities of obtaining information from digital libraries, the Scopus database was selected to link and analyse its content. Scopus provides several types of queries over the library through the standard institutional subscription. [9] Scopus API allows to retrieve information about authors and publications in a structured form, but obtain some complex information require combining several queries. [9]

The problem of sufficient speed of the entire communication during the implementation process of connection CiteViz tool with the Scopus database arose as a result of Scopus API restrictions. The low speed of communication with the Scopus database is caused by the fact that many calls are needed to get one complete record (including the relationships to other records). From the user's point of view the serial call of the API using the CiteViz tool is almost unusable due to the very long response time for retrieving data from the database.

The CiteViz tool enables sequence loading of records, whereas visualization can be processed during this data retrieval. The data is sequentially replenished and visualized as it is retrieved from a database or library. This feature was utilized for this purpose and further expanded. When many requests to Scopus API needed, there are created multiple threads, among which all the requests are equally distributed. Thus, loading occurs in a parallel manner in several threads, which makes the process of data loading significantly faster.

The previous optimization partially solved the problem of speed, but it may still take longer to get responses to queries. Although the CiteViz tool has an in-memory cache, this cache is deleted after the session when running the tool is finished. In order to build easily on the user's previous work without a long waiting time needed to download the information already retrieved, the current in-memory cache has been extended and a disk cache was added. The in-memory cache then runs in LRU (Last Recently Used) mode and the longest unused records are put into a file on the user's computer drive. After the user exits the tool, all the in-memory cache is stored in the disk cache. Consequently, once the tool has been restarted, previously downloaded records are loaded from the user's computer drive without the need to download them from the Scopus database.

Tool capabilities and functionality, and especially its connection to the Scopus database, have been tested on a wide range of scenarios. Figure 1 shows the main window - user interface of the tool with some of its basic control panels, dialog boxes, menus and controls. The window shows the citation network consisting of the 300 cited publications of the University of Hradec Kralove together with their most frequently represented subject areas. Basic controls for elaborating data are situated in the top panel. Search bar is located on the left. Controls for handling currently selected visualization view are depicted in the lower panel. Basic information about the visualized network is displayed in the popup window at the bottom left corner.

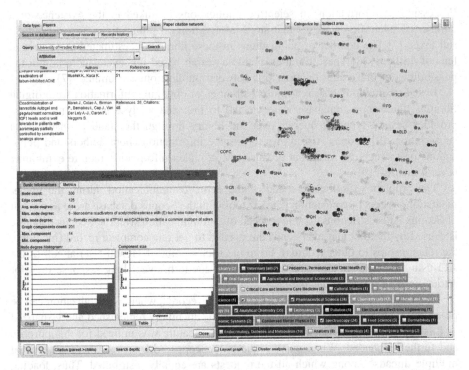

Fig. 1. The main window of the CiteViz visualization tool connected the Scopus database and depicting the citation network of the 300 most cited publications of the University of Hradec Kralove.

4 Conclusions

The possibilities of retrieving data from digital libraries for the CiteViz visualization tool developed within [6] have been explored. In consideration of all the possibilities of individual libraries it was chosen to implement the connection with the Scopus library.

During the implementation of the connection to Scopus, the speed of data retrieval proved to be the major problem. This is caused by the fact that several Scopus API calls have to be made to get one complete record including all citation relationships. Consequently, it was necessary to optimize data retrieval and the caching system in CiteViz. The optimization process resulted in using several parallel threads to retrieve the data, which shortens the communication of the visualization tool with the underlying databases. The data caching system has been expanded by the disk cache that enabled not to download repeatedly the already downloaded data even in case the work with the tool is interrupted and subsequently the tool restarted in order to continue the work.

Acknowledgements. This work and the contribution were supported by a project of Students Grant Agency (SPEV) - FIM, University of Hradec Kralove, Czech Republic. Ondrej Klapka is a student member of the research team.

References

1. Jensen, E.: Brain-based learning: the new paradigm of teaching. Corwin Press, Thousand Oaks (2008). ISBN 978-1412962568
2. Keim, D.A., Mansmann, F., Schneidewind, J., Ziegler, H.: Challenges in visual data analysis. Paper Presented at the 10th International Conference on Information Visualization, London, England, UK, 5–7 July 2006 (2006). https://doi.org/10.1109/iv.2006.31
3. Techopedia: Data Visualization: Data That Feeds Our Senses (2014). https://www.techopedia.com/2/29217/trends/big-data/when-data-visualization-works-and-when-to-avoid-it. Accessed 15 Jan 2018
4. Newman, M.E.J.: Networks: An Introduction. Oxford University Press, Oxford (2010). ISBN 978-0199206650
5. Le Grand, B.: How can data (and graph) mining techniques support research in information systems. Paper Presented at the 9th International Conference on Research Challenges in Information Science, Athens, Greece, 13–15 May 2015 (2015). https://doi.org/10.1109/rcis.2015.7128857
6. Klapka, O.: Visualization Analysis, Master Thesis (2013)
7. Dunne, C., Shneiderman, B., Gove, R., Klavans, J., Dorr, B.: Rapid understanding of scientific paper collections: integrating statistics, text analytics, and visualization. Department of Computer Science, University of Maryland. http://www.cs.umd.edu/~ben/papers/Dunne2011Rapid.pdf. Accessed 20 Jan 2018
8. Scopus. About Scopus. https://www.elsevier.com/solutions/scopus. Accessed 6 Feb 2018
9. Elsevier: Elsevier Developer Portal. https://dev.elsevier.com/. Accessed 14 Jan 2018

False-Positive Reduction in Ontology Matching Based on Concepts' Domain Similarity

Audun Vennesland[✉] and Trond Aalberg

Norwegian University of Science and Technology, Trondheim, Norway
{Audun.Vennesland,Trond.Aalberg}@ntnu.no

Abstract. In this study we explore if considering the domain similarity between concepts to be matched can contribute filter out false positive relations. This is particularly relevant in areas where the "universe of discourse" encompasses several diverse domains, such as cultural heritage. Our approach is based on an algorithm that employs the lexical resource WordNet Domains to filter out relations where the two concepts to be matched are associated with different domains. We evaluate our approach in an experiment involving Bibframe and Schema.org, two ontologies of complementary nature. The results from the evaluation show that the use of such a domain filter indeed can have a positive effect on reducing false positives while retaining true ones.

1 Introduction

Cultural heritage objects often contain rich semantic descriptions to cover their core facets, and supplementary information is often collected from various sources to include additional features relevant for their digital representation. This combined with the fact that cultural heritage is a very diverse application area leads to the use of concepts from ontologies with different levels of generality, magnitude, and spanning multiple domains, to sufficiently describe them [1]. This may lead to interoperability challenges.

Ontology matching aims to resolve such interoperability challenges, and string matching algorithms are an important component in most ontology matching systems [2]. Unfortunately, since they only focus on the string representation of concept names, without any semantic analysis, they often bring false positive relations in their computed alignment [3]. In order to optimise the quality in ontology matching it is therefore important to reduce the number of false positives from string matching operations. Our approach is based on an assumption that if two concepts are semantically similar, the domains they are associated with ought to be similar too. As an example, let us say that a string matcher computes with 96% confidence that the entity 'Content' in ontology 1 is equivalent with 'Continent' in ontology 2, which is obviously incorrect. Such a false positive relation can be filtered out if the similarity between their domains ({Metrology and Photography} vs. {Geography}) is below a given threshold.

© Springer Nature Switzerland AG 2018
E. Méndez et al. (Eds.): TPDL 2018, LNCS 11057, pp. 344–348, 2018.
https://doi.org/10.1007/978-3-030-00066-0_37

In our approach we use WordNet Domains [4], a lexical resource that offers a domain classification of Wordnet synsets (i.e. sets of synonyms for every distinct concept), to help determine domain similarity between two concepts.

2   Approach

The proposed approach to identify false positive relations is illustrated in Algorithm 1. As a preparatory step the ontologies are matched by a string matching algorithm (in our case ISub [5]) so as to produce an initial alignment (A_o) consisting of a set of relations. Each relation in this original alignment is processed in a sequence of operations.

Algorithm 1. Algorithm for filtering alignments based on domain similarity

---

Input: Original alignment A_o produced by a string matching algorithm, $minJaccard$ a threshold
    for Jaccard set similarity in the range $[0, 1]$
Output: A filtered alignment A_f where false positive relations have been removed
1: **function** $filterAlignment(A_o)$
2:   $A_f \leftarrow \emptyset$
3:   **for all** $a_i \in A_o$ **do**
4:     **if** $a_i.c_1$.equals($a_i.c_2$) **then**
5:       $A_f \leftarrow a_i$
6:     **else if** $compareConceptNamesDomains(a_i.c_1, a_i.c_2, minJaccard)$ **then**
7:       $A_f \leftarrow a_i$
8:     **else if** $fullWordRep(fullWord(a_i.c_1), fullWord(a_i.c_2), minJaccard)$ **then**
9:       $A_f \leftarrow a_i$
10:    **else if** $compoundHead(compoundWord(a_i.c_1), compoundWord(a_i.c_2), minJaccard)$ **then**
11:      $A_f \leftarrow a_i$
12:    **else if** $compareAllParts(compareAllParts(a_i.c_1), compareAllParts(a_i.c_2), minJaccard)$
      then
13:      $A_f \leftarrow a_i$
14:    **else**
15:      $a_i \notin A_f$
16:    **end if**
17:   **end for**
18:   **return** A_f
19: **end function**

The first operation (line 4) compares the two concept names for string-based equality and there is no interaction with WordNet in this operation. If the concepts are equal, they are added to the filtered alignment without further processing. If they are not, we move to the second operation. In operation 2 (`conceptNames`) on line 6, we identify the domains associated with the concept names as they are represented in the ontology, without any text processing involved. We then compare the sets of domains using Jaccard similarity of sets and state that the two concepts represent the same domain if the Jaccard score is equal to or above the $minJaccard$ parameter. Often a concept name is represented as a compound, that is, several individual words put together, for instance "TableOfContents". In operation 3 (`fullWordRep`) on line 8, a compound is split before the interaction with WordNet is performed. The remaining part of this step is similar to operation 2. In operation 4 (`compoundHead`) on line 10, we also split the compounds. However, in this step we only compare the compound head (the part that carries the basic meaning of the whole compound) of the concept names. Hence, we retrieve only the domains associated with the compound heads, and if the Jaccard score is equal to or above the $minJaccard$ threshold, we add this relation to the filtered alignment. Finally, operation 5 (`allWords`) on line 12 retrieves the domains of all "atomic" words from a compound. So if for example a relation includes the concepts "MusicNotation" and "MusicComposition" we retrieve the domains for "Music" and "Notation" for concept 1, and "Music" and "Composition" for concept 2. We merge the sets for each concept, and perform the same comparison as in previous steps using Jaccard.

3   Experimental Evaluation and Results

WordNet Domains version 3.2 is used in the experiment, and the JWNL API, version 1.4 RC2, is used for interacting with the WordNet 2.0 database. The datasets consists of two ontologies of complementary nature, that is, the ontologies to be matched are different with respect to topicality or granularity. The two ontologies are Library of Congress' Bibframe and Schema.org.

Since there is no existing reference alignment (ground truth) for these two ontologies, the original alignment produced by the string matching algorithm is inspected by the authors for true positives that are then used to form the reference alignment[1].

We compare the results of the filtered alignment with an alignment produced by the open source ontology matching system AgreementMakerLight [6], which usually ranks as a top contender of the Ontology Alignment Evaluation Initiative (OAEI) [7], an acknowledged evaluation campaign for ontology matching systems. It is run in automatic mode using a confidence threshold of 0.6.

The evaluation measures how much F-measure improvement we achieve by filtering the original alignment produced by the string matching algorithm $ISub$ [5]. The F-measure is computed as the harmonic mean of precision and recall [8]. We used a confidence threshold of 0.8 when producing the baseline alignments using $ISub$ in order to have a manageable, yet representative set of relations. The $minJaccard$ threshold is set to 0.5.

$ISub$ produced an alignment containing 93 relations between the ontologies Bibframe and Schema.org. 16 of the relations were considered correct by the manual analysis of the alignment, resulting in a precision of 17%. After applying the full WordNet Domains filter approach as described in Algorithm 1, 41 of these relations were filtered out, but all correct relations were preserved (a recall of 100%). This means that the filtered alignment consisted of 52 relations of which 16 were correct, resulting in a precision of 31%, hence a 14% increase. The F-measure was 47% for the filtered alignment, and 29% for the original alignment. While many of the true positives were relations with equal concept names, also operation 3 ($fullWordRep$) identified several true positives that were then maintained in the resulting alignment.

---

[1] Available at https://github.com/audunve/WordNetDomainsFilter.

In comparison, AML identified 12 of the correct relations in the reference alignment, having produced 21 relations in total. This results in a precision of 57%, and an F-measure of 65%.

We then analysed the quality of the alignment produced by each of the sequences of operations in the filtered approach algorithm.

This analysis revealed that all relations missed by AML were identified by sequence 1-2-3 (that is, after *fullWordRep* had been run). This alignment also included far less false positive relations than compared to when all operations were run. In summary, after the *fullWordRep* operation had been run, the alignment included all 16 correct relations of a total of 32 identified relations. This results in a precision of 50%, but with the high recall (100%) the F-measure obtained is 67%, and higher than AML. The remaining operations were too permissive, adding additional false positives to the alignment.

4   Conclusions and Further Work

The approach presented in this paper is based on filtering out false positives in alignments produced by string matching algorithms by analysing the similarity of domains associated with ontology concepts. Domain similarity is computed in a sequence of operations using a domain classification offered by the lexical resource WordNet Domains. An experimental evaluation shows that this approach improves precision and consequently F-measure compared to "unfiltered" alignments produced by the *ISub* string matching algorithm. The evaluation also indicates that the approach can compete with more sophisticated ontology matching systems. Future work includes performing more a comprehensive evaluation using larger datasets and explore how a similar filtering approach could enhance other types of alignment relations than equivalence.

References

1. Doerr, M.: Ontologies for cultural heritage. In: Staab, S., Studer, R. (eds.) Handbook on Ontologies. IHIS, pp. 463–486. Springer, Heidelberg (2009). https://doi.org/10.1007/978-3-540-92673-3_21
2. Cheatham, M., Hitzler, P.: String similarity metrics for ontology alignment. In: Alani, H. (ed.) ISWC 2013. LNCS, vol. 8219, pp. 294–309. Springer, Heidelberg (2013). https://doi.org/10.1007/978-3-642-41338-4_19
3. Po, L., Bergamaschi, S.: Automatic lexical annotation applied to the SCARLET ontology matcher. In: Nguyen, N.T., Le, M.T., Świątek, J. (eds.) ACIIDS 2010. LNCS (LNAI), vol. 5991, pp. 144–153. Springer, Heidelberg (2010). https://doi.org/10.1007/978-3-642-12101-2_16
4. Gella, S., Strapparava, C., Nastase, V.: Mapping wordnet domains, wordnet topics and Wikipedia categories to generate multilingual domain specific resources. In: LREC, pp. 1117–1121 (2014)
5. Stoilos, G., Stamou, G., Kollias, S.: A string metric for ontology alignment. In: Gil, Y., Motta, E., Benjamins, V.R., Musen, M.A. (eds.) ISWC 2005. LNCS, vol. 3729, pp. 624–637. Springer, Heidelberg (2005). https://doi.org/10.1007/11574620_45

6. Faria, D., Pesquita, C., Santos, E., Palmonari, M., Cruz, I.F., Couto, F.M.: The AgreementMakerLight ontology matching system. In: Meersman, R., et al. (eds.) OTM 2013. LNCS, vol. 8185, pp. 527–541. Springer, Heidelberg (2013). https://doi.org/10.1007/978-3-642-41030-7_38
7. Euzenat, J., Meilicke, C., Stuckenschmidt, H., Shvaiko, P., Trojahn, C.: Ontology alignment evaluation initiative: six years of experience. In: Spaccapietra, S. (ed.) Journal on Data Semantics XV. LNCS, vol. 6720, pp. 158–192. Springer, Heidelberg (2011). https://doi.org/10.1007/978-3-642-22630-4_6
8. Euzenat, J., Shvaiko, P.: Ontology Matching. Springer, Heidelberg (2013). https://doi.org/10.1007/978-3-642-38721-0

Association Rule Based Clustering of Electronic Resources in University Digital Library

Debashish Roy$^{(\boxtimes)}$ ⓘ, Chen Ding, Lei Jin, and Dana Thomas

Ryerson University, Toronto, ON M5B 2K3, Canada
{debashish.roy,cding,leijin,d1thomas}@ryerson.ca

Abstract. Library Analytics is used to analyze the huge amount of data that is collected by most colleges and universities when the library electronic resources are browsed. In this research work, we have analyzed the library usage data to accomplish the task of e-resource item clustering. We have compared different clustering algorithms and found that association-rule (ARM) based clustering is more accurate than others and it also identifies the hidden relationships between articles which are content-wise not similar. We have also shown that items in the same cluster offer a good source for recommendation.

Keywords: Association rule mining · Clustering · Hypergraphs
Recommender

1 Introduction

Nowadays, data is generated from various sources. To get profitable and attractive ideas, organizations are working to understand these data. They use different tools to identify valuable patterns hidden in the data. Most colleges and universities collect a vast amount of user interaction data through their websites. A part of this dataset comes from the library usage data. In this paper, the focus is to create an e-resource item clustering system for a university library, which can be used to facilitate the browsing task as well as recommend potentially interesting items. One of the unique challenges for the clustering task in this context is that the text information is limited. Usually, the title and the file name are the only available textual information for an e-resource item. The full text may not be easily accessible and its copyright usually belongs to the subscribed publisher instead of the university library. To overcome this challenge, we have compared the performance of different clustering algorithms that are used to cluster books/journals/articles. For clustering, we have focused mainly on content-based and ARM based approaches. We have found that ARM based clustering is able to find the relation between content-wise non-similar documents because clustering is done based on users' browsing behavior not based on the contents of the documents.

© Springer Nature Switzerland AG 2018
E. Méndez et al. (Eds.): TPDL 2018, LNCS 11057, pp. 349–353, 2018.
https://doi.org/10.1007/978-3-030-00066-0_38

The rest of this paper is organized as follows: Sect. 2 reviews and analyzes the existing research work. Section 3 explains the steps of our experiment. Section 4 discusses the results of the experiment. Finally, in Sect. 5, we conclude with a summary of results and analysis along with a future research direction.

2   Related Work

Library Analytics has been used to solve different types of problems. In [1] library usage data is used to identify the pattern of library use by the students. RAPTOR [3] is used to report e-resource usage statistics. All these research projects are good enough to provide statistics of different e-resource usage but they do not focus on applying any similarity finding technique among the e-resource items. To find similarities among documents, researchers have used different clustering algorithms. In [6] TF-IDF is used to represent the contents of the documents and K-means is used for clustering. ARM based clustering is used in [2]. Compared to these papers, our work is focused on clustering of e-resource items with limited text information. To figure out which clustering algorithm is the most effective, we have compared different clustering algorithms.

3   Experiment Design

We have used both ARM-based clustering and different content based clustering algorithms such as: K-means, FarthestFirst, Filtered, Density Based, Hierarchical and Expectation-Maximization Clustering. For ARM-based clustering, on the server logs we applied standard pre-processing steps [3] to remove unnecessary records. Then, we applied ARM to generate rules to construct a weighted hypergraph. In a hypergraph, one association rule is termed as one hyperedge and the vertices of the hypergraph correspond to the distinct documents of the library. We have used confidence as the weight for each hyperedge. Then we used HMETIS [4] to partition the hypergraph. Next, for content-based clustering, we extracted document ids and titles from the server logs. After removing the stop words we created a dictionary with all the words from the titles of all the documents and calculated Inverse document frequency (IDF) values of each word of each title. We have used the IDF-weighted vectors as input to various content-based clustering algorithms. We have used IDF because IDF assigns less weight to words that occur in more documents [5]. We didn't use the traditional TFIDF weighing scheme because usually a word occurs only once in the title and counting frequency is not that important in this case.

4   Result Analysis

To test the accuracy and the quality of different clustering results, we manually grouped the top browsed 254 e-book titles into 17 clusters with the help from the librarian. The selected titles were browsed at least 10-times. Figure 1 shows the size of the 17 manual clusters. To compare the quality of the content-based and ARM-based clusters, we calculate the matching percentage, using Eq. (4.1), between the manual clusters and the clusters generated by each clustering algorithm. In Eq. (4.1) W_i is a set of items in the i-th cluster, C_j is a set of items in the manually generated j-th cluster, k is the total number of clusters, i and j is a number between 1 and k, and max is a function to find the highest overlapping manual cluster with W_i.

$$TotalMatchQuantity = \sum_{i=1}^{k} max_j \, |W_i \cap C_j| \qquad (4.1)$$

From Fig. 2, we see that ARM based clustering i.e. HMETIS has the highest matching score and all other Content Based algorithms have not generated better clusters than the ARM based clusters. Out of all the content based clustering algorithms K-means came up with the best score. Figures 3, 4 and 5 show the size of different clusters in ARM based clusters, K-means and FarthestFirst clustering respectively. We can see distribution of items in different clusters is more balanced in the ARM based clusters. Using ARM based clustering, we have also found some hidden relationships between different documents such as: "The Hidden Factor in Climate Policy: Implicit Carbon Taxes" $->$ "Brief History of Neoliberalism". These relationships are not found by any of the content-based clustering algorithms because content-wise they are not similar. However, they are frequently accessed by the users together.

Then we used the clusters to recommend different e-books to users. Here, we have checked if e-books from the same cluster are recommended, how likely users may take the recommendation and read the book. To conduct the test, at first from the log files we downloaded all the ebrary titles for each user. There are altogether 4109 transactions available, one transaction refers to all the titles browsed by a specific user. We compared each transaction with the 17 clusters generated by different clustering algorithms. If any of the transactions overlaps with any of the clusters it means user has read at least one book from the set. To check how likely the user may visit other items in the same cluster, we group all the transactions that have overlapping of two items, three items, four items etc. For example, in case of ARM based clustering there are 372 transactions with two items' overlap, 36 transactions with three items' overlap and 5 transactions with four items' overlap. As we compare each transaction with the generated 17 clusters, we get 17 match counts. Then we find the maximum matchcount for each of the transactions. For any transaction, if the value of maximum match count is 1 then we set match percentage as 0% and if the value of maximum match count is greater than one then we set the matchcount to 100%. The idea is that if a user has browsed at least one more item from the same cluster, it is counted as a success because the recommendation based on the clustering result has been taken. We calculated average matchPercentage for all the transactions for each clustering algorithm to measure its recommendation success rate. Figure 6 shows the accuracy for different clustering algorithms. From the figure we see FarthestFirst gives the best result and ARM-based clustering has the second best result, followed by Hierarchical and K-means clustering algorithms. As we have seen in Fig. 5 FarthestFirst clustering algorithm has one big cluster, so definitely there is a higher chance to find more matches for different transactions. Similar observations can be made on K-means algorithm in Fig. 4. In this section, we have analyzed the results that are generated by the clustering algorithms and have found that: ARM-based clustering gives one of the best overall results in terms of clustering and recommendation accuracy.

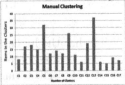

Fig. 1. Size of different clusters in manual clustering

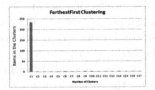

Fig. 2. Evaluation of clustering algorithms

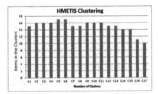

Fig. 3. Size of different clusters in HMETIS

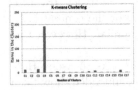

Fig. 4. Size of different clusters in K-mean

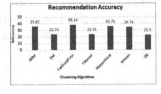

Fig. 5. Size of different clusters in farthest first

Fig. 6. Accuracy of recommendation for different clustering algorithms

5   Conclusion and Future Work

In this paper, we have worked on the problem of clustering different e-resource items in the context of library analytics. We have tested and compared different clustering algorithms based on the titles of the documents and users' browsing behavior. We have found that ARM based clustering is able to find out hidden and most accurate relationships. We have also shown that recommendation based on ARM based clustering gives one of the best results. As the future work we want to extend our analytics system to a fully functioning hybrid recommendation system, and optimize the library web site to make associated items easily accessible.

References

1. Collins, E., Stone, G.: Understanding patterns of library use among undergraduate students from different disciplines. Evid. Libr. Inf. Pract. **9**(3), 51–67 (2014)

2. Han, E., Karypis, G., Mobasher, B.: Clustering based on association rule hypergraphs. In: Workshop on Research Issues on Data Mining and Knowledge Discovery, pp. 9–13 (1997)
3. Karypis, G., Kumar, V., Smith, R., Mason, G.: Retrieval, Analysis, and Presentation Toolkit for usage of Online Resources (RAPTOR). Cardiff University. http://iam.cf.ac.uk/trac/RAPTOR. Accessed 14 Mar 2018
4. Karypis, G., Kumar, V.: A hypergraph partitioning package. Army HPC Research Center, Department of Computer Science & Engineering, University of Minnesota (1998)
5. Manning, H., Christopher, D., Raghavan, P.: Introduction to Information Retrieval, vol. 39, pp. 109–133. Cambridge University Press, Cambridge (2008)
6. Singh, V.K., Tiwari, N., Garg, S.: Document clustering using K-means, heuristic K-means and Fuzzy C-means. In: International Conference on Computational Intelligence and Communication Networks (CICN), pp. 297–301 (2011)

Hybrid Image Retrieval in Digital Libraries

A Large Scale Multicollection Experimentation of Deep Learning Techniques

Jean-Philippe Moreux[1(✉)] and Guillaume Chiron[2]

[1] Bibliothèque nationale de France, 75013 Paris, France
jean-philippe.moreux@bnf.fr
[2] L3i, University of La Rochelle, 17042 La Rochelle, France

Abstract. While digital heritage libraries historically took advantage of OCR to index their printed collections, the access to iconographic resources has not progressed in the same way, and the latter remain in the shadows. Today, it would be possible to make better use of these resources, especially by leveraging the illustrations recognized thanks to the OCR produced during the last two decades. This work presents an ETL (extract-transform-load) approach to this need, that aims to: Identify iconography wherever it may be found; Enrich the illustrations metadata with deep learning approaches; Load it all into a web app for hybrid image retrieval.

Keywords: Digital libraries · Content-based image retrieval · Deep learning
Automatic image classification · OCR · Data mining

Even though the creation of digital heritage collections began with the acquisition in image mode, several decades later to search in the content of some of these images still belongs to a more or less distant future. This apparent paradox originates in two facts: digital libraries (DLs) first focused on applying OCR to their printed materials, which renders major services in terms of information retrieval; Searching in large collections of images remains a challenge, despite the efforts of both the scientific community and GAFAs to address the underlying challenges [1]. However, the needs are very real, if one believes user surveys [2] or statistical studies of user behavior. But DLs iconographic collections are generally inadequate, given the broad spectrum of areas of knowledge and time periods surveyed by users. However, DLs are rich in many other iconographic sources. But organized in data silos that are not interoperable, most often lacking the descriptors required for image search, and exposed through text-oriented GIs. While the querying of iconographic content poses specific challenges [3], answers to various use cases, targets different knowledge domains, and finally calls for specific human-machine interactions [4, 5]. This work presents a proposal for a pragmatic solution to these two challenges, the creation of an encyclopedic heritage image database (which has never been done in DLs, to our knowledge, even if the Bayerische Staatsbibliothek[1] offers an image-based similarity search) and its querying modalities. A first section describes the initial phase of the extraction/aggregation of the

---

[1] https://bildsuche.digitale-sammlungen.de.

© Springer Nature Switzerland AG 2018
E. Méndez et al. (Eds.): TPDL 2018, LNCS 11057, pp. 354–358, 2018.
https://doi.org/10.1007/978-3-030-00066-0_39

heterogeneous data and metadata (MD) available. The next section presents the enrichments applied to the collected data, in particular the application of so-called "deep learning" techniques. Finally, a hybrid query mode is tested in a web app as a proof of concept.

1 Extract and Aggregate

A multicollection approach requires a first step of data mining in order to take into account the variability of the data available, due both to the nature of the documentary silos and to the history of the digitization policies. Our database aggregates 260k illustrations of the gallica.bnf.fr collections of images and prints related to the First World War. The data extracted from 475k pages thanks to the Gallica APIs and to SRU, OAI-PMH and IIIF protocols are stored in a XML database (basex.org).

The images collection presents particulars challenges: MD suffering from incompleteness and inconsistency; missing MD at image level like genre (picture, engraving, drawing) or color mode; portfolios exposing specific difficulties: cover and blank or text pages must be excluded, multi-illustrations pages cropped, captions extracted. For printed collections, we can leverage the OCR resources to identify illustrations a well as the text surrounding them. In the case of the daily press, the illustrations are characterized by a wide variety of genres and a large volume. It is important to note that this first step alone is worth it: even without semantic indexing of image content, it gives dedicated access to those valuable resources.

On the 530k raw illustrations extracted, we must filter the noisy ones, particularly the false detected illustrations from OCR of printed documents. The images collection presents a low noise rate ($\approx 5\%$) but it affects the quality perceived by users. For newspapers, noise varies from 10% (operator-controlled OCR) to 80% (raw OCR). Using MD and heuristics (illustration size and position, width/height ratio), the noise can be reduced (at the cost of 1 to 2% of false positives that will be handled later, Sect. 2).

2 Transform and Enrich

This step consists of transforming, enriching and aligning the MD obtained during the previous phase. Illustrations without any text descriptor are detected and their enlarged bounding box is processed with Google Cloud Vision OCR. Attempts are also made to link the illustrations to the BnF linked data service, data.bnf.fr.

Classifying Image Genres. The illustration genres are not always characterized in the catalogs (and obviously, this MD is not available for printed materials). To overcome this lack, a deep learning approach for image genres classification is implemented. We retrained the last layer of a convolutional neural network model (CNN, here Inception-v3 [6]), following a "transfer learning" method [7]. Our model has been trained on a twelve "heritage" genres ground truth (GT) of 12k illustrations produced first by bootstrapping from catalog MD and then by manual selection. Once trained, the model is evaluated: accuracy and recall are ≅90%. These results are considered to be good regarding the diversity of the training dataset (see [8] for results on a similar scenario), and performances are better with less generic models (separately trained on the images collection, F-measure rises up to 95%). Most confusions occur between engraving/ photo, line-based content (drawing, map, text) and illustrated ads, common in serials (in the newspapers set, ≈30% of the illustrations), which must be recognized to be filtered or used for scientific aims [9]. But the CNN outputs poor results because these ads can be of any graphical genre. A mixed approach (text + image) should preferably be used. The CNN model is also used to filter the unwanted illustrations that have been missed by the previous heuristic filtering step (Sect. 1), thanks to 4 noise classes (text, ornament, blank and cover pages). Recall/precision for these noise classes are highly dependent of the difficulty of the task: 98% for the images collection, 85% for the newspapers. The model can also be symmetrically applied to recall the false positives produced by the first filtering step.

Extracting Content from Images. The IBM Visual Recognition API has been used to analyze the illustrations and extract "concepts" [10] (objects, persons, colors, etc.). An evaluation is carried out on Person detection. A 4k GT is created and another evaluation is conducted on the "soldier" concept. The "person" concept has a modest recall of 55% but benefits from excellent accuracy of 98%. A decrease is observed for the more specialized "soldier" class (R = 50%; A = 80%). However, these results are to be compared with the relative silence of keyword searches: "soldier" does not exist as a concept in the bibliographic metadata and it would be necessary to write a complex keyword query like "soldier OR military officer OR gunner OR..." to obtain a 21% recall, to be compared to the 50% obtained by using the visual recognition MD only and the 70% in the hybrid scenario (visual + text descriptors), which shows the obvious interest in offering users cross-modal search [11]. Negative effects of deep learning sometimes occur: generalization may produce anachronisms; complex documents like posters are indexed with useless generic classes (machine learning techniques remain dependent on the modalities over which the training corpus has been created). The Watson API also performs face detection (R = 43%/A = 99.9%). On the same GT, a basic ResNet network gives a recall of 58% if one compromises on accuracy (92%): deep learning frameworks offers more flexibility than APIs. The enrichment indexing pipeline turns out to be complex to design and to implement. It takes into account the fact that certain genres do not need visual indexation (maps); that both text and image can be source of indexing; that this indexation may help to "unfilter" illustrations; that it may be replayed on a regular basis (using different APIs or in-house models); that new kind of content may require to retrain the model(s) and/or redesign the workflow.

3 Loading and Interacting

The BaseX database[2] is requested through REST queries. An images mosaic is fed on the fly by the Gallica IIIF server. A rudimentary faceted browsing functionality prefigures what a more ambitious user/system interaction could be[3].

Encyclopedic queries leverage the textual descriptors (metadata and OCR). While a "Georges Clemenceau" query in Gallica images collection only returns $\cong$140 results, the same query gives more than 900 illustrations with a broader spectrum of genres. The "drawing" facet can then be applied to find Clemenceau caricatures in dailies.

The conceptual classes extracted by the visual indexing overcome the silence related to the textual descriptors but also circumvent the difficulties associated with multilingual corpora, the lexical evolution or the fact that keyword-based retrieval can generate some noise. E.g, a query on "airplane" will output aerial pictures and aviator portraits whereas the "airplane" concept will do the job. Finally, the joint use of metadata and conceptual classes allows the formulation of cross-modal (or hybrid) queries: searching for visuals relating to the urban destruction following the battle of Verdun can rely on classes "street" or "ruin" and "Verdun" keyword.

4 Conclusion

The PoC source code and the database are available for the academic and digital humanities communities, which investigates more and more heritage contents for visual studies [12]. Moving towards sustainability for the illustrations MD would benefit to their reuse by information systems (like catalogs) as well as by in-house applications or end users. The IIIF Presentation API provides an elegant way to describe those illustrations, using a W3C Open Annotation attached to a Canvas in the IIIF manifest. All iconographic resources can then be operated by machine, for GLAM-specific projects, data harvesting and aggregation [13] or to the benefit of hacker/makers and social networks users. Nevertheless, the status and the management of these "new" metadata are still open questions: They are computer-generated (while catalog records are human creation) and susceptible to regular replay (AI is evolving at a frenetic pace); They can be massive (one catalog record for a newspaper title/millions of atomic data for its illustrations); An interoperability standard for expressing them is missing (IBM and Google use different taxonomies). At the same time, the maturity of modern AI techniques in image processing encourages their integration into the standard DLs toolbox. Their results, even imperfect, help to make visible and searchable large quantities of illustrations. But the industrialization of an enrichment workflow will have to cope with various challenges, mainly related to the diversity of the digital collections: neither illustration detection in documents nor deep learning for classification can generalize well on such a large spectrum of materials. Nevertheless, we can imagine that the conjunction of this abundance and a favorable technical context will open a

---

[2] https://altomator.github.io/Image_Retrieval.

[3] gallicastudio.bnf.fr.

new field of investigation for digital humanist researchers in the short term and will offer image retrieval services for all categories of users.

References

1. Datta, R., Joshi, D., Li, J., Wang J.: Image retrieval: ideas, influences, and trends of the new age. ACM Trans. Comput. Surv. (2008)
2. BnF. Enquête auprès des usagers de la bibliothèque numérique Gallica, April 2017. http://www.bnf.fr/documents/mettre_en_ligne_patrimoine_enquete.pdf
3. Picard, D., Gosselin, P.-H., Gaspard, M.-C.: Challenged in content-based image indexing of cultural heritage collections. IEEE Signal Process. Mag. Inst. Electr. Electron. Eng. 32(4), 95–102 (2015)
4. Breiteneder, C., Horst, E.: Content-based image retrieval in digital libraries. In: Proceedings of Digital Libraries Conference, Tokyo, Japan (2000)
5. Wan, G., Liu, Z.: Content-based information retrieval and digital libraries. Information Technology and Librairies, March 2008
6. Szegedy, C., Vanhoucke, V., Ioffe, S., Shlens, J., Wojna, Z.: Rethinking the Inception Architecture for Computer Vision, December 2015
7. Pan, S., Yang, Q.: A survey on transfer learning. IEEE Trans. Knowl. Data Eng. 22(10), 1345–1359 (2010)
8. Viana, M., Nguyen, Q., Babrani, M.: Document embedded images classification. In: ICDAR (2017)
9. Wevers, M., Lonij, J.: SIAMESET. KB Lab: The Hague (2017). http://lab.kb.nl/dataset/siameset
10. Karpathy, A., Fei-Fei, L.: Deep visual-semantic alignments for generating image descriptions. IEEE Trans. Pattern Anal. Mach. Intell. 39(4) (2017)
11. Wang, K., Yin, Q., Wang, W., Wu, S., Wang, L.: A comprehensive survey on cross-modal retrieval (2016). https://arxiv.org/pdf/1607.06215.pdf
12. Ginosar, S., Rakelly, K., Sachs, S., et al.: A century of portraits. A visual historical record of American high school yearbooks. In: Extreme Imaging Workshop, International Conference on Computer Vision, vol. 3 (2015)
13. Freire, N., Robson, G., Howard, J.B., Manguinhas, H., Isaac, A.: Metadata aggregation: assessing the application of IIIF and sitemaps within cultural heritage. In: Kamps, J., Tsakonas, G., Manolopoulos, Y., Iliadis, L., Karydis, I. (eds.) TPDL 2017. LNCS, vol. 10450, pp. 220–232. Springer, Cham (2017). https://doi.org/10.1007/978-3-319-67008-9_18

Grassroots Meets Grasstops: Integrated Research Data Management with EUDAT B2 Services, Dendro and LabTablet

João Rocha da Silva[(✉)] [iD], Nelson Pereira[iD], Pedro Dias, and Bruno Barros

INESC TEC, Faculdade de Engenharia, Universidade do Porto, Rua Dr. Roberto Frias, 4200-465 Porto, Portugal
joaorosilva@gmail.com, nelsonpereira1991@gmail.com,
{up201404178,up201405249}@fe.up.pt

Abstract. We present an integrated research data management (RDM) workflow that captures data from the moment of creation until its deposit. We integrated LabTablet, our electronic laboratory notebook, Dendro, our data organisation and description platform aimed at collaborative management of research data, and EUDAT's B2DROP and B2SHARE platforms. This approach combines the portability and automated metadata production abilities of LabTablet, Dendro as a collaborative RDM tool for dataset preparation, with the scalable storage of B2DROP and the long-term deposit of datasets in B2SHARE. The resulting workflow can be put to work in research groups where laboratorial or field work is central.

Keywords: Data repositories · Data curation
Research data management · Electronic laboratory notebooks
Ontologies

1 Motivation

Metadata production is often a repetitive and tedious task for researchers. However, it is essential for understanding and reusing research datasets. As such, LabTablet [1], Dendro [2] and the EUDAT CDI[1] services have been integrated to provide a seamless workflow to the researcher. The goal is to present a workflow that promoted the capture and management of data and metadata as early and continuously as possible in research projects.

2 LabTablet, an Electronic Laboratory Notebook

The LabTablet application is an electronic laboratory notebook. It takes advantage of a well-established metaphor—the laboratory notebook—where experimental contexts and settings are usually recorded in order to make sense of the

---

[1] https://www.eudat.eu/eudat-collaborative-data-infrastructure-cdi.

© Springer Nature Switzerland AG 2018
E. Méndez et al. (Eds.): TPDL 2018, LNCS 11057, pp. 359–362, 2018.
https://doi.org/10.1007/978-3-030-00066-0_40

data that is gathered. Tablets and smartphones have sensors that can be used by LabTablet during experiments or field sessions to pre-fill metadata. For example, if a researcher records a path of GPS coordinates during a field session, the Dublin Core "coverage" descriptor will be automatically suggested for that path. For researchers in different domains, distinct sets of descriptors can be used.

3   A Usage Scenario of LabTablet, Dendro and B2 Services

LabTablet, Dendro and EUDAT's B2DROP and B2SHARE services have been integrated to provide a seamless workflow to the researcher. Figure 1 shows how researchers can use these tools to support their work.

After successful authentication with their Dendro companion repository, LabTablet users can start a field session. When the application is in this mode, it continuously monitors the device's onboard sensors, turning readings into metadata values. For each field session, LabTablet creates a "Project" to package the data and metadata gathered during the session.

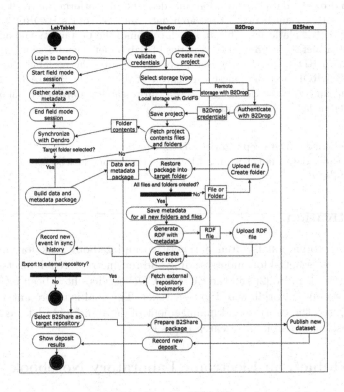

Fig. 1. The LabTablet-Dendro-EUDAT services workflow

Dendro also supports the concept of "Project", which resembles a Dropbox shared folder for members of the research group. A Project must be created before synchronization can occur between Dendro and LabTablet, as LabTablet needs a target folder to restore data and metadata at the time of synchronization. Dendro projects can have two storage types: "Local Storage", that stores all files in the Dendro server, or "B2DROP Storage", that stores all the data in EUDAT's B2DROP platform. If users opt for B2DROP storage, not only data files and folder structures but also their associated metadata will be saved and kept up to date as an RDF file in B2DROP.

To synchronize the metadata and data gathered during a session, LabTablet users select a target folder inside a Dendro project. If the project is using B2DROP storage, Dendro makes sure that the data are kept in sync in B2DROP.

In order to publish a dataset in B2SHARE, LabTablet users access the list of past synchronizations with Dendro and select the synchronized folder that they want to turn into a B2SHARE dataset. When the folder is selected, a list of possible target repositories shows up, and B2SHARE is one of the possibilities.

4   System Components and Their Integration

LabTablet is designed for low connectivity scenarios and only performs on-demand synchronization with Dendro. Figure 2 shows the main components of Dendro and LabTablet that interact with B2DROP and B2SHARE.

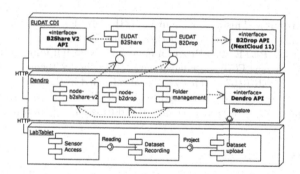

Fig. 2. Infrastructure components of the EUDAT—Dendro—LabTablet workflow

LabTablet's Sensor Access module interacts with the onboard sensors and provides readings to the Dataset Recording module, which builds metadata records based on standard schemas such as Dublin Core. The Dataset Upload module packages the data and metadata gathered by LabTablet into a zip file that is ready to be synchronized in Dendro. The Folder Management module of Dendro extracts the packages sent by external applications and loads their contents into a folder inside a Dendro Project. This module also manages the export of a folder to an external repository, such as B2SHARE.

Two NPM modules were built: `node-b2share-v2`[2], was developed to make it easy for NodeJS apps to interact with the B2SHARE v2 API. `node-b2drop`[3], allows NodeJS apps to easily interact with B2DROP. Dendro uses these modules to interact with both B2Services.

5   Conclusions and Future Work

The B2DROP platform is effective at managing files and folders, but its metadata production and representation capabilities are limited when compared to those available in Dendro, which integrates data storage and semantic metadata production in a single interface. Dendro also supports different ontologies instead of per-community metadata descriptor sets, as seen in B2SHARE. To pass on metadata to B2DROP, Dendro needs to produce RDF files that are updated and placed into B2DROP storage, but this metadata is not visible as such by other applications, being handled like any other files.

The Dendro interface and metadata layer can be considered as an alternative in the workflows supported by the B2DROP service, which provides a scalable storage backend but is not designed for metadata production. Thus, we are planning a comparison between B2DROP and Dendro as data preparation platforms prior to deposit in B2SHARE, from the point of view of those research groups in the long-tail of science.

Acknowledgements. This work is financed by the ERDF - European Regional Development Fund through the Operational Programme for Competitiveness and Internationalization - COMPETE 2020 Programme and by National Funds through the Portuguese funding agency, FCT - Fundação para a Ciência e a Tecnologia within project POCI-01-0145-FEDER-016736. This work is also funded in the context of an extended pilot by EUDAT and carried out by INESC TEC. We also thank Daan Broeder and Damien Lecarpentier for the assistance provided during our collaboration with EUDAT.

References

1. Amorim, R.C., Castro, J.A., Silva, J.R., Ribeiro, C.: LabTablet: semantic metadata collection on a multi-domain laboratory notebook. In: Closs, S., Studer, R., Garoufallou, E., Sicilia, M.-A. (eds.) MTSR 2014. CCIS, vol. 478, pp. 193–205. Springer, Cham (2014). https://doi.org/10.1007/978-3-319-13674-5_19
2. Silva, J.R., Ribeiro, C., Lopes, J.C.: Ranking Dublin Core descriptor lists from user interactions: a case study with Dublin Core Terms using the Dendro platform. Int. J. Digital Libr. (2018). ISSN 1432-1300. https://doi.org/10.1007/s00799-018-0238-x

---

[2] https://www.npmjs.com/package/@feup-infolab/node-b2share-v2.
[3] https://www.npmjs.com/package/@feup-infolab/node-b2drop.

Linked Publications and Research Data: Use Cases for Digital Libraries

Fidan Limani[✉], Atif Latif, and Klaus Tochtermann

ZBW - Leibniz Information Center for Economics, Kiel, Germany
{f.limani,a.latif,k.tochtermann}@zbw.eu

Abstract. Linking publications to research data is becoming important for a more complete research picture. Siloes of publication and data collections within institutions hamper this realization. We explore few cases that result from linking scholarly resources in a digital library setting.

Keywords: Research publications · Research data · Digital libraries

1 Introduction

Research data (RD) is experiencing a more prominent presence in research. Already emerging as a 1st class research citizen, there is a need to consider different aspects in order to maximize its research impact. Starting from metadata provision, to services development, it is necessary to handle this new tenant of scholarly communication, similar to what has been done with respect to research publications. The outcome is expected to provide a richer scholarly communication experience, where research publications, data, and other research-related artifacts provide a more comprehensive research picture.

Digital libraries (DL) are already considering extending their services by adding RD to their catalogues and services (see Borgman [1], for example). Establishing links between publications and data is one of the building blocks of a holistic scholarly view – and a prerequisite for realizing relevant scenarios – something we aim at in this paper: Specifically, we explore capabilities – in the form of use cases – that these links bring with resources from the domain of economics in the context of a DL.

2 Motivation

There are two cases that motivate this work:

Publications and RD from the Same Research Work. We often find that publication and RD collections are siloed even within the same institution. This prohibits readers from a more complete view of research outcomes. This use case links publications and data – part of the same research work/project/etc.

E. Méndez et al. (Eds.): TPDL 2018, LNCS 11057, pp. 363–367, 2018.
https://doi.org/10.1007/978-3-030-00066-0_41

– for a more complete research experience. Examples include recommending relevant data for a publication in a DL collection, or measuring the usability of a dataset in a research domain, to name a few.

Relevant Publication-Data Links. This use case focuses on identifying data that support a publication, or provide more context via publications that reference data, for resources that are not part of the same research work (publication or data). The use case identifies resources based on different criteria, such as research subject, publication date, publisher, etc.

3 Related Work

Cross-linking scholarly resources is already in the focus of multiple initiatives. From encoding link semantics, to applying linking technologies, solutions that rely on ontologies, metadata standards, persistent identifiers, user-provided classification terms, and more to further structure this linkage (see [2–4]).

Projects with different scopes also exist that establish links between scholarly resources. RMap relies on Semantic Web[1] and Linked Data[2] to model the relationships between such resources, extending beyond publications and data (see Hanson et al. [5] for more). Similarly, ResearchObject[3] relies on mechanisms to represent scholarly resources as a bundle to be accessible as a whole, with the final result published according to FAIR[4] principles, with accessibility support for both humans and machines.

Standardization efforts that propose solutions at a general level also exist. Such is the case with the Scholix Framework[5]. Driven by the RDA/WDS Publishing Data Services Working Group[6] and its partners, it represents a set of guidelines that foster interoperability between scholarly resources. The Data-Literature Interlinking (DLI) Service represents an implementation instance adhering to these guidelines, and currently offers more than 8 M links (see Burton et al. [6,7]).

4 Methodology

In this section we present the methodology – dataset and metadata selection, and the workflow that supports our approach.

As mentioned earlier in the paper, often there are silos of research resource collections (publications, RD, software code, etc.) even within institutions. Such is the case with the two collections selected for this work, operated by a single organization:

---

[1] https://www.w3.org/standards/semanticweb/.
[2] https://www.w3.org/standards/semanticweb/data.
[3] http://www.researchobject.org/.
[4] https://www.force11.org/group/fairgroup/fairprinciples.
[5] http://www.scholix.org/home.
[6] https://rd-alliance.org/groups/rdawds-publishing-data-services-wg.html.

- Journal Data Archive (JDA): targets RD from journals in the domain of economics. Researchers can upload different types of datasets, such as raw data, scripts or implementation code, etc. The current collection of JDA includes 70 datasets from different economics journals.
- EconBiz is a publications portal that focuses on the domain of economics. It supports many types of publications that researchers can store. With well over 10 M publications across participating databases and its set of services, it offers a great support to researchers in finding relevant publications.

4.1   Describing Research Resources: Metadata

For our problem domain we consider a common set of metadata for both collections. We take this metadata set into consideration during metadata harvesting and matching operations. Following is a brief description of these elements.

- EconBiz metadata elements: The typical descriptive metadata, such as title, creator, publication data, publisher, and resource type, among few others, are present. There is also a metadata element that specifies the subject of the resource, which is based on a thesaurus term – the STW[7].
- JDA metadata elements: We see the usual descriptive metadata elements in this collection (among few others) as well. Analogous to the previous collection, there is one element that specifies the subject of the resource, but this time based on terms from the JEL[8]. It is important to note that these terms are already mapped to the ones used for publications in EconBiz.

Having this terminology mapping is an opportunity to identify relevant resources across both collections, either by direct matching, or by narrowing/broadening search criteria, depending on the results.

4.2   Approach Workflow

Harvest Collections: Both collections provide REST interfaces; JDA entries are stored as JSON files, whereas the EconBiz search is conducted on-the-fly for every JDA entry, with only the highest-ranking result considered for matching.

Establish Links: For every dataset entry from the JDA collection, we search EconBiz for the same publication title (Use case 1). In case such a link cannot be established (the original publication that used the data is not hosted on EconBiz), we use other subject categorization attributes in both collections to conduct search (Use case 2). Other search scenarios are also possible: filtering candidate links based on publication year, publisher, access policy (open or closed), etc., are all viable refinements of this part.

Link Re-use: Once established, there is great potential for sharing the publication-data links. Adopting Scholix Framework principles, or Linked Open

---

[7] http://zbw.eu/stw/version/latest/about.
[8] https://www.aeaweb.org/econlit/jelCodes.php.

Data as a publication medium, or any other method mentioned in the related work, are just few of the options that could further increase the impact of the links.

On the technical side of things, our (object-oriented) model maps and implements the metadata described above – via harvesting both data collections based on their corresponding REST APIs – and stores identified links in a relational database.

5   Results

Resources Linking: For the first use case (publications-data linking) our approach matched 68 out of 70 JDA entries (around 97%) to publications in EconBiz. For the second use case, we successfully searched for RD based on certain criteria. For example, finding RD that correspond to Germany, retrieves 8 such resources from the JDA collection. One could as easily track RD publication by country, institution, publisher; specify the research domain and publication date, or uniquely address certain dataset, to name but a few scenarios.

Link Re-use: The links themselves represent a valuable asset to external parties; sharing the links according to the Scholix Framework principles is a way to increase the exposure of linked resources beyond the scope of local repository.

6   Conclusion and Future Work

In this paper we explore the role of publications-to-data links in a DL setting. Two types of use cases clearly show the benefits that result from establishing such links. In the future we plan to broaden the scope of both data selection (clearly our approach focused on exploring potential use cases, not having the largest link collection), and enrichment of links themselves in order to develop new, more complete use cases.

References

1. Borgman, C.L.: The conundrum of sharing research data. J. Assoc. Inf. Sci. Technol. **63**(6), 1059–1078 (2012)
2. Mayernik, M.S., Phillips, J., Nienhouse, E.: Linking publications and data: challenges, trends, and opportunities. D-Lib Mag. **22**(5/6), 11 (2016)
3. Aalbersberg, I.J., Dunham, J., Koers, H.: Connecting scientific articles with research data: new directions in online scholarly publishing. Data Sci. J. **12**, WDS235–WDS242 (2013)
4. Nüst, D., et al.: Opening the publication process with executable research compendia. D-Lib Mag. **23**(1/2) (2017). http://www.dlib.org/dlib/january17/nuest/01nuest.html
5. Hanson, K.L., Di Lauro, T., Donoghue, M.: The RMap project: capturing and preserving associations amongst multi-part distributed publications. In: Proceedings of the 15th ACM/IEEE-CS Joint Conference on Digital Libraries, pp. 281–282. ACM, June 2015

6. Burton, A., et al.: The Scholix framework for interoperability in data-literature information exchange. D-Lib Mag. **23**(1/2) (2017). http://www.dlib.org/dlib/january17/burton/01burton.html
7. Burton, A., et al.: The data-literature interlinking service: towards a common infrastructure for sharing data-article links. Program **51**(1), 75–100 (2017). https://doi.org/10.1108/PROG-06-2016-0048

The Emergence of Thai OER to Support Open Education

Titima Thumbumrung[✉] and Boonlert Aroonpiboon

National Science and Technology Development Agency,
111 Thailand Science Park, Phahonyothin Road,
Khlong Nueng, Khlong Luang 12120, Pathum Thani, Thailand
{titima, boonlert}@nstda.or.th

Abstract. This paper aims to present the practical work of the development of Open Educational Resources (OER) in a developing country, Thailand, to support open education and lifelong learning in the society. Thai OER is an ongoing project under the Online Learning Resources for Distance Learning project in the Celebration of the Auspicious Occasion of Her Royal Highness Princess Maha Chakri Sirindhorn's 60th Birthday Anniversary on the 2nd April 2015. It is developed by the collaborative efforts of multiple stakeholders in the country to share educational materials via the Internet under an open licensing agreement. This is to reduce the cost, access and usage barriers of students, teachers and learners, especially disadvantaged and disabled children and young people who lack opportunities to access education and knowledge. The materials, provided in Thai OER, cover a range of topics in different fields, especially Thai local and indigenous knowledge, and in different formats for all users. This paper also presents the benefits of Thai OER for different levels and major challenges to develop and adopt OER in a developing country, Thailand.

Keywords: Open educational resources · Open education · Lifelong learning

1 Background and Emergence of Thai OER

A decade ago, educational resources such as books, articles and audio had been curated in libraries. With the development of digital technologies, a large number of educational resources can be access anywhere and at any time through the Internet and digital devices. However, access to these resources is usually limited to registered students and users within specific institutions.

In the early 2000s, academic institutions such as Massachusetts Institute of Technology [1], University of Tubingen and Open University [2] broke barriers to access and use of educational resources by opening access to their resources to another institution as Open Educational Resources (OER). Commonwealth of Learning's OER global report 2017 suggested increasing support for OER [3].

Despite continuous efforts to reform the education system in Thailand for almost two decades, there were still issues that need to be tackled. For instance, disadvantaged and disabled children and young people lacked opportunities to access education and knowledge. Also, rural schools lacked teachers and educational materials. These issues

© Springer Nature Switzerland AG 2018
E. Méndez et al. (Eds.): TPDL 2018, LNCS 11057, pp. 368–372, 2018.
https://doi.org/10.1007/978-3-030-00066-0_42

impacted on the development of human resources for the country's development and global competitiveness.

For more than twenty years Her Royal Highness Princess Maha Chakri Sirindhorn gave her support to Thai government organizations to implement projects that use technology to enhance the quality of Thailand's educational system. In 2014, the Online Learning Resources for Distance Learning project was set up under the Initiative of Her Royal Highness Princess Maha Chakri Sirinhorn. It aimed to develop a nationwide online system which provided free and flexible educational materials and online courses to students, teachers and learners all over the country by using Information and Communication Technology (ICT). It was implemented by the National Science and Technology Development Agency (NSTDA), the Ministry of Science and Technology, with funding support from the Office of Basic Education, the Ministry of Education. In 2015, Thai OER (https://oer.learn.in.th) was started under the Online Learning Resources for Distance Learning project. It aimed (i) to provide quality educational resources to students, teachers and learners anywhere and at any time by free of charge; (ii) to encourage knowledge production and knowledge sharing by the collaboration of different organizations and use of ICT; and (iii) to concretely encourage academic morality and ethics through open license.

2 Collections and Technical Sides of Thai OER

Thai OER provides more than 40 different types of educational materials, such as, books (in different formats for different types of users), course materials, images, clip art, audio files and video clips. These materials cover a range of topics in science and technology, humanities and social sciences to support education and lifelong learning. Thai OER also collects and shares learning materials about Thai local knowledge such as herbs, traditional clothing and indigenous crafts. Therefore, in Thai OER users are not only able to access knowledge on classroom curricula but are also able to access Thai local content and indigenous knowledge.

NSTDA, the host of Thai OER, collaborates with different institutions in Thailand to share their digital collections via Thai OER. For instance, NSTDA worked with the National Archives of Thailand to digitize a collection of century-old royal photographic glass-plate negatives. The collection was last December designated as a "Memory of the World" by the United Nations Educational, Scientific and Cultural Organisation (UNESCO). Thai OER was selected by the National Archives of Thailand to share the collection for digital use. Presently, there are approximately 85,000 educational materials shared in Thai OER under a Creative Commons license.

Thai OER was developed by using Fedora Repository, an open source system for digital content management and dissemination. A survey of OER projects located at other institutions indicated the use of Fedora Repository. According to the open source system, another institution can download, install, use, modify and distribute the system free of charge. Fedora REST API was used to facilitate interoperability with client applications. Drupal was tested and used as a user interface because it was highly adaptable and customizable.

To make the resources on Thai OER globally discoverable, the selection of an internationally acceptable metadata standard used in OER was important. Thai OER adopted the OER Commons Metadata framework [4] which based on the Dublin Core Metadata standard and widely used by a number of OER sites. The metadata fields of Thai OER and a guideline for metadata annotation are provided at Thai OER website. OAI-PMH was used to facilitate the diffusion of metadata between Thai OER and other institutions. Each digital content in Thai OER was kept in Fedora and its set of descriptive metadata were indexed and searched by Apache Solr. 4 Virtual Machines was used to provide functionality of a physical computer. The Thai OER system could handle approximately 3,000 concurrent requests. Presently, the system is replicated by eight organizations in Thailand.

3 Benefits of Thai OER

Thai OER creates benefits for different levels in the country. Firstly, Thailand has its central OER repository to collect and disseminate quality teaching and learning resources, especially Thai local and indigenous knowledge which serves as a source of identity of Thais. However, this knowledge is quickly disappearing because of many factors such as vast modernization and the drastic change of economic framework.

Secondly, Thai OER encourages social movement and cross-community collaboration in education to support open and flexible learning opportunities in the country. It encourages creators to share their educational resources with other people broadly under an open licensing agreement. Today, more than 88 organizations in both the public and private sectors, and more than 1,700 individuals from different fields willingly share their educational materials through Thai OER under a Creative Commons license. These organizations include universities, research centers, museums, archives, libraries, governmental departments, temples and associations.

Thirdly, students, teachers and learners throughout the country who have access to the Internet can access quality educational resources anywhere and at any time, free of charge. As of April 2018, a total of 2,758,881 users accessed Thai OER and 5,011,300 educational items were downloaded, up from 416 users and 432 educational items in April 2015, when the Thai OER website was launched.

Finally, open educational materials, open licenses and copyright awareness are promoted to more than 5,000 students and teachers through a number of seminars and training. This helps students and teachers make better decisions when using copyrighted work in their teaching and study activities to reduce copyright infringement risks among them.

4 Challenges for OER Development and Adaptation in Thailand

Introducing Thai OER to academics and institutions for four years through a number of seminars and training, there are four significant challenges relevant to OER development and adaptation in the developing country, Thailand.

Firstly, more Thais now have internet access through their mobile devices. However, some Thais, especially in rural and in distant areas, still lack ICT access, bandwidth and connectivity. This situation impacts OER access and use. Therefore, ICT infrastructure development and offline solutions need to be considered.

Secondly, the investment of resources in broadband and connectivity, hardware and software, is still lacking. Most public educational institutions have insufficient resources to create effective and high quality educational resources. Therefore, there is a need to explore new business models and funding sources to solve this issue.

Furthermore, there are many social challenges related to the Thai OER project and movement. For instance, there is a lack of policies and top management support that focus on OER to encourage staff members to participate in OER activities. Policies and top management support play an important role, as funding and participation are attached to policies, and therefore coherent policies and strategies for adopting OER are required. Students, teachers and learners lack digital skills to find, access, evaluate, select, use, modify, share and integrate OER into their teaching and learning activities. Institutions have concerns about sharing their knowledge and intellectual property or losing control of materials that have income-earning potential. This mindset inhibits the practices of OER movement.

Finally, knowledge and awareness about copyright, open licensing and fair use of most students, teachers and learners are still limited. For instance, most students and teachers have a misconception that they use will be fair use if they use attribution or a disclaimer identifying the original creator. Also, they think that anything, especially on the internet, without a copyright notice is not protected.

5   Conclusion

A number of countries around the world, including Thailand, have joined OER. Thai OER was initiated to reduce gaps in educational and knowledge access and usage of students, teachers and learners in the country, especially disadvantaged and disabled children and young people. It also aimed to support open education and lifelong learning in the society. Some of the key strategies for Thai OER development included: using the open source software to create the system; collaborating with both private and public organizations and individuals to create and share educational resources; capitalizing on the fact that the organizations and individuals could host their resources on public sites through an open licensing agreement; and encouraging the organizations and individuals to share their resources to co-create the Thailand's central OER repository together. However, the major challenges of developing and implementing OER in the developing country, Thailand, included: providing effective and sufficient resources for OER activities; developing digital skills and understanding on intellectual property licensing; encouraging knowledge sharing culture; and especially developing policies that focus on OER at national and institutional levels.

References

1. D'Antoni, S.: Open educational resources: reviewing initiatives and issues. Open Learn. **24** (1), 3–10 (2009)
2. Commonwealth of Learning: Understanding Open Educational Resource. Commonwealth of Learning, British Columbia (2015)
3. Commonwealth of Learning: Open Educational Resources global report 2017. Commonwealth of Learning, British Columbia (2017)
4. OER Commons Metadata. https://iskme.zendesk.com/hc/en-us/articles/115001935503-Metadata. Accessed 05 Apr 2018

Fair Play at Carlos III University of Madrid Library

Belén Fernández-del-Pino Torres[(✉)] [iD]
and Teresa Malo-de-Molina y Martín-Montalvo

Universidad Carlos III de Madrid, Leganés, Spain
{belen.fernandez-delpino,teresa.malo}@uc3m.es

Abstract. Our purpose is to show projects held at Carlos III University Library related to FAIR principles. As time passes the Library evolves from a traditional library to "as open as possible, as closed as necessary" Digital Library.

Keywords: Author disambiguation · Digital libraries · Geolocation
Identifiers · Linked data · Metadata · Open data · Open science
Open source · ORCID · Repositories · Research data management
Library research services · University libraries

1 Introduction

"Fair play" has been part of our objectives before the concept was established and defined by Wilkinson [1] as FAIR principles: "FAIR Principles put specific emphasis on enhancing the ability of machines to automatically find and use the data, in addition to supporting its reuse by individuals". Likewise, Sipos [2] affirmed: "FAIR: an acronym that recently became inevitable for anyone involved in research data management or in any of the initiatives relating to the European Open Science Cloud". Gergely continues clarifying that "The FAIR principles precede implementation choices, and do not enforce or recommend any specific technology, standard, or implementation-solution. The principles are also not a standard or a specification. They establish a concise and measurable set that can act as a common denominator across institutes, across data and service providers and across disciplines. This means that they can be used as a guide to help data and tool owners to evaluate if their data, tools and services are findable, accessible, interoperable, and reusable". For example, Gergely summarizes "to make data Findable (F), should: have a persistent identifier and be described by metadata".

2 Carlos III University of Madrid FAIR Projects

When in 2006 Carlos III University of Madrid (UC3M) [3] offered to their researchers community the Institutional Repository *e-Archivo* [4], made the first step related with FAIR principles: indeed every item included is Findable (F) because of the handle assigned and also thanks to the Dublin Core metadata used. Furthermore, since its aim is to preserve and disseminate the intellectual production resulting from UC3M Academic and Research activity, and to offer Open Access to such documents. *e-Archivo* complies also with being Accessible (A) as described by Kraft [5] like: "Data and metadata should be archived long-term and made available in such a way that they can be easily retrieved by machines and humans, or be used locally with the help of standard communication protocols". The fact that *e-Archivo* is harvested by OpenAire [6], RePEc [7], Recolecta [8], etc. make it proven Interoperable (I) as Angelina explains: "Data should be available in such a format that it can be exchanged, interpreted and combined in a (semi) automated manner with other data, to be carried out by man and machine operations". *e-Archivo* collection includes mainly Doctoral Theses, Articles, Books and Chapters, Reports, Conference Proceedings, Datasets, Preprints, Working Papers, etc. doing the items easily Reusable (R) as Angelina describes: "A good description of data and metadata ensures that the data can be re-used for future research and are comparable to other compatible sources. It must be possible to cite the data properly, and the conditions under which the data can be re-used should be presented in a way that is easy for man and machines to understand".

At this point, the reader may think that, by the moment, nothing new has been presented, but this project from 2006 was the first stone for our actual ecosystem towards Open Science, described by FOSTER [9] as "the movement to make scientific research, data and dissemination accessible to all levels of an inquiring society". In fact, *e-Archivo* launched the beginning of a roadmap for cultural change introducing the OA Philosophy among the Library Staff, the Researchers and all the stakeholders, which really was a challenge... and continues to be so.

However, meanwhile, UC3M Library began to explore new services related with the repository; therefore, thanks to the incipient collaboration with UC3M Research Service, the integration with the CRIS was achieved. The target was to facilitate to UC3M researchers the deposit of OA publications via self-archiving from the CRIS, as well as easily acceptance of the deposit license related to each publication. This connection assured interoperability between the two platforms and supposed an asset to gain researchers confidence. This project was fully described by Rasero and Poveda [10] in 2012.

ORCID implementation in CRIS during 2014 helped to strengthen the researchers' confidence challenge assuring disambiguation and helping Authority management.

Nevertheless, the more relevant step was the creation in 2016 of a research discovery tool: *UC3M - Research Portal* [11], which allows the visualization of the information related to the research activity hosted in UC3M CRIS (including *e-Archivo*) and UC3M website. We opted to use VIVO, an Open Source platform. The data includes research structures and projects, bibliographic references of publications and other research outputs, such as patents, software and Doctoral thesis, In case the bibliographic references contents the URI (handle or DOI), full text of the publications could be reachable. This project made all UC3M research activity completely FAIR and the high access rates confirms a significant impact and great repercussion on research community.

Almost at the same time, the Library was preparing to offer services for data management related with H2020 requirements and other funders. In this particular activity, we realized that our offer would be wider if we cooperated among the public universities in the Autonomous Region of Madrid and the UNED consortium: Consorcio Madroño. We started contacting with the Digital Curation Centre (DCC) and adapting DMPonline to the Spanish-speaking researchers by creating PGDonline [12], a tool to create, review, and share data management plans with all the information and documentation in Spanish, available and free for any researcher, regardless of the institution to which he/she belongs.

The data management services includes *e-CienciaDatos* [13], a multidisciplinary data repository that hosts the scientific datasets from Consorcio Madroño researchers. It is based in Dataverse, hosted at the consortium, but managed by each university. Registered at DataCite, re3data.org, OpenAIRE compatible data provider and with DOI included in every dataset. The main challenge was to obtain datasets, but as time passes, we are gaining reputation and attracting researchers' interest.

Lately, we have started to explore how to increase visibility of some items or projects located at UC3M repositories, and we found two possible solutions: create a webpage and add the Geospatial Metadata to offer data geolocation. The result is the *Federico-Tena World Trade Historical Database* [14]. We consider remarkable the development of this project because, first, it has allowed us to expand the number of datasets included in *e-cienciaDatos*. In addition, it provides a splendid way to display the data. On top of that, the know-how and experience would be exportable to other projects and university research groups.

3 Future and Conclusions

During all these years, we have learned that cooperation is crucial; not only within the same institution, but between different institutions and departments. That is why we are always open to new collaborations and trying to keep up to date in new trends and requirements, as for example, how to obtain and expose standardized usage metrics related with the different research evaluation and assessment processes. In fact, UC3M Library is already participating in the *Pilot for Usage Metrics for OpenAIRE Content Providers* [15].

Adding FAIR values to services and collections sometimes is easier than you might think; small steps may bring great results and may help the university to go far. UC3M Library will keep exploring and playing fair.

References

1. Wilkinson, M., et al.: The FAIR guiding principles for scientific data management and stewardship. Sci. Data **3**, 160018–160027 (2016). https://doi.org/10.1038/sdata.2016.18
2. Sipos, G.: What is FAIR? EGI newsletter Inspired, no. 28 (2017). https://www.egi.eu/about/newsletters/what-is-fair/
3. Carlos III University of Madrid (Spain), Public University created in 1989, with more than 21.000 Students and about 200 Academic Staff. UC3M is part of YERUN Network of Young European Research Universities. UC3M. https://www.uc3m.es/Home. Accessed 11 Apr 2018

4. e-Archivo. https://e-archivo.uc3m.es/. Accessed 11 Apr 2018

5. Kraft, A.: The FAIR Data Principles for Research Data. TIB-Blog (2017). https://blogs.tib.eu/wp/tib/2017/09/12/the-fair-data-principles-for-research-data/#infos

6. OpenAIRE compatible data providers: e-Archivo. https://www.openaire.eu/search/data-providers#text:e-archivo. Accessed 11 Apr 2018

7. RePEc (Research Papers in Economics). https://edirc.repec.org/spain.html. Accessed 11 Apr 2018

8. RECOLECTA [Open Science Harvester] is a platform that gathers all the Spanish scientific repositories together in one place and provides services to repository managers, researchers and decision-makers. https://www.recolecta.fecyt.es/. Accessed 11 Apr 2018

9. FOSTER. https://www.fosteropenscience.eu/foster-taxonomy/open-science. Accessed 11 Apr 2018

10. Rasero, V., Poveda, A.: Integración del Sistema de Gestión de la Investigación y el Repositorio Institucional. e-Archivo (2012). http://hdl.handle.net/10016/20697

11. UC3M Research Portal. https://researchportal.uc3m.es. Accessed 11 Apr 2018

12. InvestigaM: PaGoDa. http://www.consorciomadrono.es/en/investigam/crear-su-pgd/. Accessed 11 Apr 2018

13. e-CienciaDatos. https://edatos.consorciomadrono.es/. Accessed 11 Apr 2018

14. Federico, G., Tena Junguito, A.: Federico-Tena World Trade Historical Database: World Trade (2018). e-cienciaDatos, V2. https://doi.org/10.21950/jkzfdp

15. Principe, P.: Infrastructures for open science in Europe: the power of repositories. In: Ponencia presentada en el Congreso Ecosistemas del Conocimiento Abierto (ECA 2017), Salamanca (2017). http://hdl.handle.net/10366/135640

Supporting Description of Research Data: Evaluation and Comparison of Term and Concept Extraction Approaches

Cláudio Monteiro[1]([⊠]) [iD], Carla Teixeira Lopes[1,2] [iD], and João Rocha Silva[1,2] [iD]

[1] Faculty of Engineering, University of Porto,
Rua Dr. Roberto Frias, 4200-465 Porto, Portugal
{claudio.monteiro,ctl}@fe.up.pt, joaorosilva@gmail.com
[2] INESC TEC, Faculty of Engineering, University of Porto,
Rua Dr. Roberto Frias, 4200-465 Porto, Portugal

Abstract. The importance of research data management is widely recognized. Dendro is an ontology-based platform that allows researchers to describe datasets using generic and domain-specific descriptors from ontologies. Selecting or building the right ontologies for each research domain or group requires meetings between curators and researchers in order to capture the main concepts of their research. Envisioning a tool to assist curators through the automatic extraction of key concepts from research documents, we propose 2 concept extraction methods and compare them with a term extraction method. To compare the three approaches, we use as ground truth an ontology previously created by human curators.

Keywords: Term extraction · Ontology learning
Research data management

1 Introduction

Research data requires contextual information in order to be accessed and interpreted, making metadata essential for their reuse. At the same time, metadata can vary greatly across research domains, making it hard to produce comprehensive and accurate metadata [1]. Dendro was created taking this problem into consideration by allowing users to use descriptors from different ontologies in metadata records. This enables descriptors from domain-specific ontologies (e.g.: Data Documentation Initiative) to be mixed with generic ones (e.g.: Dublin Core or Friend of a Friend), depending on the requirements of the research project [5].

The process of selecting or creating the ontologies starts with a meeting between the researchers, experts in their fields, and a curator, who shares knowledge regarding ontologies and research data management (RDM). After this, the curator analyses the interview results as well as some of the publications of the researchers, and selects a list of descriptors related to the research domain, which will then be validated by the researcher before being added to Dendro.

© Springer Nature Switzerland AG 2018
E. Méndez et al. (Eds.): TPDL 2018, LNCS 11057, pp. 377–380, 2018.
https://doi.org/10.1007/978-3-030-00066-0_44

Envisioning a tool to assist curators in descriptor selection, in this work we apply and compare techniques used in ontology learning systems to the textual materials that they have available at the end of the first interviews.

In recent years several ontology learning tools have been developed. These use different methods to extract terms, concepts and relations. State of the art systems include Text2Onto [3] which uses the Gate framework during preprocessing and allows the user to mix different term extraction methods such as TF-IDF and C/NC-value. Another system is OntoGain [6] which uses both OpenNLP and WordNet Java Library (JWNL) during preprocessing and C/NC-value for term extraction.

2   Concept Extraction Approaches

In order to assist the creation or selection of ontologies for Dendro, we have implemented different methods from ontology learning tasks. During these tasks, we perform three distinct steps, namely, preprocessing, term extraction and a query to DBpedia and/or Linked Open Vocabularies (LOV) with terms validated by the curator, to fetch potentially relevant concepts. An overview of the process can be seen in Fig. 1. In the tool we envision to implement, the curators will be able to validate the output after term extraction and at the end of the process. After term extraction, they filter the relevant terms and can manually add terms to the proposed list before the concept extraction stage.

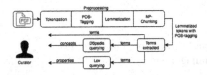

Fig. 1. Overview of our approaches

The **preprocessing** stage, essential in every ontology learning system, is handled by CoreNLP, which is used for sentence parsing, tokenization, Part-of-speech tagging and lemmatization. After, we apply linguistic filters that will allow us to extract noun phrases, which are required in the next phase. We have used two different filters, Noun+Noun and (Adj—Noun)+Noun [4].

For **term extraction** we selected C-Value, an hybrid approach for the extraction of multi-worded terms that mixes both linguistic and statistical information [4]. Part of the linguistic component has already been done in the preprocessing phase, the other consists of applying a stopword list, which can later be edited by the curator for each research domain. This list of stopwords is used to exclude words that are unlikely to be useful terms. During the statistical part of the method, a termhood value is assigned to each candidate in order to rank it in the output list. This component takes into consideration four different characteristics: total frequency in the corpus, total frequency as part of a longer term, the number of these longer terms and their length (in number of words). For example, the term "basal cell carcinoma" is a part of the longer candidates "cystic basal cell carcinoma" and "adenoid cystic basal cell carcinoma" which means that, during computation, it will require all four characteristics previously mentioned, while the third term is not a part of a bigger term and will only require its own frequency for computation [4].

Curators are usually not experts in the research domain being analyzed. By querying **DBpedia** with each term extracted they may get a description for them which will help decide if the term is useful or not. This is also why we allow the curator to add new entries to the list extracted in the previous phase. We are currently querying DBpedia in order to associate a concept to each extracted term. But concepts are usually resources, and since our main objective is to suggest properties as candidate descriptors, we have decided to also make queries to LOV. Using LOV's own score system we propose curators a "starting" descriptor, letting them decide to accept it as is of specialize it further—as a sub-property, for example.

3   Evaluation

As a preliminary assessment of our approaches, we used a scenario based on vehicle simulation where an ontology containing 12 descriptors already exists and was approved by researchers in that area [2]. As input, we have used the same five documents that were used during the manual process by the curators.

To evaluate the effectiveness, we computed the precision and recall for both linguistic filters of C-Value based on an exact match of the complete extracted term/concept (e.g.: vehicle frontal area), its 2 last words (e.g.: frontal area) and its last word (e.g.: area). We included the 2 last methods because we noted that often the output of our approaches was very specific.

3.1   Results

The results for both C-value filters can be seen in Fig. 2. As expected the (Adj—Noun)+Noun linguistic filter offers an increase in recall due to the extra terms allowed. With the Noun+Noun filter, 8 out of 281 had an exact match with the ground truth. With the (Adj—Noun)+Noun, this proportion was 10 out of 398.

Regarding DBpedia, comparing the descriptor and the complete DBPedia label, we were only able to associate a concept for about 25% of the wanted descriptors, while almost reaching 50% recall if using only the last 2 words.

With LOV we were able to achieve around 75% only using the last word, since in some cases LOV does not offer descriptors as specific as the ones in the ontology being used, with examples such as, "mass" instead of "vehicle mass" or "ratio" instead of "gear ratio".

As expected the scores are better when using shorter multi-word terms since the terms are more specific when using more words.

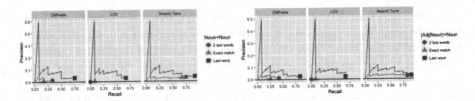

Fig. 2. Precision vs recall graphs

4   Conclusions and Future Work

In this paper, we presented an ontology learning tool built for the Dendro platform and a preliminary evaluation of its performance. The main idea is not to create an ontology from start to finish, but to assist curators during the process.

We were pleased with our results, especially for recall, but we agree that there is work to be done to improve precision. Apart from using the method that only used the last word during the comparison, we were able to reach a precision of around 6%. As for recall, we were able to reach over 90% comparing only with the extracted terms and 50% when querying DBpedia, both of these when using the (Adj—Noun)+Noun filter. These results highlight how challenging this retrieval task is, due to the high diversity and specificity of the descriptors being retrieved.

As for future work we will start by experimenting other term extraction methods to improve the precision figures. Also, DBpedia sometimes does not return any results from the queries, so we will look into reformulating the latter.

References

1. Amorim, R.C., Castro, J.A., da Silva, J.R., Ribeiro, C.: A comparative study of platforms for research data management: interoperability, metadata capabilities and integration potential. In: Rocha, A., Correia, A.M., Costanzo, S., Reis, L.P. (eds.) New Contributions in Information Systems and Technologies. AISC, vol. 353, pp. 101–111. Springer, Cham (2015). https://doi.org/10.1007/978-3-319-16486-1_10
2. Castro, J.A., Perrotta, D., Amorim, R.C., da Silva, J.R., Ribeiro, C.: Ontologies for research data description: a design process applied to vehicle simulation. In: Garoufallou, E., Hartley, R.J., Gaitanou, P. (eds.) MTSR 2015. CCIS, vol. 544, pp. 348–354. Springer, Cham (2015). https://doi.org/10.1007/978-3-319-24129-6_30
3. Cimiano, P., Mädche, A., Staab, S., Völker, J.: Ontology learning. In: Staab, S., Studer, R. (eds.) Handbook on Ontologies. IHIS, pp. 245–267. Springer, Heidelberg (2009). https://doi.org/10.1007/978-3-540-92673-3_11
4. Frantzi, K.T., Ananiadou, S., Tsujii, J.: The *C-value/NC-value* method of automatic recognition for multi-word terms. In: Nikolaou, C., Stephanidis, C. (eds.) ECDL 1998. LNCS, vol. 1513, pp. 585–604. Springer, Heidelberg (1998). https://doi.org/10.1007/3-540-49653-X_35
5. Rocha, J., Ribeiro, C., Lopes, J.: Ranking Dublin Core descriptor lists from user interactions: a case study with Dublin Core Terms using the Dendro platform. Int. J. Digital Libr. (2018). https://doi.org/10.1007/s00799-018-0238-x
6. Wong, W., Liu, W., Bennamoun, M.: Ontology learning from text. ACM Comput. Surv. **44**(4), 1–36 (2012)

Anonymized Distributed PHR Using Blockchain for Openness and Non-repudiation Guarantee

David Mendes[1], Irene Rodrigues[1]([✉]), César Fonseca[2], Manuel Lopes[3], José Manuel García-Alonso[4], and Javier Berrocal[4]

[1] LISP, ECT, Universidade de Évora, Évora, Portugal
dmendes@uevora.pt, ipr@di.uevora.pt
[2] Departamento de Enfermagem, Universidade de Évora, Évora, Portugal
cfonseca@uevora.pt
[3] Rede Nacional de Cuidados Continuados, Ministério da Saúde, Lisbon, Portugal
mjl@uevora.pt
[4] Universidad de Extremadura, Cáceres, Spain
{jgaralo,jberolm}@unex.es

Abstract. We introduce our solution developed for data privacy, and specifically for cognitive security that can be enforced and guaranteed using blockchain technology in SAAL (Smart Ambient Assisted Living) environments. Using our proposal the access to a patient's clinical process resists tampering and ransomware attacks that have recently plagued the HIS (Hospital Information Systems) in various countries. One important side effect of this data infrastructure is that it can be accessed in open form, for research purposes for instance, since no individual re-identification or group profiling is possible by any means.

Keywords: Blockchain · Data privacy · Interoperability · Open access

1 Introduction

In the realm of clinical information storage and maintenance one of the most hazardous situations that have been developing lately are the ransomware attacks and sensitive information breaches that are frightening the Hospital and National Health Information Services all around the world. Some new forms of data (actually information and knowledge) storage are in need that can circumvent this problem urgently for the adherence to health information processing that is emergent in these times of Artificial Intelligence and Big Data Analytics dawn. We suggest a decentralized structure that show characteristics that prevent, by design, all these problems and is not vulnerable to these kind of threats while

This work was supported by 4IE project (0045-4IE-4-P) funded by the Interreg V-A España-Portugal (POCTEP) 2014–2020 program and by LISP, Laboratório de Informática, Sistemas e Paralelismo ref: UID/CEC/4668/2016.

E. Méndez et al. (Eds.): TPDL 2018, LNCS 11057, pp. 381–385, 2018.
https://doi.org/10.1007/978-3-030-00066-0_45

promoting security in the edge-computing era [1]. We define an abstraction that we call ICP (Individual Care Process) a knowledge item that collects comprehensive information about an individual's health and care history. Any of the stakeholders may, in accordance with the fulfillment of the necessary authorizations for access to clinical data, consult and change this data. The distributed technology that allows us to guarantee this type of access while maintaining the privacy and confidentiality of the data is Blockchain, in which the different actors maintain the ledger of all the transactions. We can visualize the ICP as the ledger for all events related to the health/care process of a citizen. Blockchain technology ensures that only the owner of the private authentication key can authorize the manipulation of the sensitive data of your ICP. A physician (or other clinical staff) bound by professional secrecy may have access to diagnostic data, therapy or medical history provided that they are authenticated under the eIDAS but may safeguard some specific diagnostic or outcome data as enforced by the upcoming GDPR.

2   Methods

Cognitive Security Impact Evaluation. It has become utterly important that data protection be not only concerned with data in isolated terms but with the cognitive power that systems can extract from data when taken aggregated. Nowadays data owners can infer cognitive relations when in possession of disparate data chunks. Individual profiling as well as Group profiling, are currently major privacy concerns, and to avoid them a special attention has to be provided to Cognitive Security [4,10]. This kind of concern has lead in European Union to the enforcement of General Data Protection Regulation that is effective in all EU countries from May 25 of 2018. In wireless networks like those present in AAL environments special concerns have to be taken has illustrated in [4] and particularly in Smart Environments [2,5,10,12,13] as already predicted by [4].

Blockchain Data Privacy and Protection. It is necessary for the operation of the comprehensive ICP (Individual Care Process) to keep the information coming from many sources that can change without central control, but with the need to keep a record of all immutable state transitions. The distributed technology that allows us to ensure this type of access and data confidentiality is the Blockchain [5–9], in which the different actors maintain the ledger of every healthcare transaction [12]. We can visualize the ICP as the ledger of all events related to the process of health/care of a citizen. Blockchain technology ensures that only the owner of the private authentication key may authorize the handling of sensitive data from his/her CPAIP. Access to data, which a particular healthcare provider may have access to be encapsulated in the ICP itself by prior informed consent and it is possible to maintain a high level of granularity based on these consents [11,13]. It is important to note the use of DLA (Distributed Ledger Algorithms) algorithms that require only little computational power while maintaining an adequate level of Justice in the transactions order. These algorithms are deeply studied to support DLT (Distributed Ledger Technologies) and already available that we will use in our solution. Specifically it is implemented the DLA that use BFT (Byzantine Fault Tolerance) [3] based on the Hyperledger project of the Linux Foundation [8]. With these algorithms, even the IoT gateways, based on smartphones, may act on the ledger while ensuring absolute authenticity and privacy of the ICP [9].

Personal Rights and Information Protection. We obtain the necessary notifications and authorizations for collecting and processing the data (including specific authorizations and the necessary approvals, if applicable) and the free and fully informed consent of the research participants.

3   Blockchain

Blockchain is a shared, distributed ledger that facilitates the process of recording transactions and tracking assets in a business network. Transactions can be verified and recorded through the consensus of all parties involved. A blockchain requires each individual participant – or node – to hold a copy of the record. The blockchain architecture gives participants the ability to share a ledger that is updated, through peer-to-peer replication, every time a transaction occurs. Blockchain is particularly valuable at increasing the level of trust among network participants. Blockchain can hold the complete medical history for each patient, with multiple granularities of control by the patient, doctors, regulators, hospitals, insurers, and so on, providing a secure mechanism to record and maintain a comprehensive medical history for every patient. Blockchain provides the validation that the healthcare industry needs, and it delivers that service in a way all parties can trust.

Distributed Ledger Algorithms. It is important to use Distributed Ledger Algorithms (DLA) algorithms that only require small computational power and maintain an adequate level of justice in the transaction order. These algorithms are deeply studied to support the technologies already available from DLT (Distributed Ledger Technologies) that we will use in our solution. Specifically, DLAs implementing "Byzantine Fault Tolerance" [8] such as the Linux Foundation's Hyperledger project [8] are implemented. With these algorithms, the implemented Smartphone-based SAAL (Smart Ambient Assisted Living) IoT gateways can act on the ledger while guaranteeing the authenticity and absolute privacy of the ICP, even in IoT [9].

Byzantine Fault Tolerance. To ensure the consensus of the transactions needed for building the Hyperledger blockchain is used an algorithm based in Byzantine Fault Tolerance system. To ensure that a transaction is accepted as valid, 2f+1 valid signatures from distinct peers are needed. If some error occurs in a peer, due to an invalid message or timeout, then a transaction to the next peer in the chain is sent. In non-failure cases, a client submits a transaction to a leader peer. That peer verifies the transaction and signs it. It then broadcasts to the remaining 2f+1 validating peers. To detect failures, when a peer sends a transaction, it is given a timeout for receiving an answer. If that timeout is reached then a new transaction is made for an additional peer in the chain. The process is repeated until reaching 2f+1 valid signatures. At that time, the transaction is considered valid and a broadcast with that signed transaction is made to all peers [4].

4 Conclusions

We present the usage of Blockchain technology as a means to achieve unsurpassed security in health records bookkeeping. While completely tamper proof, we indicate the algorithms which usage can lead to a fair, democratic maintenance of the ledger while being low computational power consumers. This characteristic enables the usability by low computing power device like those present in the AAL environments. The level of safety perceived by monitored patients in these domiciled or institutionalized environments is very high while their health information is guaranteed to be at no risk. The ability of preserving the anonymity of the information structure, both individually as in group, allows the usage of this kind of proposed data support for research and open book keeping purposes.

References

1. Ahmed, A., Ahmed, E.: A survey on mobile edge computing. In: 10th International Conference on Intelligent Systems and Control (ISCO), pp. 1–8, January 2016
2. Asano, S., Yashiro, T., Sakamura, K.: Device collaboration framework in IoT-aggregator for realizing smart environment. In: TRON Symposium, December 2016
3. Castro, M., Liskov, B.: Practical byzantine fault tolerance and proactive recovery. ACM Trans. Comput. Syst. **20**(4), 398–461 (2002)
4. Greenstadt, R., Beal, J.: Cognitive security for personal devices. In: Proceedings of the 1st ACM Workshop on Workshop on AISec, AISec 2008, pp. 27–30. ACM, New York (2008)
5. Holler, J., Tsiatsis, V., Mulligan, C., Karnouskos, S., Boyle, D.: From Machine-to-Machine to the Internet of Things: Introduction to a New Age of Intelligence. Elsevier Science (2014)
6. Ichikawa, D., Kashiyama, M., Ueno, T.: Tamper-resistant mobile health using blockchain technology. JMIR Mhealth Uhealth **5**(7), e111 (2017)
7. Iroha. Hyperledger Iroha. Accessed 29 Aug 2017
8. Jacobovitz, O.: Blockchain for identity management. Technical report, The Lynne and William Frankel Center for Computer Science Department of Computer Science, Ben-Gurion University, Beer Sheva, Israel, December 2016. Technical Report #16-02
9. Jain, S., Kajal, A.: Effective analysis of risks and vulnerabilities in internet of things. Int. J. Comput. Corp. Res. **5**(2) (2015)
10. Liang, X., Shetty, S., Tosh, D., Kamhoua, C., Kwiat, K., Njilla, L.: Provchain: a blockchain-based data provenance architecture in cloud environment with enhanced privacy and availability. In: 2017 17th IEEE/ACM International Symposium on Cluster, Cloud and Grid Computing (CCGRID), pp. 468–477, May 2017

11. NASDAQ. Byzantine Fault Tolerance. Accessed 13 Apr 2018
12. Pramanik, M.I., Lau, R.Y., Demirkan, H., Azad, M.A.K.: Smart health: big data enabled health paradigm within smart cities. Expert Syst. Appl. **87**, 370–383 (2017)
13. RegEU. Regulation EU No 910/2014 of the European parliament and of the council of 23 July 2014 on electronic identification and trust services for electronic transactions in the internal market and repealing directive 1999/93/EC (eIDAS regulation) (2014). European union: 4459

Tutorials

Linked Data Generation from Digital Libraries

Anastasia Dimou[✉], Pieter Heyvaert, and Ben Demeester

Ghent University, Ghent, Belgium

Knowledge acquisition, modeling and publishing are important in digital libraries with large heterogeneous data sources. The process of extracting, structuring, and organizing knowledge from one or multiple data sources is required to construct knowledge-intensive systems and services for the Semantic Web. This way, the processing of large and originally semantically heterogeneous data sources is enabled and new knowledge is captured. Thus, offering existing data as Linked Data increases its shareability, extensibility and reusability. However, using Linking Data, as a means to represent knowledge, has proven to be easier said than done!

During this tutorial, we will elaborate the importance of semantically annotating data and how existing technologies facilitate the generation of their corresponding Linked Data: We will (i) introduce the [R2]RML[1,2] language(s) to generate Linked Data derived from different heterogeneous data sources, e.g., tabular data in databases, hierarchical data in XML published as Open Data or in JSON derived from a Web API; and (ii) show to non-Semantic Web experts how to annotate their data with the RMLEditor[3] which, thanks to its innovative user interface, allows all underlying Semantic Web technologies to be invisible to the end users. In the end, participants, independently of their knowledge background, will model, annotate and publish some Linked Data on their own!

The goal is to show that domain-experts can easily model the knowledge as Linked Data without being aware of Semantic Web technologies or dependent on Semantic Web experts. By the end of this tutorial, knowledge management or domain experts, data specialists and publishers should know how to model the knowledge that appears in their data as Linked Data, as well as how to annotate their data to generate and publish them as Linked Data.

The tutorial is organized as follows.
In the first session, the participants follow the introduction to Linked Data and Semantic Web and the presentation of exemplary tools that allow them to semantically annotate and publish Linked Data. In the second session, the participants follow the tutorial organizers as they introduce the tools to semantically annotate some sample data and publish them. Thus, there is less time to experiment on their own with the tool chain and data.

---

[1] http://rml.io/.
[2] https://www.w3.org/TR/r2rml/.
[3] http://rml.io/RMLeditor.html.

© Springer Nature Switzerland AG 2018
E. Méndez et al. (Eds.): TPDL 2018, LNCS 11057, p. 389, 2018.
https://doi.org/10.1007/978-3-030-00066-0

Research the Past Web Using Web Archives

Daniel Gomes(✉), Daniel Bicho, and Fernando Melo

FCT: Arquivo.pt, Lisbon, Portugal

The Web is the largest source of public information ever built. However, 80% of the web pages disappear or are changed to a different content within 1 year. Web archives provide services and tools that preserve and enable access to information published online since 1996.

The main objectives of this tutorial provided by the Arquivo.pt team are to:

- Motivate to the pertinence of web archiving, present use cases and share recommendations to create preservable websites for future access;
- Introduce tools to create and explore web archives such as: *oldweb.today*, *Memento Time Travel Portal*, *Arquivo.pt*, *robustify.js*, *ArchiveReady.com*, *webrecorder.io* or *brozzler*;
- Present methods and technologies to develop web applications that automatically access and process information preserved in web archives, for instance using the *Wayback Machine*, *Memento Time Travel protocol* or *Arquivo.pt API*.

Daniel Gomes started Arquivo.pt (the Portuguese web-archive) and currently leads this public service. He obtained his Ph.D in Computer Science in 2007 with a thesis focused on the design of large-scale systems for the processing of web data. He is a researcher in web archiving and web-based information systems since 2001.

Daniel Bicho has 8 years of experience in computer engineering and holds a degree in Telecommunications and Computers engineering. Currently is finishing his Master thesis in the field of Computer Vision, focusing at image classification using Deep Neural Network techniques. He is responsible for operating the crawling system of Arquivo.pt.

Fernando Melo is a software developer and researcher at Arquivo.pt. He obtained his Master degree in Computer Science with a thesis that addressed how to automatically perform the georeferencing of textual documents. He is currently applying and developing Big Data techniques to enable large-scale processing of web-archived content. Fernando Melo participated on the development of the Application Programming Interfaces provided by Arquivo.pt.

© Springer Nature Switzerland AG 2018
E. Méndez et al. (Eds.): TPDL 2018, LNCS 11057, p. 390, 2018.
https://doi.org/10.1007/978-3-030-00066-0

Europeana Hands-On Session

Hugo Manguinhas[1](✉) and Antoine Isaac[1,2]

[1] Europeana, The Hague, The Netherlands
[2] Vrije Universiteit Amsterdam, Amsterdam, The Netherlands

The Europeana REST API allows you to build applications that use the wealth of Europeana collections drawn from the major libraries, museums, archives, and galleries across Europe. The Europeana collections contain over 54 million cultural heritage items, from books and paintings to 3D objects and audiovisual material, that celebrate over 3,500 cultural institutions across Europe.

Over the past couple of years, the Europeana REST API has grown beyond its initial scope as set out in September 2011, into a wide range of specialized APIs. At the moment, we offer several APIs that you can use to not only get the most out of Europeana but also to contribute back.

This tutorial session will walk you through the wide range of APIs that Europeana now offers, followed by an hands-on session where you will be able to experience first hand what you can do with it.

Hugo Manguinhas is Product Owner for APIs at Europeana Foundation. His main focus is to shape the vision of the APIs and ensure a sustainable and consistent development of the products. He is also involved in the elaboration of requirements and specifications that contribute to the further development of the Europeana Data Model. Prior to Europeana, Hugo has been involved in several European Projects as a Researcher working at the Lisbon Technical University and INESC-ID Portugal in subjects such as interoperability, linked open data and semantic technologies from which he takes his enthusiasm for his current position at Europeana.

Antoine Isaac (Europeana Foundation) works as scientific coordinator for Europeana. He has been researching and promoting the use of Semantic Web and Linked Data technology in culture since his PhD studies at Paris-Sorbonne and the Institut National de l'Audiovisuel. He has especially worked on the representation and interoperability of collections and their vocabularies. He has served in other related W3C efforts, for example on SKOS, Library Linked Data, Data on the Web Best Practices, Data Exchange. He co-chairs the Technical Working Group of the *RightsStatements.org* initiative.

E. Méndez et al. (Eds.): TPDL 2018, LNCS 11057, p. 391, 2018.
https://doi.org/10.1007/978-3-030-00066-0

Author Index

Printed in the United States
By Bookmasters